EXCLUSIVE
PROCESSES
at High Momentum Transfer

EXCLUSIVE PROCESSES
at High Momentum Transfer

May 15 – 18, 2002
Jefferson Lab, Newport News, Virginia

editors

Anatoly Radyushkin
Old Dominion University and
Jefferson Lab, USA

Paul Stoler
Rensselaer Polytechnic Institute, USA

World Scientific
New Jersey • London • Singapore • Hong Kong

Published by

World Scientific Publishing Co. Pte. Ltd.

5 Toh Tuck Link, Singapore 596224

USA office: Suite 202, 1060 Main Street, River Edge, NJ 07661

UK office: 57 Shelton Street, Covent Garden, London WC2H 9HE

British Library Cataloguing-in-Publication Data

A catalogue record for this book is available from the British Library.

EXCLUSIVE PROCESSES AT HIGH MOMENTUM TRANSFER

ISBN 981-238-255-0

Printed in Singapore by World Scientific Printers (S) Pte Ltd

PROGRAM ADVISORY PANEL

ORGANIZING COMMITTEE

C. Carlson (William & Mary)
J.M. Laget (Saclay)
G. Miller (Seattle)
A. Radyushkin (Old Dominion & Jefferson Lab)
P. Stoler (Rensselaer)
B. Wojtsekhowski (Jefferson Lab)

PREFACE

These are the proceedings of the Workshop on Exclusive Processes at High Momentum Transfer, which was held 15–18 May, 2002 at the Thomas Jefferson National Accelerator Facility (Jefferson Lab) in Newport News, Virginia, USA. The workshop was sponsored by Jefferson Lab, CEA-Saclay, College of William & Mary, EC Research Network ESOP, National Science Foundation, Old Dominion University, Rensselaer Polytechnic Institute Rutgers University, and University of Illinois.

The workshop was open to all interested scientists, and embraced both theorists and experimentalists. The Organizing Committee thanks the speakers and participants at the conference for the stimulating atmosphere that prevailed. We trust the written versions of the talks presented in the volume will reproduce some of that feeling.

Thanks are also due to the people who worked hard to organize the conference and make it a success. The Conference Services group at Jefferson Lab was indispensible. Allow us to give special thanks to Ruth Bizot and Linda Ceraul, and even more special thanks to Mary Fox for all her efforts before, during, and since the conference.

Paul Stoler
Anatoly Radyushkin
August 2002

CONTENTS

SESSION PRESENTATIONS

PERSPECTIVES ON EXCLUSIVE PROCESSES IN QCD

S. J. BRODSKY

Stanford Linear Accelerator Center
2575 Sand Hill Road
Menlo Park, CA 94025, USA
E-mail: sjbth@slac.stanford.edu

Hard hadronic exclusive processes are now at the forefront of QCD studies, particularly because of their role in the interpretation of exclusive hadronic B decays, processes which are essential for determining the CKM phases and the physics of CP violation. Perturbative QCD and its factorization properties at high momentum transfer provide an essential guide to the phenomenology of exclusive amplitudes at large momentum transfer—the leading power fall-off of form factors and fixed-angle cross sections, the dominant helicity structures, and their color transparency properties. The reduced amplitude formalism provides an extension of the perturbative QCD predictions to exclusive nuclear amplitudes. The hard scattering subprocess T_H controlling the leading-twist amplitude is only evaluated in the QCD perturbative domain where the propagator virtualities are above the separation scale. A critical question is the momentum transfer required such that leading-twist perturbative QCD contributions dominate. I review some of the contentious theoretical issues and empirical challenges to Perturbative QCD based analyses, such as the magnitude of the leading-twist contributions, the role of soft and higher twist QCD mechanisms, the effects of non-zero orbital angular momentum, the possibility of single-spin asymmetries in deeply virtual Compton scattering, the role of hidden color in nuclear wavefunctions, the behavior of the ratio of Pauli and Dirac nucleon form factors, the origin of anomalous J/ψ decays, the apparent breakdown of color transparency in quasi-elastic proton-proton scattering, and the measurement of hadron and photon wavefunctions in diffractive dijet production.

1. Introduction

Exclusive processes provide a unique window for viewing QCD processes and hadron dynamics at the amplitude level[1]. Hadronic exclusive processes are closely related to exclusive hadronic B decays, processes which are essential for determining the CKM phases and the physics of CP violation. The universal light-front wavefunctions which control hard exclusive processes such as form factors, deeply virtual Compton scattering, high momentum transfer photoproduction, and two-photon processes, are also required for computing exclusive heavy hadron decays[2,3,4,5], such as

$B \to K\pi$, $B \to \ell\nu\pi$, and $B \to Kp\bar{p}$ [6]. The same physics issues, including color transparency, hadron helicity rules, and the question of dominance of leading-twist perturbative QCD mechanisms enter in both realms of physics. New tests of theory and comprehensive measurements of hard exclusive amplitudes can be carried out for electroproduction at Jefferson Laboratory and in two-photon collisions at CLEO, Belle, and BaBar [7]. The perturbative QCD approach to exclusive processes is now facing a number of strong empirical challenges. New data from Jefferson Laboratory [8] for the ratio of Pauli and Dirac form factors of the proton appears to be at variance with QCD expectations. This has led to a new focus on the range of validity of leading-twist perturbative QCD predictions and the necessity to have better theoretical control on higher-twist contributions. The Pauli form factor is particularly interesting, since it measures spin-orbit $\vec{S} \cdot \vec{L}$ couplings and thus the presence of orbital angular momentum in the proton light-front wavefunction. The new Jefferson Laboratory results appear to call into question hadron helicity conservation [9,10], a key feature of the leading-twist predictions. It is often claimed that the leading-twist predictions for the spacelike pion and proton form factors strongly underestimate their empirical magnitudes. The assumed relation between diffractive dijet production to the shape of the projectile light-front wavefunction has also been questioned. I will give my perspective on these challenges to theory in this report. QCD mechanisms for exclusive processes are illustrated in Figs. 1 and 2.

2. Perturbative QCD and Exclusive Processes

There has been considerable progress analyzing exclusive and diffractive reactions at large momentum transfer from first principles in QCD. Rigorous statements can be made on the basis of asymptotic freedom and factorization theorems which separate the underlying hard quark and gluon subprocess amplitude from the nonperturbative physics of the hadronic wavefunctions. The leading-power contribution to exclusive hadronic amplitudes such as quarkonium decay, heavy hadron decay, and scattering amplitudes where hadrons are scattered with large momentum transfer can often be factorized as a convolution of distribution amplitudes $\phi_H(x_i, \Lambda)$ and hard-scattering quark/gluon scattering amplitudes T_H integrated over

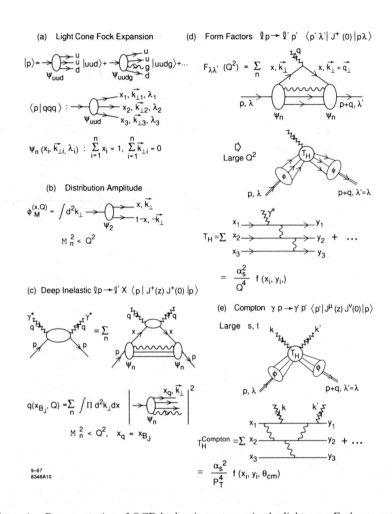

Figure 1. Representation of QCD hadronic processes in the light-cone Fock expansion.
(a) The valence uud and $uudg$ contributions to the light-cone Fock expansion for the
proton. (b) The distribution amplitude $\phi(x,Q)$ of a meson expressed as an integral
over its valence light-cone wavefunction restricted to $q\bar{q}$ invariant mass less than Q.
(c) Representation of deep inelastic scattering and the quark distributions $q(x,Q)$ as
probabilistic measures of the light-cone Fock wavefunctions. The sum is over the Fock
states with invariant mass less than Q. (d) Exact representation of spacelike form factors
of the proton in the light-cone Fock basis. The sum is over all Fock components. At large
momentum transfer the leading-twist contribution factorizes as the product of the hard
scattering amplitude T_H for the scattering of the valence quarks collinear with the initial
to final direction convoluted with the proton distribution amplitude. (e) Leading-twist
factorization of the Compton amplitude at large momentum transfer.

4

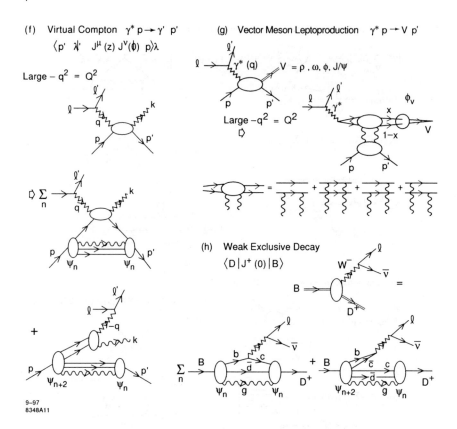

(f) Virtual Compton $\gamma^* p \to \gamma' p'$
$\langle p' \ \lambda' \ J^\mu(z) J^\nu(\phi) \ p \rangle \lambda$

Large $-q^2 = Q^2$

$\Downarrow \sum_n$

$+$

9–97
8348A11

(g) Vector Meson Leptoproduction $\gamma^* p \to V p'$

$V = \rho, \omega, \phi, J/\psi$

Large $-q^2 = Q^2$
$\Downarrow$

(h) Weak Exclusive Decay
$\langle D | J^+(0) | B \rangle$

$\sum_n$

Figure 2. (f) Representation of deeply virtual Compton scattering in the light-cone Fock expansion at leading twist. Both diagonal $n \to n$ and off-diagonal $n + 2 \to n$ contributions are required. (g) Diffractive vector meson production at large photon virtuality Q^2 and longitudinal polarization. The high energy behavior involves two gluons in the t channel coupling to the compact color dipole structure of the upper vertex. The bound-state structure of the vector meson enters through its distribution amplitude. (h) Exact representation of the weak semileptonic decays of heavy hadrons in the light-cone Fock expansion. Both diagonal $n \to n$ and off-diagonal pair annihilation $n + 2 \to n$ contributions are required.

the light-cone momentum fractions of the valence quarks[11]:

$$\mathcal{M}_{\text{Hadron}} = \int \prod \phi_H^{(\Lambda)}(x_i, \lambda_i) T_H^{(\Lambda)} dx_i . \tag{1}$$

Here $T_H^{(\Lambda)}$ is the underlying quark-gluon subprocess scattering amplitude in which each incident and final hadron is replaced by valence quarks with collinear momenta $k_i^+ = x_i p_H^+$, $\vec{k}_{\perp i} = x_i \vec{p}_{\perp H}$. The invariant mass of all intermediate states in T_H is evaluated above the separation scale $\mathcal{M}_n^2 >$

Λ^2. The essential part of the hadronic wavefunction is the distribution amplitude [11], defined as the integral over transverse momenta of the valence (lowest particle number) Fock wavefunction; *e.g.* for the pion

$$\phi_\pi(x_i, Q) \equiv \int d^2k_\perp \, \psi_{q\bar{q}/\pi}^{(Q)}(x_i, \vec{k}_{\perp i}, \lambda) \tag{2}$$

where the separation scale Λ can be taken to be order of the characteristic momentum transfer Q in the process. It should be emphasized that the hard scattering amplitude T_H is evaluated in the QCD perturbative domain where the propagator virtualities are above the separation scale.

The leading power fall-off of the hard scattering amplitude as given by dimensional counting rules follows from the nominal scaling of the hard-scattering amplitude: $T_H \sim 1/Q^{n-4}$, where n is the total number of fields (quarks, leptons, or gauge fields) participating in the hard scattering[12,13]. Thus the reaction is dominated by subprocesses and Fock states involving the minimum number of interacting fields. In the case of $2 \to 2$ scattering amplitudes, this implies $\frac{d\sigma}{dt}(AB \to CD) = F_{AB \to CD}(t/s)/s^{n-2}$. In the case of form factors, the dominant helicity conserving amplitude obeys $F(t) \sim (1/t)^{n_H - 1}$ where n_H is the minimum number of fields in the hadron H. The full predictions from PQCD modify the nominal scaling by logarithms from the running coupling and the evolution of the distribution amplitudes. In some cases, such as large angle $pp \to pp$ scattering, there can be "pinch" contributions[14] when the scattering can occur from a sequence of independent near-on shell quark-quark scattering amplitudes at the same CM angle. After inclusion of Sudakov suppression form factors, these contributions also have a scaling behavior close to that predicted by constituent counting.

The constituent counting rules[12,13] were originally derived in 1973 before the development of QCD, in anticipation that the underlying theory of hadron physics would be renormalizable and close to a conformal theory. The factorizable structure of hard exclusive amplitudes in terms of a convolution of valence hadron wavefunctions times a hard scattering quark scattering amplitude was also proposed. Upon the discovery of the asymptotic freedom in QCD, there was a systematical development of the theory of hard exclusive reactions, including factorization theorems, counting rules, and evolution equations for the hadronic distribution amplitudes[15,16,17,18]. In a remarkable recent development, Polchinski and Strassler[19] have derived the constituent counting rules using string duality, mapping features of gravitational theories in higher dimensions (AdS_5) to physical QCD in ordinary 3+1 space-time.

The distribution amplitudes which control leading-twist exclusive amplitudes at high momentum transfer can be related to the gauge-invariant Bethe-Salpeter wavefunction at equal light-cone time $\tau = x^+$. The logarithmic evolution of the hadron distribution amplitudes $\phi_H(x_i, Q)$ with respect to the resolution scale Q can be derived from the perturbatively-computable tail of the valence light-cone wavefunction in the high transverse momentum regime. The DGLAP evolution of quark and gluon distributions can also be derived in an analogous way by computing the variation of the Fock expansion with respect to the separation scale. Other key features of the perturbative QCD analyses are: (a) evolution equations for distribution amplitudes which incorporate the operator product expansion, renormalization group invariance, and conformal symmetry[11,20,21,22,23]; (b) hadron helicity conservation which follows from the underlying chiral structure of QCD[9]; (c) color transparency, which eliminates corrections to hard exclusive amplitudes from initial and final state interactions at leading power and reflects the underlying gauge theoretic basis for the strong interactions[24] and (d) hidden color degrees of freedom in nuclear wavefunctions, which reflect the color structure of hadron and nuclear wavefunctions[25]. There have also been recent advances eliminating renormalization scale ambiguities in hard-scattering amplitudes via commensurate scale relations[26] which connect the couplings entering exclusive amplitudes to the α_V coupling which controls the QCD heavy quark potential.

3. The Pion Form Factor

The pion spacelike form factor provides an important illustration of the perturbative QCD formalism. The proof of factorization begins with the exact Drell-Yan-West representation[27,28,29] of the current in terms of the light-cone Fock wavefunctions (see Section 7.) The integration over the momenta of the constituents of each wavefunction can be divided into two domains $\mathcal{M}_n^2 < \Lambda^2$ and $\mathcal{M}_n^2 > \Lambda^2$, where $\mathcal{M}_n^2$ is the invariant mass of the n-particle state. Λ plays the role of a separation scale. In practice, it can be taken to be of order of the momentum transfer.

Consider the contribution of the two-particle Fock state. The argument of the final state pion wavefunction is $k_\perp + (1 - x)q_\perp$. First take $k_\perp$ small. At high momentum transfer where

$$\mathcal{M}^2 \sim \frac{(1 - x)^2 q_\perp^2}{x(1 - x)} = \frac{Q^2(1 - x)}{x} > \Lambda^2, \tag{3}$$

one can iterate the equation of motion for the valence light-front wavefunction using the one gluon exchange kernel. Including all of the hard scattering domains, one can organize the result into the factorized form:

$$F_\pi(Q^2) = \int_0^1 dx \int_0^1 dy \phi_\pi(y,\Lambda) T_H(x,y,Q^2) \phi_\pi(x,\Lambda), \qquad (4)$$

where T_H is the hard-scattering amplitude $\gamma^*(q\bar{q}) \to (q\bar{q})$ for the production of the valence quarks nearly collinear with each meson, and $\phi_M(x,\Lambda)$ is the distribution amplitude for finding the valence q and $\bar{q}$ with light-cone fractions of the meson's momentum, integrated over invariant mass up to Λ. The process independent distribution amplitudes contain the soft physics intrinsic to the nonperturbative structure of the hadrons. Note that T_H is non-zero only if $\frac{(1-x)Q^2}{x} > \Lambda^2$ and $\frac{(1-y)Q^2}{y} > \Lambda^2$. In this hard-scattering domain, the transverse momenta in the formula for T_H can be ignored at leading power, so that the structure of the process has the form of hard scattering on collinear quark and gluon constituents: $T_H(x,y,Q^2) = \frac{16\pi C_F \alpha_s(Q^{*2})}{(1-x)(1-y)Q^2}(1+\mathcal{O}(\alpha_s))$ and thus[15,16,17,11,18,30,31,32,33]

$$F_\pi(Q^2) = \frac{16\pi C_F \alpha_s(Q^{*2})}{Q^2} \int_0^{\widehat{x}} dx \frac{\phi_\pi(x,\Lambda)}{(1-x)} \int_0^{\widehat{y}} dy \frac{\phi_\pi(y,\Lambda)}{(1-y)}, \qquad (5)$$

to leading order in $\alpha_s(Q^{*2})$ and leading power in $1/Q$. Here $C_F = 4/3$ and Q^* can be taken as the BLM scale[34]. The endpoint regions of integration $1 - x < \frac{\Lambda^2}{Q^2} = 1 - \widehat{x}$ and $1 - y < \frac{\Lambda^2}{Q^2} = 1 - \widehat{y}$ are to be explicitly excluded in the leading-twist formula. However, since the integrals over x and y are convergent, one can formally extend the integration range to $0 < x < 1$ and $0 < y < 1$ with an error of higher twist. This is only done for convenience – the actual domain only encompasses the off-shell regime. The contribution from the endpoint regions of integration, $x \sim 1$ and $y \sim 1$, are power-law and Sudakov suppressed and thus contribute corrections at higher order in $1/Q$ [16,17,11]. The contributions from non-valence Fock states and corrections from fixed transverse momentum entering the hard subprocess amplitude are higher twist, *i.e.*, power-law suppressed. Loop corrections involving hard momenta give next-to-leading-order (NLO) corrections in α_s.

It is sometimes assumed that higher twist terms in the LC wave function, such as those with $L_z \neq 0$, have flat distributions at the $x \to 0, 1$ endpoints. This is difficult to justify since it would correspond to bound state wavefunctions which fall-off in transverse momentum but have no fall-off at large k_z. After evolution to $Q^2 \to \infty$, higher twist distributions can evolve eventually to constant behavior at $x = 0, 1$; however, the wave-functions are in practice only being probed at moderate scales. In fact, if the higher twist terms are evaluated in the soft domain, then there is no evolution at all. A recent analysis by Beneke[35] indicates that the $1/Q^4$ contribution to the pion form factor is only logarithmically enhanced even if the twist-3 term is flat at the endpoints. It is also possible that contributions from the twist three $q\bar{q}g$ light-front wavefunctions may well cancel even this enhancement.

Thus perturbative QCD can unambiguously predict the leading-twist behavior of exclusive amplitudes. These contributions only involve the truncated integration domain of x and $k_\perp$ momenta where the quark and gluon propagators and couplings are perturbative; by definition the soft regime is excluded. The central question is then whether the PQCD leading-twist prediction can account for the observed leading power-law fall-off of the form factors and other exclusive processes. Assuming the pion distribution amplitude is close to its asymptotic form, one can predict the normalization of exclusive amplitudes such as the spacelike pion form factor $Q^2 F_\pi(Q^2)$. Next-to-leading order predictions are available which incorporate higher order corrections to the pion distribution amplitude as well as the hard scattering amplitude[21,36,37,38]. The natural renormalization scheme for the QCD coupling in hard exclusive processes is $\alpha_V(Q)$, the effective charge defined from the scattering of two infinitely-heavy quark test charges. Assuming $\alpha_V(Q^*) \simeq 0.4$ at the BLM scale Q^*, the QCD LO prediction appears to be smaller by approximately a factor of 2 compared to the presently available data extracted from pion electroproduction experiments[34]. However, the extrapolation from spacelike t to the pion pole in electroproduction may be unreliable, in the same sense that lattice gauge theory extrapolations to $m_\pi^2 \to 0$ are known to be nonanalytic. Thus it is not clear that there is an actual discrepancy between perturbative QCD and experiment. It would be interesting to develop predictions for the transition form factor $F_{q\bar{q}\to\pi}(t, q^2)$ that is in effect measured in electroproduction.

Compton scattering is a key test of the perturbative QCD approach[39,40,41]. A detailed recalculation of the helicity amplitudes and differential cross section for proton Compton scattering at fixed angle has been carried out recently by Brooks and Dixon[41] at leading-twist and at leading order in α_s. They use contour deformations to evaluate the singular integrals in the light-cone momentum fractions arising from pinch contributions. The shapes and scaling behavior predicted by perturbative QCD agree well with the existing data[42]. In order to reduce uncertainties associated with α_s and the three-quark wave function normalization, Brooks and Dixon have normalized the Compton cross section using the proton elastic form factor. The theoretical predictions for the ratio of Compton scattering to electron-proton scattering is about an order of magnitude below existing experimental data. However, this discrepancy of a factor of 3 in the relative normalization of the amplitudes could be attributed to the fact that the number of diagrams contributing to the Compton amplitude at next-to-leading order (α_s^3) is much larger in Compton scattering compared to the proton form factor.

A debate has continued on whether processes such as the pion and proton form factors and elastic Compton scattering $\gamma p \to \gamma p$ might be dominated by higher twist mechanisms until very large momentum transfers[43,44,45,46,47]. For example, if one assumes that the light-cone wavefunction of the pion has the form $\psi_{\text{soft}}(x, k_\perp) = A \exp\left(-b\frac{k_\perp^2}{x(1-x)}\right)$, then the Feynman endpoint contribution to the overlap integral at small $k_\perp$ and $x \simeq 1$ will dominate the form factor compared to the hard-scattering contribution until very large Q^2. However, this form for $\psi_{\text{soft}}(x, k_\perp)$ does not fall-off at all for $k_\perp = 0$ and $k_z \to -\infty$. A soft QCD wavefunction would be expected to be exponentially suppressed in this regime, as in the BHL model $\psi_n^{\text{soft}}(x_i, k_{\perp i}) = A \exp\left(-b \sum_i^n [\frac{\vec{k}_\perp^2 + m^2}{x}]_i\right)$ [48]. The endpoint contributions are also suppressed by a QCD Sudakov form factor[49], reflecting the fact that a near-on-shell quark must radiate if it absorbs large momentum. If the endpoint contribution dominates proton Compton scattering, then both photons will interact on the same quark line in a local fashion and the amplitude is real, in strong contrast to the QCD predictions which have a complex phase structure. The perturbative QCD predictions[39,40,41] for the Compton amplitude phase can be tested in virtual Compton scattering by interference with Bethe-Heitler processes[50]. Recently the "handbag" approach to Compton scattering[46,47] has been applied to $\bar{p}p \to \gamma\gamma$ at large energy[51]. In this case, one assumes that the process occurs via the exchange of a diquark with light-cone momentum fraction $x \sim 0$, so that the hard subprocess is $\bar{q}q \to \gamma\gamma$ where the quarks annihilate with the full energy of the baryons and nearly on-shell. The critical question is whether the proton wavefunction has significant support when the massive diquark has zero light-front momentum fraction, since the diquark light-cone kinetic energy and the bound state wavefunction become far-off shell $k_F^2 \sim -(m^2 + k_\perp^2)/x \to -\infty$ in this domain.

It is interesting to compare the calculation of a meson form factors in QCD with the calculations of form factors of bound states in QED. The analog to a soft wavefunction is the Schrödinger-Coulomb solution $\psi_{1s}(\vec{k}) \propto (1 + \vec{p}^2/(\alpha m_{\text{red}})^2)^{-2}$, and the full wavefunction, which incorporates transversely polarized photon exchange, differs by a factor $(1 + \vec{p}^2/m_{\text{red}}^2)$. Thus the leading-twist dominance of form factors in QED occurs at relativistic scales $Q^2 > m_{\text{red}}^2$ [1].

4. Perturbative QCD Calculation of Baryon Form Factors

The baryon form factor at large momentum transfer provides another important example of the application of perturbative QCD to exclusive processes. Away from possible special points in the x_i integrations (which are suppressed by Sudakov form factors) baryon form factors can be written to leading order in $1/Q^2$ as a convolution of a connected hard-scattering amplitude T_H convoluted with the baryon distribution amplitudes. The Q^2-evolution of the baryon distribution amplitude can be derived from the operator product expansion of three quark fields or from the gluon exchange kernel. Taking into account the evolution of the baryon distribution amplitude, the nucleon magnetic form factors at large Q^2, has the form[11,17,9]

$$G_M(Q^2) \to \frac{\alpha_s^2(Q^2)}{Q^4} \sum_{n,m} b_{nm} \left(\log \frac{Q^2}{\Lambda^2} \right)^{\gamma_n^B + \gamma_n^B} \left[1 + \mathcal{O}\left(\alpha_s(Q^2), \frac{m^2}{Q^2} \right) \right] \quad .(6)$$

where the γ_n^B are computable anomalous dimensions[52] of the baryon three-quark wave function at short distance, and the b_{mn} are determined from the value of the distribution amplitude $\phi_B(x, Q_0^2)$ at a given point Q_0^2 and the normalization of T_H. Asymptotically, the dominant term has the minimum anomalous dimension. The contribution from the endpoint regions of integration, $x \sim 1$ and $y \sim 1$, at finite $k_\perp$ is Sudakov suppressed [16,17,11]; however, the endpoint region may play a significant role in phenomenology.

The proton form factor appears to scale at $Q^2 > 5$ GeV^2 according to the PQCD predictions. See Fig. 3. Nucleon form factors are approximately described phenomenologically by the well-known dipole form $G_M(Q^2) \simeq 1/(1 + Q^2/0.71 \text{ GeV}^2)^2$ which behaves asymptotically as $G_M(Q^2) \simeq (1/Q^4)(1 - 1.42 \text{ GeV}^2/Q^2 + \cdots)$. This suggests that the corrections to leading twist in the proton form factor and similar exclusive processes involving protons become important in the range $Q^2 < 1.4$ GeV^2.

The shape of the distribution amplitude controls the normalization of the leading-twist prediction for the proton form factor. If one assumes that the proton distribution amplitude has the asymptotic form: $\phi_N = Cx_1x_2x_3$, then the convolution with the leading order form for T_H gives zero! If one takes a non-relativistic form peaked at $x_i = 1/3$, the sign is negative, requiring a crossing point zero in the form factor at some finite Q^2. The broad asymmetric distribution amplitudes advocated by Chernyak and Zhitnitsky[54,55] gives a more satisfactory result. If one assumes a constant value of $\alpha_s = 0.3$, and $f_N = 5.3 \times 10^{-3}$GeV^2, the leading order prediction

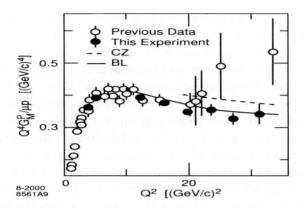

Figure 3. Predictions for the normalization and sign of the proton form factor at high Q^2 using perturbative QCD factorization and QCD sum rule predictions for the proton distribution amplitude (From Ji et al.[53]) The curve labelled BL has arbitrary normalization and incorporates the fall-off of two powers of the running coupling. The dotted line is the QCD sum rule prediction of given by Chernyak and Zhitnitsky[54,55]. The results are similar for the model distribution amplitudes of King and Sachrajda[56], and Gari and Stefanis[57].

is below the data by a factor of ≈ 3. However, since the form factor is proportional to $\alpha_s^2 f_N^2$, one can obtain agreement with experiment by a simple renormalization of the parameters. For example, if one uses the central value of Ioffe's determination $f_N = 8 \times 10^{-3} \mathrm{GeV}^2$, then good agreement is obtained[58]. The normalization of the proton distribution amplitude is also important for estimating the proton decay rate[59]. The most recent lattice results[60] suggest a significantly larger normalization for the required proton matrix elements, 3 to 5 times larger than earlier phenomenological estimates. One can also use PQCD to predict ratios of various baryon and isobar form factors assuming isospin or $SU(3)$-flavor symmetry for the basic wave function structure. Results for the neutral weak and charged weak form factors assuming standard $SU(2) \times U(1)$ symmetry can also be derived[61].

A useful technique for obtaining the solutions to the baryon evolution equations is to construct completely antisymmetric representations as a polynomial orthonormal basis for the distribution amplitude of multi-quark bound states. In this way one obtain a distinctive classification of nucleon (N) and Delta (Δ) wave functions and the corresponding Q^2 dependence which discriminates N and Δ form factors. More recently Braun and collaborators have shown how one can use conformal symmetry to classify the eigensolutions of the baryon distribution amplitude[23]. They identify a new

'hidden' quantum number which distinguishes components in the $\lambda = 3/2$
distribution amplitudes with different scale dependence. They are able to
find analytic solution of the evolution equation for $\lambda = 3/2$ and $\lambda = 1/2$
baryons where the two lowest anomalous dimensions for the $\lambda = 1/2$ operators (one for each parity) are separated from the rest of the spectrum
by a finite 'mass gap'. These special states can be interpreted as baryons
with scalar diquarks. Their results may support Carlson's solution[62] to the
puzzle that the proton to Δ form factor falls faster[63] than other $p \to N^*$
amplitudes if the Δ distribution amplitude has a symmetric $x_1 x_2 x_3$ form.

In a remarkable new development, Pobylitsa et al.[64] have shown how to
compute transition form factors linking the proton to nucleon-pion states
which have minimal invariant mass W. A new soft pion theorem for high
momentum transfers allows one to compute the three quark distribution
amplitudes for the near threshold pion states from a chiral rotation. The
new soft pion results are in a good agreement with the SLAC electroproduction data for $W^2 < 1.4$ GeV^2 and $7 < Q^2 < 30.7$ GeV^2.

5. Hadron Helicity Conservation

Hadron helicity conservation (HHC) is a QCD selection rule concerning
the behavior of helicity amplitudes at high momentum transfer, such as
fixed CM scattering. Since the convolution of T_H with the light-cone wavefunctions projects out states with $L_z = 0$, the leading hadron amplitudes
conserve hadron helicity[9,10]. Thus the dominant amplitudes are those in
which the sum of hadron helicities in the initial state equals the sum of
hadron helicities in the final state; other helicity amplitudes are relatively
suppressed by an inverse power in the momentum transfer.

In the case of electron-proton scattering, hadron helicity conservation
states that the proton helicity-conserving form factor (which is proportional
to G_M) dominates over the proton helicity-flip amplitude (proportional to
$G_E/\sqrt{\tau}$) at large momentum transfer. Here $\tau = Q^2/4M^2, Q^2 = -q^2$. Thus
HHC predicts $G_E(Q^2)/\sqrt{\tau}G_M(Q^2) \to 0$ at large Q^2. The new data from
Jefferson Laboratory[8] which shows a decrease in the ratio $G_E(Q^2)/G_M(Q^2)$
is not itself in disagreement with the HHC prediction.

The leading-twist QCD motivated form $Q^4 G_M(Q^2) \simeq \mathrm{const}/Q^4 \ln Q^2 \Lambda^2$
provides a good guide to both the time-like and spacelike proton form factor data at $Q^2 > 5$ GeV^2 [65]. However, the Jefferson Laboratory data[8]
appears to suggest $Q F_2(Q^2)/F_1(Q^2) \simeq \mathrm{const}$, for the ratio of the proton's Pauli and Dirac form factors in contrast to the nominal expectation
$Q^2 F_2(Q^2)/F_1(Q^2) \simeq \mathrm{const}$ expected (modulo logarithms) from PQCD. It

should however be emphasized that a PQCD-motivated fit is not precluded. For example, Hiller, Hwang and I [66] have noted that the form

$$\frac{F_2(Q^2)}{F_1(Q^2)} = \frac{\mu_A}{1 + (Q^2/c)\ln^b(1 + Q^2/a)} \tag{7}$$

with $\mu_A = 1.79$, $a = 4m_\pi^2 = 0.073$ GeV^2, $b = -0.5922$, $c = 0.9599$ GeV^2 also fits the data well. The extra logarithmic factor is not unexpected for higher twist contributions. This fit is shown in Fig. 4. The fitted form is consistent with hadron helicity conservation. The predictions for the time-like domain using simple crossing of the above form is shown by the dotted lines.

The study of time-like hadronic form factors using e^+e^- colliding beams can provide very sensitive tests of HHC, since the virtual photon in $e^+e^- \to \gamma^* \to h_A \bar{h}_B$ always has spin ± 1 along the beam axis at high energies. Angular momentum conservation implies that the virtual photon can "decay" with one of only two possible angular distributions in the center of momentum frame: $(1+\cos^2\theta)$ for $|\lambda_A - \lambda_B| = 1$ and $\sin^2\theta$ for $|\lambda_A - \lambda_B| = 0$ where λ_A and λ_B are the helicities of the outgoing hadrons. Hadronic helicity conservation, as required by QCD, greatly restricts the possibilities. It implies that $\lambda_A + \lambda_B = 0$. Consequently, angular momentum conservation requires $|\lambda_A| = |\lambda_B| = 1/2$ for baryons, and $|\lambda_A| = |\lambda_B| = 0$ for mesons; thus the angular distributions for any sets of hadron pairs are now completely determined at leading twist: $\frac{d\sigma}{d\cos\theta}(e^+e^- = B\bar{B}) \propto 1 + \cos^2\theta$ and $\frac{d\sigma}{d\cos\theta}(e^+e^- = M\bar{M}) \propto \sin^2\theta$. Verifying these angular distributions for vector mesons and other higher spin mesons and baryons would verify the vector nature of the gluon in QCD and the validity of PQCD applications to exclusive reactions.

It is usually assumed that a heavy quarkonium state such as the J/ψ always decays to light hadrons via the annihilation of its heavy quark constituents to gluons. However, as Karliner and I [67] have shown, the transition $J/\psi \to \rho\pi$ can also occur by the rearrangement of the $c\bar{c}$ from the J/ψ into the $|q\bar{q}c\bar{c}\rangle$ intrinsic charm Fock state of the ρ or π. On the other hand, the overlap rearrangement integral in the decay $\psi' \to \rho\pi$ will be suppressed since the intrinsic charm Fock state radial wavefunction of the light hadrons will evidently not have nodes in its radial wavefunction. This observation provides a natural explanation of the long-standing puzzle why the J/ψ decays prominently to two-body pseudoscalar-vector final states in conflict with HHC, whereas the ψ' does not. If the intrinsic charm explanation is correct, then this mechanism will complicate the analysis of virtually all

14

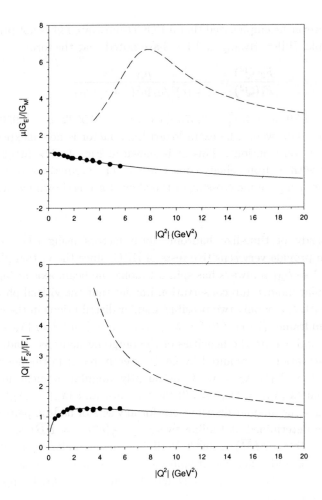

Figure 4. Perturbative QCD motivated fit[66] to the spacelike form factor ratios and G_E/G_M and QF_2/F_1. The fit is described in the text. The data are from Jefferson Laboratory[8]. Predictions for the time-like form factor ratios are shown as dotted curves.

heavy hadron decays such as $J/\psi \to p\bar{p}$. In addition, the existence of intrinsic charm Fock states, even at a few percent level, provides new, competitive decay mechanisms for B decays which are nominally CKM-suppressed[68]. For example, the weak decays of the B-meson to two-body exclusive states consisting of strange plus light hadrons, such as $B \to \pi K$, are expected to be dominated by penguin contributions since the tree-level $b \to su\bar{u}$ decay is CKM suppressed. However, higher Fock states in the B wave function containing charm quark pairs can mediate the decay via a CKM-favored $b \to sc\bar{c}$ tree-level transition. The presence of intrinsic charm in the b meson can be checked by the observation of final states containing three charmed quarks, such as $B \to J/\psi D\pi$ [69].

6. Other Applications

There are a large number of measured exclusive reactions in which the empirical power law fall-off predicted by dimensional counting and PQCD appears to be accurate over a large range of momentum transfer. The ap-

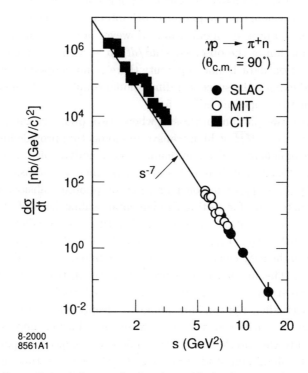

Figure 5. Comparison of photoproduction data with the dimensional counting power-law prediction. The data are summarized in Anderson *et al.*[70]

proach to scaling of $s^7 d\sigma/dt(\gamma p \to \pi^+ n)$ shown in Fig. 5 appears to indicate that leading-twist PQCD is applicable at momentum transfers exceeding a few GeV. If anything, the scaling appears to work too well, considering that one expects logarithmic deviations from the running of the QCD coupling and logarithmic evolution of the hadron distribution amplitudes. The deviations from scaling at lower energies[71] are interesting and can be attributed to s-channel resonances or perhaps heavy quark threshold effects, merging into the fixed-angle scaling in a similar way as one observes the approach to leading-twist Bjorken-scaling behavior in deep inelastic scattering via quark-hadron duality[72]. The absence of significant corrections to leading-twist scaling suggests that the running coupling is effectively frozen at the kinematics relevant to the data. If higher-twist soft processes are conspiring to mimic leading-twist scaling $s^7 d\sigma/dt(\gamma p \to \pi^+ n)$, then we would have the strange situation of seeing two separate kinematic domains of s^7 scaling of the photoproduction cross section. It has been argued[44,73] that the Compton amplitude is dominated by soft end-point contributions of the proton wavefunctions where the two photons both interact on a quark line carrying nearly all of the proton's momentum. However, a corresponding soft end-point explanation of the observed $s^7 d\sigma/dt(\gamma p \to \pi^+ n)$ scaling of the pion photoproduction data is not apparent; there is no endpoint contribution which could explain the success of dimensional counting in large-angle pion photoproduction apparent in Fig. 5.

Exclusive two-photon processes where two photons annihilate into hadron pairs $\gamma\gamma \to H\overline{H}$ at high transverse momentum provide highly valuable probes of coherent effects in quantum chromodynamics. For example, in the case of exclusive final states at high momentum transfer and fixed θ_{cm} such as $\gamma\gamma \to p\bar{p}$ or meson pairs, photon-photon collisions provide a time-like microscope for testing fundamental scaling laws of PQCD and for measuring distribution amplitudes. Counting rules predict asymptotic fall-off $s^4 d\sigma/dt \sim f(t/s)$ for meson pairs and $s^6 d\sigma/dt \sim f(t/s)$ for baryon pairs. Hadron-helicity conservation predicts dominance of final states with $\lambda_H + \lambda_{\overline{H}} = 0$. The angular dependence reflects the distribution amplitudes. One can also study $\gamma^*\gamma \to$ hadron pairs in $e^{\pm}e^-$ collisions as a function of photon virtuality, the time-like analog of deeply virtual Compton scattering which is sensitive to the two hadron distribution amplitude. One can also study the interference of the time-like Compton amplitude with the bremsstrahlung amplitude $e^{\pm}e \to BBe^{\pm}e^-$. where a time-like photon produces the pair. The $e^{\pm}$ asymmetry measures the relative phase of the time-like hadron form factor with that of the virtual Compton amplitude.

The PQCD predictions for the two-photon production of charged pions and kaons is insensitive to the shape of the meson distribution amplitudes. In fact, the ratio of the $\gamma\gamma \to \pi^+\pi^-$ and $e^+e^- \to \mu^+\mu^-$ amplitudes at large s and fixed θ_{CM} can be predicted since the ratio is nearly insensitive to the running coupling and the shape of the pion distribution amplitude:

$$\frac{\frac{d\sigma}{dt}(\gamma\gamma \to \pi^+\pi^-)}{\frac{d\sigma}{dt}(\gamma\gamma \to \mu^+\mu^-)} \sim \frac{4|F_\pi(s)|^2}{1 - \cos^2\theta_{c.m.}}. \tag{8}$$

The comparison of the PQCD prediction for the sum of $\pi^+\pi^-$ plus K^+K^- channels with recent CLEO data[76] is shown in Fig. 6. Results for separate pion and kaon channels have been given by the TPC/2γ collaboration[74].

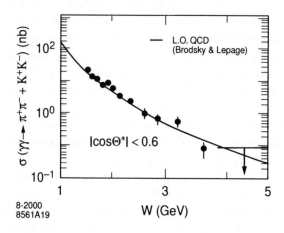

8-2000
8561A19

Figure 6.   Comparison of the sum of $\gamma\gamma \to \pi^+\pi^-$ and $\gamma\gamma \to K^+K^-$ meson pair production cross sections with the parameter-free perturbative QCD prediction of Brodsky and Lepage[75]. The data are from the CLEO collaboration[76].

The angular distribution of meson pairs is also predicted by PQCD at large momentum transfer. The CLEO data for charged pion and kaon pairs show a clear transition to the angular distribution predicted by PQCD for $W = \sqrt{s}_{\gamma\gamma} > 2$ GeV. Similarly in $\gamma\gamma \to p\bar{p}$ one can see a dramatic change in the fixed angle distribution as one enters the hard scattering domain. It is clearly important to measure the two-photon production of neutral pions and $\rho^+\rho^-$ cross sections in view of their strong sensitivity to the shape of meson distribution amplitudes. Furthermore, the ratio of $\pi^+\pi^-$ to $\pi^0\pi^0$ cross sections is highly sensitive to the production dynamics. The ratio $\frac{\sigma(\gamma\gamma \to \pi^0\pi^0)}{\sigma(\gamma\gamma \to \pi^+\pi^-)}$ at fixed angles which is very small in PQCD, and of order 1 in soft handbag models.

An interesting contribution to $K^+ p \to K^+ p$ scattering comes from the exchange of the common u quark. The quark interchange amplitude for $A + B \to C + D$ can be written as a convolution of the four light-cone wavefunctions multiplied by a factor $\Delta^- = P_A^- + P_B^- - \sum_i k_i^-$, the inverse of the central propagator[77]. The interchange amplitude is consistent with constituent counting rule scaling, and often provides a phenomenologically accurate representation of the $\theta_{c.m.}$ angular distribution at large momentum transfer. For example, the angular distribution of processes such as $K^+ p \to K^+ p$ appear to follow the predictions based on hard scattering diagrams based on quark interchange, e.g., $T_H((u_1 \bar{s})(u_2 u_3 d) \to (u_2 \bar{s})(u_1 u_3 d)$ [77]. This mechanism also provides constraints on Regge intercepts $\alpha_R(t)$ for meson exchange trajectories at large momentum transfer[78]. An extensive review of this phenomenology is given in the review by Sivers et al.[79]

One of the most interesting areas of exclusive processes is to amplitudes where the nuclear wavefunction has to absorb large momentum transfer. For example, the helicity-conserving deuteron form factor is predicted to scale as $F_d(Q^2) \propto (Q^2)^{-5}$ reflecting the minimal six quark component of nuclear wavefunction. The deuteron form factor at high Q^2 is sensitive to wavefunction configurations where all six quarks overlap within an impact separation $b_{\perp i} < \mathcal{O}(1/Q)$. The leading power-law fall off predicted by QCD is $F_d(Q^2) = f(\alpha_s(Q^2))/(Q^2)^5$, where, asymptotically[80,25], $f(\alpha_s(Q^2)) \propto \alpha_s(Q^2)^{5+2\gamma}$. In general, the six-quark wavefunction of a deuteron is a mixture of five different color-singlet states. The dominant color configuration at large distances corresponds to the usual proton-neutron bound state. However, at small impact space separation, all five Fock color-singlet components eventually acquire equal weight, i.e., the deuteron wavefunction evolves to 80% "hidden color"[25]. The relatively large normalization of the deuteron form factor observed at large Q^2 hints at sizable hidden-color contributions[81]. Hidden color components can also play a predominant role in the reaction $\gamma d \to J/\psi p n$ at threshold if it is dominated by the multi-fusion process $\gamma g g \to J/\psi$. In the case of nuclear structure functions beyond the single nucleon kinematic limit, $1 < x_{bj} < A$, the nuclear light-cone momentum must be transferred to a single quark, requiring quark-quark correlations between quarks of different nucleons in a compact, far-off-shell regime. This physics is also sensitive to the part of the nuclear wavefunction which contains hidden-color components in distinction from a convolution of separate color-singlet nucleon wavefunctions. One also sees the onset of the predicted perturbative QCD scaling behavior for exclusive nuclear amplitudes such as deuteron photodisintegration $(n = 1 + 6 + 3 + 3 = 13)$ $s^{11} \frac{d\sigma}{dt}(\gamma d \to p n) \sim$ constant at fixed CM angle.

The measured deuteron form factor and the deuteron photodisintegration cross section appear to follow the leading-twist QCD predictions at large momentum transfers in the few GeV region[82,83]. To first approximation, the proton and neutron share the deuteron's momentum equally. Since the deuteron form factor contains the probability amplitudes for the proton and neutron to scatter from $p/2$ to $p/2 + q/2$; it is natural to define the reduced deuteron form factor[80,25]

$$f_d(Q^2) \equiv \frac{F_d(Q^2)}{F_{1N}\left(\frac{Q^2}{4}\right) F_{1N}\left(\frac{Q^2}{4}\right)}. \tag{9}$$

The effect of nucleon compositeness is removed from the reduced form factor. QCD then predicts the scaling

$$f_d(Q^2) \sim \frac{1}{Q^2}; \tag{10}$$

i.e. the same scaling law as a meson form factor. This scaling is consistent with experiment for $Q \gtrsim 1$ GeV. In the case of deuteron photodisintegration $\gamma d \to pn$ the amplitude requires the scattering of each nucleon at $t_N = t_d/4$. The perturbative QCD scaling is[84]

$$\frac{d\sigma}{d\Omega_{c.m.}}(\gamma d \to np) = \frac{1}{\sqrt{s(s - M_d^2)}} \frac{F_n^2(t_d/4) F_p^2(t_d/4) f_{red}^2(\theta_{c.m})}{p_\perp^2}. \tag{11}$$

The predicted scaling of the reduced photodisintegration amplitude $f_{red}(\theta_{c.m.}) \simeq$ const is also consistent with experiment[84,82,83]. See Fig. 7.

The postulate that the QCD coupling has an infrared fixed-point provides an understanding of the applicability of conformal scaling and constituent counting rules to physical QCD processes[12,13]. The general success of dimensional counting rules implies that the effective coupling $\alpha_V(Q^*)$ controlling the gluon exchange propagators in T_H are frozen in the infrared, since the effective momentum transfers Q^* exchanged by the gluons are often a small fraction of the overall momentum transfer[34]. In this case, the pinch contributions are suppressed by a factor decreasing faster than a fixed power[12]. The effective coupling $\alpha_\tau(s)$ extracted from τ decays displays a flat behavior at low mass scales[87].

The field of analyzable exclusive processes has been expanded to a wide range of QCD processes, such as electroweak decay amplitudes, highly virtual diffractive processes such as $\gamma^* p \to \rho p$ [88,89], and semi-exclusive processes such as $\gamma^* p \to \pi^+ X$ [101] where the π^+ is produced in isolation at large p_T. An important new application is the recent analysis of hard exclusive B decays by Beneke *et al.*[2] and Keum *et al.*[3] Deeply virtual Compton amplitude $\gamma^* p \to \gamma p$ has emerged as one of the most important exclusive QCD

20

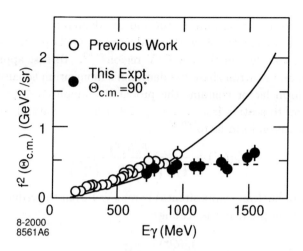

Figure 7. Comparison of deuteron photodisintegration data with the scaling prediction which requires $f^2(\theta_{cm})$ to be at most logarithmically dependent on energy at large momentum transfer. The data in are from Belz *et al.*[85] The solid curve is a nuclear physics prediction[86].

reactions[90,91,92,93]. The process factorizes into a hard amplitude representing Compton scattering on the quark times skewed parton distributions. The resulting skewed parton form factors can be represented as diagonal and off-diagonal convolutions of light-cone wavefunctions, as in semileptonic B decay[94]. New sum rules can be constructed which correspond to gravitons coupling to the quarks of the proton[90]. It is possible that the handbag approximation to DVCS may be modified by corrections to the quark propagator similar to those which appear in the final state interaction corrections to deep inelastic scattering[95,96]. In particular, one can expect that the propagator corrections will give single-spin asymmetries correlating the spin of the proton with the normal to the production plane in DVCS [97].

The hard diffraction of vector mesons $\gamma^* p \to V^0 p$ at high Q^2 and high energies for longitudinally polarized vector mesons factorizes into a skewed parton distribution times the hard scale $\gamma^* g \to g V^0$ amplitude, where the physics of the vector meson is contained in its distribution amplitude[88,21,98]. The data appears consistent with the s, t and Q^2 dependence predicted by theory. Ratios of these processes for different mesons are sensitive to the ratio of $1/x$ moments of the V^0 distribution amplitudes.

The two-photon annihilation process $\gamma^* \gamma \to$ hadrons, which is measurable in single-tagged $e^+ e^- \to e^+ e^-$ hadrons events provides a semi-local probe of even charge conjugation $C =^+$ hadron systems $\pi^0, \eta^0, \eta', \eta_c, \pi^+\pi^-$, etc. The $\gamma^*\gamma \to \pi^+\pi^-$ hadron pair process is related to virtual Compton scattering on a pion target by crossing. Hadron pair production is of particular interest since the leading-twist amplitude is sensitive to the $1/x - 1/(1 - x)$ moment of the two-pion distribution amplitude coupled to two valence quarks[21,99]. This type of measurement can also constrain the parameters of the effective chiral theory, which is needed for example to constrain the hadronic light-by-light contribution to the muon magnetic moment[100].

One can also study hard "semi-exclusive" processes[101] of the form $A + B \to C + Y$ which are characterized by a large momentum transfer between the particles A and C and a large rapidity gap between the final state particle C and the inclusive system Y. Such reactions are in effect generalizations of deep inelastic lepton scattering, providing novel currents which probe specific quark distributions of the target B at fixed momentum fraction and novel spin-dependent parton distributions.

7. Exact Formulae for Exclusive Processes

The natural formalism for describing the hadronic wavefunctions which enter exclusive and diffractive amplitudes is the light-cone Fock representation obtained by quantizing the theory at fixed "light-cone" time $\tau = t + z/c$ [102]. For example, the proton state has the Fock expansion

$$|p\rangle = \sum_n \langle n|p\rangle \, |n\rangle$$

$$= \psi_{3q/p}^{(\Lambda)}(x_i, \vec{k}_{\perp i}, \lambda_i) \, |uud\rangle \qquad (12)$$

$$+ \psi_{3qg/p}^{(\Lambda)}(x_i, \vec{k}_{\perp i}, \lambda_i) \, |uudg\rangle + \cdots$$

representing the expansion of the exact QCD eigenstate on a non-interacting quark and gluon basis. The probability amplitude for each such n-particle state of on-mass shell quarks and gluons in a hadron is given by a light-cone Fock state wavefunction $\psi_{n/H}(x_i, \vec{k}_{\perp i}, \lambda_i)$, where the constituents have longitudinal light-cone momentum fractions $x_i = k_i^+/p^+ = (k_i^0 + k_i^z)/(p^0 + p^z)$, $\sum_{i=1}^n x_i = 1$, relative transverse momentum $\vec{k}_{\perp i}$, $\sum_{i=1}^n \vec{k}_{\perp i} = \vec{0}_\perp$, and helicities λ_i. The effective lifetime of each $n-$ parton configuration in the laboratory frame is $\frac{2P_{lab}}{\mathcal{M}_n^2 - M_p^2}$ where $\mathcal{M}_n^2 = \sum_{i=1}^n (k_{\perp i}^2 + m_i^2)/x_i < \Lambda^2$ is the off-shell invariant mass and Λ is a global ultraviolet regulator. A crucial feature of the light-cone formalism is the fact that the form of the $\psi_{n/H}^{(\Lambda)}(x_i, \vec{k}_{\perp i}, \lambda_i)$ is invariant under longitudinal boosts; i.e., the light-cone wavefunctions expressed in the relative coordinates x_i and $k_{\perp i}$ are independent of the total momentum P^+, $\vec{P}_\perp$ of the hadron. Angular momentum conservation also has a precise meaning in the light-front representation: for each n–particle Fock state,

$$J^z = \sum_{i=1}^{n} S_i^z + \sum_i^{n-1} L_i^z \tag{13}$$

since there are only $n - 1$ relative orbital angular momenta[96].

Matrix elements of space-like local operators for the coupling of photons, gravitons and the deep inelastic structure functions can all be expressed as overlaps of light-cone wavefunctions with the same number of Fock constituents. This is possible since one can choose the special frame $q^+ = 0$ [27,28] for space-like momentum transfer and take matrix elements of "plus" components of currents such as J^+ and T^{++}. Since the physical vacuum in light-cone quantization coincides with the perturbative vacuum, no contributions to matrix elements from vacuum fluctuations occur[102]. In the light-cone formalism one can identify the Dirac and Pauli form factors from the light-cone spin-conserving and spin-flip vector current matrix elements of the J^+ current[29]: $\left\langle P + q, \uparrow \left| \frac{J^+(0)}{2P^+} \right| P, \uparrow \right\rangle = F_1(q^2)$, $\left\langle P + q, \uparrow \left| \frac{J^+(0)}{2P^+} \right| P, \downarrow \right\rangle = -\frac{(q^1 - iq^2)}{2M} F_2(q^2)$. More explicitly, the Pauli form factor can be calculated from the expression

$$-(q^1 - iq^2)\frac{F_2(q^2)}{2M} = \sum_a \int \frac{d^2\vec{k}_\perp dx}{16\pi^3} \sum_j e_j \, \psi_a^{\uparrow *}(x_i, \vec{k}'_{\perp i}, \lambda_i) \, \psi_a^{\downarrow}(x_i, \vec{k}_{\perp i}, \lambda_i) , \tag{14}$$

where the summation is over all contributing Fock states a and struck constituent charges e_j. The arguments of the final-state light-cone wavefunction are[27,28] $\vec{k}'_{\perp i} = \vec{k}_{\perp i} + (1 - x_i)\vec{q}_\perp$ for the struck constituent and $\vec{k}'_{\perp i} = \vec{k}_{\perp i} - x_i \vec{q}_\perp$ for each spectator. The Pauli form factor couples Fock states differing by one unit of orbital angular momentum, since the initial and final states have opposite total spin J_z and the constituent spins are unchanged in the overlap formula. In effect, one is measuring the spin-orbit $\vec{S} \cdot \vec{L}$ in the light-front formalism. Thus the F_2 form factor and the anomalous moment directly measure relative angular momentum in the proton. This is also true for the E generalized parton distribution determined in DVCS and the single-spin asymmetries measured in DIS. In these cases, the quark contributions are weighted by the square of the quark charges.

In the ultra-relativistic limit where the radius of the system is small compared to its Compton scale $1/M$, the anomalous magnetic moment must vanish[103]. The light-cone formalism is consistent with this theorem. The anomalous moment coupling $B(0)$ to a graviton vanishes for any composite system. This remarkable result, first derived by Okun and Kobzarev[104,105,106,107], follows directly from the Lorentz boost properties of the light-cone Fock representation[96].

The overlap formula for the form factors is invariant under $\vec{q}_\perp \to -\vec{q}_\perp$. Thus at large momentum transfer one obtains an expansion of form factors in powers of $1/q^2$ modulo logarithms. This also can be seen from a twist expansion of the operator product expansion[108].

Exclusive semi-leptonic B-decay amplitudes involving time-like currents such as $B \to A\ell\bar{\nu}$ can also be evaluated exactly in the light-cone framework[109,110]. In this case, the $q^+ = 0$ frame cannot be used, and the time-like decay matrix elements require the computation of both the diagonal matrix element $n \to n$ where parton number is conserved and the off-diagonal $n+1 \to n-1$ convolution such that the current operator annihilates a $q\bar{q}'$ pair in the initial B wavefunction. See Fig. 2(h). A similar result holds for the light-cone wavefunction representation of the deeply virtual Compton amplitude[94]. This feature will carry over to exclusive hadronic B-decays, such as $B^0 \to \pi^- D^+$. In this case the pion can be produced from the coalescence of a $d\bar{u}$ pair emerging from the initial higher particle number Fock wavefunction of the B. The D meson is then formed from the remaining quarks after the internal exchange of a W boson.

In principle, a precise evaluation of the hadronic matrix elements needed for B-decays and other exclusive electroweak decay amplitudes requires knowledge of all of the light-cone Fock wavefunctions of the initial and final state hadrons. In the case of model gauge theories such as QCD(1+1) [111] or collinear QCD [112] in one-space and one-time dimensions, the complete evaluation of the light-cone wavefunction is possible for each baryon or meson bound-state using the DLCQ method[113,112]. It would be interesting to use such solutions as a model for physical B-decays.

There are now real prospects of computing the hadron wavefunctions and distribution amplitudes from first principles in QCD as exemplified by the computation[114] of the pion distribution amplitude using a combination of DLCQ and the transverse lattice methods and recent results from traditional lattice gauge theory[115]. Instanton models predict a pion distribution amplitude close to the asymptotic form[116]. A new result for the proton distribution amplitude treating nucleons as chiral solitons has recently been derived by Diakonov and Petrov[117]. Dyson-Schwinger models[118] of hadronic Bethe-Salpeter wavefunctions can also be used to predict light-cone wavefunctions and hadron distribution amplitudes by integrating over the relative k^- momentum.

8. Color Transparency

Each hadron entering or emitted from a hard exclusive reaction initially emerges with high momentum and small transverse size $b_\perp = \mathcal{O}(1/\widetilde{Q})$. A fundamental feature of gauge theory is that soft gluons decouple from the small color-dipole moment of the compact fast-moving color-singlet wavefunction configurations of the incident and final-state hadrons. The transversely compact color-singlet configurations can effectively persist over a distance of order $\ell_{\mathrm{Ioffe}} = \mathcal{O}(E_{\mathrm{lab}}/Q^2)$, the Ioffe coherence length. Thus if we study hard quasi-elastic processes in a nuclear target such as $eA \to e'p'(A-1)$ or $pA \to p'(A-1)$, the outgoing and ingoing hadrons will have minimal absorption in a nucleus. The diminished absorption of hadrons produced in hard exclusive reactions implies additivity of the nuclear cross section in nucleon number A and is the theoretical basis for the "color transparency" of hard quasi-elastic reactions[24,119,120,121]. In contrast, in conventional Glauber scattering, one predicts strong, nearly energy-independent initial and final state attenuation. Similarly, in hard diffractive processes such as $\gamma^*(Q^2)p \to \rho p$ [88] only the small transverse configurations $b_\perp \sim 1/Q$ of the longitudinally polarized vector meson distribution amplitude is involved. Its hadronic interactions as it exits the nucleus will be minimal, and thus the $\gamma^*(Q^2)N \to \rho N$ reaction can occur coherently throughout a nuclear target in reactions without absorption or shadowing. Evidence for color transparency in such reactions has been reported by Fermilab experiment E665[122].

The most convincing demonstration of color transparency has been reported by the E791 group at FermiLab[123] in measurements of diffractive dissociation of a high energy pions to high transverse momentum dijets; $\pi A \to jet\ jet\ A$; the forward diffractive amplitude is observed to grow in proportion to the total number of nucleons in the nucleus, in strong contrast to standard Glauber theory which predicts that only the front surface of the nucleus should be effective.

There is also evidence for the onset of color transparency in large angle quasi-elastic pp scattering in nuclear targets[124,125,126], in the regime $6 < s < 25$ GeV^2, indicating that small wavefunction configurations are indeed controlling this exclusive reaction at moderate momentum transfers. However at $p_{\mathrm{lab}} \simeq 12$ GeV, $E_{cm} \simeq 5$ GeV, color transparency dramatically fails. It is noteworthy that in the same energy range, the normal-normal spin asymmetry A_{NN} in elastic $pp \to pp$ scattering at $\theta_{cm} = 90^0$ increases dramatically to $A_{NN} \simeq 0.6$ – it is about four times more probable that the protons scatter with helicity normal to the scattering plane than anti-normal[127].

The unusual spin and color transparency effects seen in elastic proton-proton scattering at $E_{CM} \sim 5$ GeV and large angles could be related to the charm threshold and the effects of a $|\,uuduudc\bar{c}\rangle$ resonance which would appear as in the $J = L = S = 1$ pp partial wave[128,129]. The intermediate state $|uuduudc\bar{c}\rangle$ has odd intrinsic parity and couples to the $J = S = 1$ initial state, thus strongly enhancing scattering when the incident projectile and target protons have their spins parallel and normal to the scattering plane. A similar enhancement of A_{NN} is observed at the strangeness threshold. The physical protons coupling at the charm threshold will have normal Glauber interactions, thus explaining the anomalous change in color transparency observed at the same energy in quasi-elastic pp scattering. A crucial test of the charm hypothesis is the observation of open charm production near threshold with a cross section of order of $1\mu b$ [128,129]. A similar cross section is expected for the second threshold for open charm production from $p\bar{p} \to$ charm $p\bar{p}$. An alternative explanation of the color transparency and spin anomalies in pp elastic scattering has been postulated by Ralston, Jain, and Pire[130,120]. The oscillatory effects in the large-angle $pp \to pp$ cross section and spin structure are postulated to be due to the interference of Landshoff pinch and perturbative QCD amplitudes. In the case of quasi-elastic reactions, the nuclear medium absorbs and filters out the non-compact pinch contributions, leaving the additive hard contributions unabsorbed. It is clearly important that these two alternative explanations be checked by experiment.

In general, one can expect strong effects whenever heavy quarks are produced at low relative velocity with respect to each other or the other quarks in the reaction since the QCD van der Waals interactions become maximal in this domain. The opening of the strangeness and charm threshold in intermediate states can become most apparent in large angle reactions such as pp scattering and pion photoproduction since the competing perturbative QCD amplitudes are power-suppressed. Charm and bottom production near threshold such as J/ψ photoproduction is also sensitive to the multi-quark, gluonic, and hidden-color correlations of hadronic and nuclear wavefunctions in QCD since all of the target's constituents must act coherently within the small interaction volume of the heavy quark production subprocess[131]. Although such multi-parton subprocess cross sections are suppressed by powers of $1/m_Q^2$, they have less phase-space suppression and can dominate the contributions of the leading-twist single-gluon subprocesses in the threshold regime.

9. Self-Resolved Diffractive Reactions and Light Cone Wavefunctions

Diffractive multi-jet production in heavy nuclei provides a novel way to measure the shape of the LC Fock state wavefunctions and test color transparency. For example, consider the reaction[132,133,134] $\pi A \rightarrow \mathrm{Jet}_1 + \mathrm{Jet}_2 + A'$ at high energy where the nucleus A' is left intact in its ground state. The transverse momenta of the jets balance so that $\vec{k}_{\perp i} + \vec{k}_{\perp 2} = \vec{q}_\perp < R^{-1}{}_A$. The light-cone longitudinal momentum fractions also need to add to $x_1 + x_2 \sim 1$ so that $\Delta p_L < R_A^{-1}$. The process can then occur coherently in the nucleus. Because of color transparency, the valence wavefunction of the pion with small impact separation, will penetrate the nucleus with minimal interactions, diffracting into jet pairs[132]. The $x_1 = x$, $x_2 = 1 - x$ dependence of the di-jet distributions will thus reflect the shape of the pion valence light-cone wavefunction in x; similarly, the $\vec{k}_{\perp 1} - \vec{k}_{\perp 2}$ relative transverse momenta of the jets gives key information on the derivative of the underlying shape of the valence pion wavefunction[133,134,135]. The diffractive nuclear amplitude extrapolated to $t = 0$ should be linear in nuclear number A if color transparency is correct. The integrated diffractive rate should then scale as $A^2/R_A^2 \sim A^{4/3}$ as verified by E791 for 500 GeV incident pions on nuclear targets[123]. The measured momentum fraction distribution of the jets[136] is consistent with the shape of the pion asymptotic distribution amplitude, $\phi_\pi^{\mathrm{asympt}}(x) = \sqrt{3} f_\pi x(1 - x)$. Data from CLEO[137] for the $\gamma \gamma^* \rightarrow \pi^0$ transition form factor also favor a form for the pion distribution amplitude close to the asymptotic solution[17,11] to the perturbative QCD evolution equation.

The diffractive dissociation of a hadron or nucleus can also occur via the Coulomb dissociation of a beam particle on an electron beam (e.g. at HERA or eRHIC) or on the strong Coulomb field of a heavy nucleus (e.g. at RHIC or nuclear collisions at the LHC)[135]. The amplitude for Coulomb exchange at small momentum transfer is proportional to the first derivative $\sum_i e_i \frac{\partial}{\partial k_{Ti}} \psi$ of the light-cone wavefunction, summed over the charged constituents. The Coulomb exchange reactions fall off less fast at high transverse momentum compared to pomeron exchange reactions since the light-cone wavefunction is effective differentiated twice in two-gluon exchange reactions. It is also interesting to study diffractive tri-jet production using proton beams $pA \rightarrow \mathrm{Jet}_1 + \mathrm{Jet}_2 + \mathrm{Jet}_3 + A'$ to determine the fundamental shape of the 3-quark structure of the valence light-cone wavefunction of the nucleon at small transverse separation[133].

There has been an important debate whether diffractive jet production faithfully measures the light-front wavefunctions of the projectile. Braun et *al.*[138] and Chernyak[139] have argued that one should systematically iterate the gluon exchange kernel from all sources, including final state interactions. Thus if the hard momentum exchange which produces the high transverse momentum di-jets occurs in the final state, then the x and $k_\perp$ distributions will reflect the gluon exchange kernel, not the pion's wavefunction. However, it should be noted that the measurements of pion diffraction by the E791 experiment[136] are performed on a platinum target. Only the part of the pion wavefunction with small impact separation can give the observed color transparency; *i.e.*, additivity of the amplitude on nuclear number. Thus the nucleus automatically selects events where the jets are produced at high transverse momentum in the initial state before the pion reaches the nucleus[132].

The debate[140,138,139] concerning the nature of diffractive dijet dissociation also applies to the simpler analysis of diffractive dissociation via Coulomb exchange. The one-photon exchange matrix element can be identified with the spacelike electromagnetic form factor for $\pi \to q\bar{q}$; $\langle \pi; P - q | j^+(0) | q\bar{q}; P \rangle$. Here the state $| q\bar{q} \rangle$ is the eigenstate of the QCD Hamiltonian; it is effectively an 'out' state. If we choose the $q^+ = 0$ frame where $q^2 = -\vec{q}_\perp^2$, then the form factor is exactly the overlap integral in transverse momentum of the pion and $\bar{q}q$ LCWFs summed over Fock States. The form factor vanishes at $Q^2 = 0$ because it is the matrix element of the total charge operator and the pion and jet-jet eigenstates are orthogonal. The $n = 2$ contribution to the form factor is the convolution $\psi_\pi(x, k_\perp - (1-x)q_\perp)$ with $\psi_{\bar{q}q}(x, k_\perp)$. This can be expanded at small q^2 in terms of the transverse momentum derivatives of the pion wavefunction. The final-state wavefunction represents an outgoing wave of free quarks with momentum $y, \ell_\perp$ and $1 - y, -\ell_\perp$. To first approximation the wavefunction $\psi_{\bar{q}q}(x, k_\perp)$ peaks strongly at $x = y$ and $k_\perp = \ell_\perp$. Using this approximation, the form factor at small Q^2 is proportional to the derivative of the pion light-cone wavefunction $[e_q(1-x) - e_{\bar{q}}x] \frac{\partial}{dk_\perp} \psi_\pi(x, k_\perp)$ evaluated at $x = y$ and $k_\perp = \ell_\perp$. One can also consider corrections to the final state wavefunction from gluon exchange. However, the final quarks are already moving in the correct direction at zeroth order, so these corrections would be expected to be of higher order.

10. Conclusions

Perturbative QCD provides an important guide to high momentum transfer exclusive processes. The theory involves fundamental details of hadron structure at the amplitude level. The hadron wavefunctions required for these perturbative QCD analyses are also relevant for computing exclusive heavy hadron decays.

The leading-twist contributions to exclusive amplitudes derive from the kinematic regime where the quarks and gluons propagators are evaluated in the perturbative regime. There are many successes of the perturbative approach, including important checks of color transparency and hadron helicity conservation. The successes of perturbative QCD scaling for exclusive processes at presently accessible momentum transfers can be understood if the effective QCD coupling is approximately constant at the momentum transfers scales relevant to the hard scattering amplitudes. The Sudakov suppression of the long-distance contributions is strengthened if the coupling is frozen because it involves the exponentiation of a double logarithmic series.

In this review I have argued that the new Jefferson Laboratory measurements of the ratio of proton form factors are not necessarily incompatible with the perturbative QCD predictions. I have also argued that the apparent discrepancy of theory with the normalization of the spacelike pion form factor may be due to the difficulty of extrapolating electroproduction data from the off-shell regime to the pion pole.

Further experimental studies, particularly measurements of electroproduction at Jefferson Laboratory and the study of two-photon exclusive channels at CLEO and the B-factories have the potential of providing critical information on the hadron wavefunctions as well as testing the dominant dynamical processes at short distances. Testing quantum chromodynamics to high precision in exclusive processes is not easy. Virtually all QCD processes are complicated by the presence of dynamical higher twist effects, including power-law suppressed contributions due to multi-parton correlations, intrinsic transverse momentum, and finite quark masses. Many of these effects are inherently nonperturbative in nature and require detailed knowledge of hadron wavefunctions themselves. New systematic approaches to higher twist contributions are required, such as the recent development of effective field theories[141,142].

Diffractive dijet production on nuclei has provided a compelling demonstration of color transparency and because of the color filtering effect of the nuclear target has yielded strong empirical constraints on the shape of pion distribution amplitude. I have argued that these "self-resolving" diffractive processes can also provide direct experimental information on the light-cone wavefunctions of the photon and proton in terms of their QCD degrees of freedom, as well as the composition of nuclei in terms of their nucleon and mesonic degrees of freedom.

11. Acknowledgements

I am grateful to Anatoly Radyushkin and Paul Stoler for their kind invitation to this very interesting and provocative workshop. I also thank Volodya Braun, Carl Carlson, Markus Diehl, Lyonya Frankfurt, Haiyan Gao, Susan Gardner, Gudrun Hiller, John Hiller, Paul Hoyer, Dae Sung Hwang, Peter Kroll, Jerry Miller, Stephane Peigne, Mark Strikman, and Christian Weiss for helpful conversations.

References

1. For reviews, see S. J. Brodsky and G. P. Lepage, SLAC-PUB-4947 *In *A.H. Mueller, (ed): Perturbative Quantum Chromodynamics, 1989, p. 93-240,* and S. J. Brodsky, SLAC-PUB-8649 *In *Shifman, M. (ed.): At the frontier of particle physics, vol. 2* 1343-1444.*

2. M. Beneke, G. Buchalla, M. Neubert and C. T. Sachrajda, Nucl. Phys. B **591**, 313 (2000) [arXiv:hep-ph/0006124].

3. Y. Y. Keum, H. N. Li and A. I. Sanda, Phys. Rev. D **63**, 054008 (2001) [arXiv:hep-ph/0004173].

4. A. Szczepaniak, E. M. Henley and S. J. Brodsky, Phys. Lett. B **243**, 287 (1990).

5. A review of QCD analyses of exclusive *B* decays is given in S. J. Brodsky, [arXiv:hep-ph/0104153].

6. C. K. Chua, W. S. Hou and S. Y. Tsai, [arXiv:hep-ph/0204185].

7. For a review of QCD tests in photon-photon collisions see, S. J. Brodsky, in *Proc. of the e^+e^- Physics at Intermediate Energies Conference* ed. Diego Bettoni, eConf **C010430**, W01 (2001) [arXiv:hep-ph/0106294].

8. M. K. Jones [Jefferson Lab Hall A Collaboration], Nucl. Phys. A **699**, 124 (2002).

9. S. J. Brodsky and G. P. Lepage, Phys. Rev. **D24**, 2848 (1981).

10. V. Chernyak, [arXiv:hep-ph/9906387].

11. G. P. Lepage and S. J. Brodsky, Phys. Rev. D **22**, 2157 (1980).

12. S. J. Brodsky and G. R. Farrar, Phys. Rev. D **11**, 1309 (1975).

13. V. A. Matveev, R. M. Muradian and A. N. Tavkhelidze, Lett. Nuovo Cim. **7**, 719 (1973).

14. P. V. Landshoff, Phys. Rev. D **10**, 1024 (1974).

15. S. J. Brodsky and G. P. Lepage, SLAC-PUB-2294 *Workshop on Current Topics in High Energy Physics,* Cal Tech., Pasadena, Calif., Feb 13-17, 1979.

16. G. P. Lepage and S. J. Brodsky, Phys. Rev. Lett. **43**, 545 (1979).

17. G. P. Lepage and S. J. Brodsky, Phys. Rev. Lett. **B 87**, 359 (1979).

18. A. V. Efremov and A. V. Radyushkin, Theor. Math. Phys. **42** (1980) 97;

19. J. Polchinski and M. J. Strassler, Phys. Rev. Lett. **88**, 031601 (2002) [arXiv:hep-th/0109174].

20. S. J. Brodsky, Y. Frishman, G. P. Lepage and C. Sachrajda, Phys. Lett. **91B**, 239 (1980).

21. D. Muller, Phys. Rev. D **51**, 3855 (1995) [arXiv:hep-ph/9411338].

22. P. Ball and V. M. Braun, Nucl. Phys. B **543**, 201 (1999) [arXiv:hep-ph/9810475].
23. V. M. Braun, S. E. Derkachov, G. P. Korchemsky and A. N. Manashov, Nucl. Phys. **B553**, 355 (1999) [arXiv:hep-ph/9902375].
24. S. J. Brodsky and A. H. Mueller, Phys. Lett. **B206**, 685 (1988).
25. S. J. Brodsky, C. R. Ji and G. P. Lepage, Phys. Rev. Lett. **51**, 83 (1983).
26. S. J. Brodsky and H. J. Lu, Phys. Rev. D **51**, 3652 (1995) [arXiv:hep-ph/9405218].
27. S. D. Drell and T. Yan, Phys. Rev. Lett. **24**, 181 (1970).
28. G. B. West, Phys. Rev. Lett. **24**, 1206 (1970).
29. S. J. Brodsky and S. D. Drell, Phys. Rev. D **22**, 2236 (1980).
30. V. L. Chernyak, A. R. Zhitnitsky and V. G. Serbo, JETP Lett. **26**, 594 (1977).
31. V. L. Chernyak, V. G. Serbo and A. R. Zhitnitsky, Sov. J. Nucl. Phys. **31**, 552 (1980).
32. G. R. Farrar and D. R. Jackson, Phys. Rev. Lett. **43**, 246 (1979).
33. A. Duncan and A. H. Mueller, Phys. Rev. **D21**, 1636 (1980).
34. S. J. Brodsky, C. R. Ji, A. Pang and D. G. Robertson, Phys. Rev. D **57**, 245 (1998) [arXiv:hep-ph/9705221].
35. M. Beneke, [arXiv:hep-ph/0202056].
36. D. Muller, D. Robaschik, B. Geyer, F. M. Dittes and J. Horejsi, Fortsch. Phys. **42**, 101 (1994) [arXiv:hep-ph/9812448].
37. B. Melic, B. Nizic and K. Passek, Phys. Rev. D **60**, 074004 (1999) [arXiv:hep-ph/9802204].
38. A. Szczepaniak, A. Radyushkin and C. Ji, Phys. Rev. **D57**, 2813 (1998) [arXiv:hep-ph/9708237].
39. A. S. Kronfeld and B. Nizic, Phys. Rev. **D44**, 3445 (1991).
40. P. A. Guichon and M. Vanderhaeghen, Prog. Part. Nucl. Phys. **41**, 125 (1998) [arXiv:hep-ph/9806305].
41. T. C. Brooks and L. J. Dixon, Phys. Rev. D **62**, 114021 (2000) [arXiv:hep-ph/0004143].
42. M. A. Shupe *et al.*, Phys. Rev. D **19**, 1921 (1979).
43. N. Isgur and C. H. Llewellyn Smith, Phys. Lett. **B217**, 535 (1989).
44. A. V. Radyushkin, Phys. Rev. **D58**, 114008 (1998) hep-ph/9803316.
45. J. Bolz and P. Kroll, Z. Phys. **A356**, 327 (1996) hep-ph/9603289.
46. M. Diehl, T. Feldmann, R. Jakob and P. Kroll, Eur. Phys. J. C **8**, 409 (1999) [arXiv:hep-ph/9811253].
47. H. W. Huang, P. Kroll and T. Morii, Eur. Phys. J. C **23**, 301 (2002) [arXiv:hep-ph/0110208].
48. G. P. Lepage, S. J. Brodsky, T. Huang and P. B. Mackenzie, and S. J. Brodsky, T. Huang and G. P. Lepage, *In *Banff 1981, Proceedings, Particles and Fields 2*, 143-199*.
49. H. N. Li and G. Sterman, Nucl. Phys. B **381**, 129 (1992).
50. S. J. Brodsky, F. E. Close and J. F. Gunion, Phys. Rev. **D6**, 177 (1972).
51. C. Weiss, [arXiv:hep-ph/0206295].
52. M. E. Peskin, Phys. Lett. **B88**, 128 (1979).
53. C-R Ji, A. F. Sill and R. M. Lombard-Nelsen, Phys. Rev. **D36**, 165 (1987).

54. V. L. Chernyak and I. R. Zhitnitsky, Nucl. Phys. **B246**, 52 (1984).
55. V. L. Chernyak, A. A. Ogloblin and I. R. Zhitnitsky, Z. Phys. **C42**, 583 (1989).
56. I. D. King and C. T. Sachrajda, Nucl. Phys. **B279**, 785 (1987).
57. M. Gari and N. Stefanis, Phys. Lett. **B175**, 462 (1986), M. Gari and N. Stefanis, Phys. Lett. **187B**, 401 (1987).
58. N. G. Stefanis, Eur. Phys. J. **C7**, 1 (1999) [arXiv:hep-ph/9911375].
59. S. J. Brodsky, J. R. Ellis, J. S. Hagelin and C. T. Sachrajda, Nucl. Phys. B **238**, 561 (1984).
60. Y. Kuramashi [JLQCD Collaboration], [arXiv:hep-ph/0103264].
61. S. J. Brodsky, G. P. Lepage and S. A. Zaidi, Phys. Rev. **D23**, 1152 (1981).
62. C. E. Carlson, Phys. Rev. **D34**, 2704 (1986).
63. P. Stoler, Phys. Rept. **226**, 103 (1993).
64. P. V. Pobylitsa, V. Polyakov and M. Strikman, Phys. Rev. Lett. **87**, 022001 (2001) [arXiv:hep-ph/0101279].
65. M. Ambrogiani *et al.* [E835 Collaboration], Phys. Rev. D **60**, 032002 (1999).
66. S. Brodsky, D. S. Hwang, and J. Hiller, in preparation.
67. S. J. Brodsky and M. Karliner, Phys. Rev. Lett. **78**, 4682 (1997), hep-ph/9704379.
68. S. J. Brodsky and S. Gardner, Phys. Rev. D **65**, 054016 (2002) [arXiv:hep-ph/0108121].
69. C. H. Chang and W. S. Hou, Phys. Rev. D **64**, 071501 (2001).
70. R. L. Anderson *et al.*, Phys. Rev. Lett. **30**, 627 (1973)
71. H. J. Besch, F. Krautschneider, K. P. Sternemann and W. Vollrath, Z. Phys. C **16**, 1 (1982).
72. W. Melnitchouk, Nucl. Phys. A **699**, 278 (2002) [arXiv:hep-ph/0106262].
73. M. Diehl, P. Kroll and C. Vogt, Phys. Lett. B **532**, 99 (2002) [arXiv:hep-ph/0112274].
74. J. Boyer *et al.*, Phys. Rev. Lett. **56**, 207 (1980); TPC/Two Gamma Collaboration (H. Aihara *et al.*), Phys. Rev. Lett. **57**, 404 (1986).
75. S. J. Brodsky and G. P. Lepage, Phys. Rev. **D24**, 1808 (1981).
76. H. Paar *et al.*, CLEO collaboration (to be published).
77. J. F. Gunion, S. J. Brodsky and R. Blankenbecler, Phys. Rev. **D8**, 287 (1973).
78. R. Blankenbecler, S.J. Brodsky, J. F. Gunion and R. Savit, Phys. Rev. **D8**, 4117 (1973).
79. D. Sivers, S. J. Brodsky and R. Blankenbecler, Phys. Rept. **23**, 1 (1976).
80. S. J. Brodsky and B. T. Chertok, Phys. Rev. **D14**, 3003 (1976).
81. G. R. Farrar, K. Huleihel and H. Zhang, Phys. Rev. Lett. **74**, 650 (1995).
82. R. J. Holt, Phys. Rev. **C41**, 2400 (1990).
83. C. Bochna *et al.* [E89-012 Collaboration], Phys. Rev. Lett. **81**, 4576 (1998) [arXiv:nucl-ex/9808001].
84. S. J. Brodsky and J. R. Hiller, Phys. Rev. **C28**, 475 (1983).
85. J. E. Belz *et al.*, Phys. Rev. Lett. **74**, 646 (1995).
86. T. S. Lee, CONF-8805140-10.
87. S. J. Brodsky, S. Menke, and J. Rathman (in preparation).
88. S. J. Brodsky, L. Frankfurt, J. F. Gunion, A. H. Mueller and M. Strikman, Phys. Rev. **D50**, 3134 (1994), [arXiv:hep-ph/9402283].

32

89. J. C. Collins, Phys. Rev. D **57**, 3051 (1998) [Erratum-ibid. D **61**, 019902 (2000)] [arXiv:hep-ph/9709499].

90. X. Ji, Phys. Rev. **D55**, 7114 (1997), [arXiv:hep-ph/9609381].

91. A. V. Radyushkin, Phys. Rev. **D56**, 5524 (1997) [hep-ph/9704207].

92. M. Diehl, T. Feldmann, R. Jakob and P. Kroll, Phys. Lett. **B460**, 204 (1999) [arXiv:hep-ph/9903268].

93. M. Diehl, T. Feldmann, R. Jakob and P. Kroll, Eur. Phys. J. **C8**, 409 (1999), [arXiv:hep-ph/9811253].

94. S. J. Brodsky, M. Diehl and D. S. Hwang, Nucl. Phys. B **596**, 99 (2001) [arXiv:hep-ph/0009254].

95. S. J. Brodsky, P. Hoyer, N. Marchal, S. Peigne and F. Sannino, Phys. Rev. D **65**, 114025 (2002) [arXiv:hep-ph/0104291].

96. S. J. Brodsky, D. S. Hwang, B. Q. Ma and I. Schmidt, Nucl. Phys. B **593**, 311 (2001) [arXiv:hep-th/0003082].

97. S. J. Brodsky, D. S. Hwang and I. Schmidt, Phys. Lett. B **530**, 99 (2002) [arXiv:hep-ph/0201296].

98. J. C. Collins, L. Frankfurt and M. Strikman, Phys. Rev. **D56**, 2982 (1997) [arXiv:hep-ph/9611433].

99. M. Diehl, T. Gousset, and B. Pire, [arXiv:hep-ph/0003233].

100. M. Ramsey-Musolf and M. B. Wise, [arXiv:hep-ph/0201297].

101. S. J. Brodsky, M. Diehl, P. Hoyer and S. Peigne, Phys. Lett. B **449**, 306 (1999) [arXiv:hep-ph/9812277].

102. For an extensive review and further references see S. J. Brodsky, H. Pauli and S. S. Pinsky, Phys. Rept. **301**, 299 (1998), [arXiv:hep-ph/9705477].

103. S. J. Brodsky and F. Schlumpf, Phys. Lett. **B 329** (1994) 111.

104. L. Okun and I. Yu. Kobzarev, JETP **16** 1343 (1963).

105. X. D. Ji, Phys. Rev. Lett. **78**, 610 (1997) [arXiv:hep-ph/9603249].

106. X. D. Ji, Phys. Rev. D **58**, 056003 (1998) [arXiv:hep-ph/9710290].

107. O. V. Teryaev, [arXiv:hep-ph/9904376].

108. V. M. Braun, A. Lenz, N. Mahnke and E. Stein, Phys. Rev. D **65**, 074011 (2002) [arXiv:hep-ph/0112085].

109. S. J. Brodsky and D. S. Hwang, Nucl. Phys. **B543**, 239 (1999), [arXiv:hep-ph/9806358.]

110. C. R. Ji and H. M. Choi, Fizika **B8**, 321 (1999).

111. K. Hornbostel, S. J. Brodsky, and H. C. Pauli, Phys. Rev. **D41** 3814 (1990).

112. F. Antonuccio and S. Dalley, Phys. Lett. **B348**, 55 (1995); Phys. Lett. **B376**, 154 (1996); Nucl. Phys. **B461**, 275 (1996).

113. H.-C. Pauli and S. J. Brodsky, Phys. Rev. **D32** (1985) 1993 and 2001.

114. S. Dalley, Nucl. Phys. Proc. Suppl. **90**, 227 (2000) [arXiv:hep-ph/0007081]. See also M. Burkardt and S. K. Seal, Phys. Rev. **D65**, 034501 (2002).

115. L. Del Debbio, M. Di Pierro, A. Dougall and C. Sachrajda, Nucl. Phys. Proc. Suppl. **83-84**, 235 (2000) [arXiv:hep-lat/9909147].

116. V. Y. Petrov, M. V. Polyakov, R. Ruskov, C. Weiss and K. Goeke, Phys. Rev. **D59**, 114018 (1999) [arXiv:hep-ph/9807229].

117. D. Diakonov and V. Y. Petrov, [arXiv:hep-ph/0009006].

118. M. B. Hecht, C. D. Roberts and S. M. Schmidt, [arXiv:nucl-th/0008049].

119. L. L. Frankfurt and M. I. Strikman, Phys. Rept. **160**, 235 (1988).

120. P. Jain, B. Pire and J. P. Ralston, Phys. Rept. **271**, 67 (1996) [arXiv:hep-ph/9511333].

121. For a review of the experimental status of color transparency, see K. Griffioen, these proceedings.

122. M. R. Adams *et al.* [E665 Collaboration], 470-GeV," Phys. Rev. Lett. **74**, 1525 (1995).

123. E. M. Aitala *et al.* [E791 Collaboration], Phys. Rev. Lett. **86**, 4773 (2001) [arXiv:hep-ex/0010044].

124. A. S. Carroll *et al.*, Phys. Rev. Lett. **61**, 1698 (1988).

125. Y. Mardor *et al.*, Phys. Lett. **B437**, 257 (1998) [arXiv:nucl-ex/9710002].

126. A. Leksanov *et al.*, Phys. Rev. Lett. **87**, 212301 (2001) [arXiv:hep-ex/0104039].

127. G. R. Court *et al.*, Phys. Rev. Lett. **57**, 507 (1986).

128. S. J. Brodsky and G. F. de Teramond, Phys. Rev. Lett. **60**, 1924 (1988).

129. G. F. de Teramond, R. Espinoza and M. Ortega-Rodriguez, Phys. Rev. **D58**, 034012 (1998) [arXiv:hep-ph/9708202].

130. J. P. Ralston and B. Pire, Phys. Rev. Lett. **57**, 2330 (1986).

131. S. J. Brodsky, E. Chudakov, P. Hoyer and J. M. Laget, Phys. Lett. B **498**, 23 (2001) [arXiv:hep-ph/0010343].

132. G. Bertsch, S. J. Brodsky, A. S. Goldhaber, and J. F. Gunion, Phys. Rev. Lett. **47**, 297 (1981).

133. L. Frankfurt, G. A. Miller, and M. Strikman, Phys. Lett. **B304**, 1 (1993), [arXiv:hep-ph/9305228].

134. L. Frankfurt, G. A. Miller and M. Strikman, Found. Phys. **30**, 533 (2000) [arXiv:hep-ph/9907214].

135. S. Brodsky, M. Diehl, P. Hoyer, and S. Peigne, in preparation.

136. E. M. Aitala *et al.* [E791 Collaboration], Phys. Rev. Lett. **86**, 4768 (2001) [arXiv:hep-ex/0010043].

137. J. Gronberg *et al.* [CLEO Collaboration], Phys. Rev. **D57**, 33 (1998), hep-ex/9707031.

138. V. M. Braun, D. Y. Ivanov, A. Schafer and L. Szymanowski, [arXiv:hep-ph/0204191].

139. V. L. Chernyak, [arXiv:hep-ph/0206144].

140. N. N. Nikolaev, W. Schafer and G. Schwiete, Phys. Rev. D **63**, 014020 (2001) [arXiv:hep-ph/0009038].

141. C. W. Bauer, S. Fleming, D. Pirjol and I. W. Stewart, Phys. Rev. D **63**, 114020 (2001) [arXiv:hep-ph/0011336].

142. M. Beneke, A. P. Chapovsky, M. Diehl and T. Feldmann, [arXiv:hep-ph/0206152].

THE UPGRADE OF CEBAF TO 12 GEV: PHYSICS MOTIVATIONS AND TECHNICAL ASPECTS

B. A. MECKING AND L. S. CARDMAN

Physics Division, Thomas Jefferson National Accelerator Facility,
12000 Jefferson Avenue, Newport News, Virginia 23606, U.S.A.
E-mail: mecking@jlab.org, cardman@jlab.org

The Continuous Electron Beam Accelerator Facility, CEBAF, makes use of electron and photon beams with an energy up to 6 GeV to investigate the electromagnetic structure of mesons, nucleons, and nuclei. We discuss the physics motivation for upgrading the facility to a maximum energy of 12 GeV and some of the key technical aspects of the upgrade.

1. Introduction

There are important physics questions that cannot be addressed now due to the lack of an electron accelerator with the combination of high energy, duty-cycle, and beam power required to study rare phenomena. Of particular importance are:

- clarifying the origin of quark confinement by searching for $q\bar{q}$ systems with excited flux tubes
- determining the longitudinal momentum distribution of the valence quarks in the nucleon via deep-inelastic inclusive scattering
- measuring quark–quark correlations via exclusive processes (by exploiting the recently developed formalism of the Generalized Parton Distributions).

To address these issues, plans have been developed for increasing the energy of the CEBAF accelerator to 12 GeV, and upgrading the experimental equipment to provide capabilities matched to the higher beam energy. These plans have been endorsed by the Nuclear Science Advisory Committee, NSAC.

2. The Status of CEBAF Today

The Southeastern Universities Research Association (SURA), a university consortium, operates the Thomas Jefferson National Accelerator Facility, also called Jefferson Lab (or JLab), for the U.S. Department of Energy. The main research instrument at JLab is the Continuous Electron Beam Accelerator Facility, CEBAF, a superconducting electron accelerator capable of delivering three electron beams with independent currents and independent, but correlated energies for simultaneous experiments in three halls, labeled Halls A, B, and C. The beam has a maximum energy of 5.8 GeV, a maximum power of 1 MW, and 100% duty-cycle. The halls contain complementary equipment to cover a wide range of physics problems. Hall A has two high resolution magnetic spectrometers for experiments that require the precise reconstruction of the mass of an unobserved hadronic final state. Hall B is equipped with a toroidal multi-gap spectrometer (the CEBAF Large Acceptance Spectrometer, CLAS) for the detection of several loosely correlated particles in the hadronic final state. Hall C includes two general-purpose magnetic spectrometers: the High Momentum Spectrometer (HMS) and the Short Orbit Spectrometer (SOS), as well as the infrastructure and space for mounting a variety of specialized experiments. Examples of recent experiments that also serve to highlight the capabilities of the accelerator and the present equipment in the halls are: the measurement of the ratio of the electric and magnetic form factors of the proton [1], the study of the strange quark content of the nucleon via parity violation in elastic $\vec{e} - p$ scattering [2], the separation of the three form factors of the deuteron via combining the tensor polarization of the recoiling deuteron in elastic $e - d$ scattering with the existing unpolarized measurements [3], and the measurement of the quadrupole strength in the $N \to \Delta$ transition [4,5].

3. The Physics of CEBAF at 12 GeV

The physics motivation for the 12 GeV upgrade has been described in detail in a White Paper [6]. Some important examples presented in that document are discussed in the following sections.

3.1. *Spectroscopy of Gluonic Excitations*

Exploratory numerical solutions of QCD ("lattice QCD") for distant static heavy quarks suggest [7] that confinement is caused by the formation of a string-like chromoelectric flux tube. Models for light quarks [8] predict that

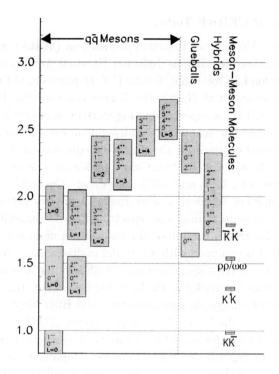

Figure 1. Level diagram showing masses (in GeV) of the conventional meson nonets and of expected glueballs and hybrids; also shown are the thresholds for molecular (2-meson) states. L refers to the angular momentum between the quarks, and each J^{PC} refers to a nonet of mesons.

a new form of hadronic matter exists in the mass region around 2 GeV in which the gluonic degree-of-freedom of a quark-antiquark system is excited. For some of these new states the vibrational quantum numbers of the flux tube, when added to those of the quarks, produce J^{PC} combinations not allowed for ordinary $q\bar{q}$ states. These unusual states (such as 0^{+-}, 1^{-+}, and 2^{+-}) are called *exotic hybrid mesons* [9]. Their masses, orderings, and decay mechanisms will provide experimental information on the mechanism that produces the flux tube.

Figure 1 shows a level diagram giving the range of masses for the conventional $q\bar{q}$ nonets, estimates of the masses of the lightest glueballs and hybrids, and thresholds for possible "molecular" meson-meson bound states.

Most of the searches for exotic mesons have been conducted with pion beams. The disadvantage of that approach is that the production of exotic combinations is suppressed because the spins of the quarks in the probe are antiparallel ($S = 0$) while the spins of the quarks in the exotic hybrid states are parallel ($S = 1$). In contrast, a photon can easily fluctuate into a vector meson with parallel quark spins. When the flux tube in this $S = 1$ system is excited, both ordinary and exotic J^{PC} are possible. This makes photon beams particularly suitable for the production of exotic hybrids [8].

Experimental prerequisites to search for exotic hybrids in the 1.5 to 3 GeV mass region are a high-flux beam of linearly polarized photons with an energy of about 9 GeV, and a large acceptance detector with hermetic coverage for multi-particle final states to facilitate the extraction of exotic waves via a partial-wave analysis.

3.2. *Momentum Distributions of the Valence Quarks*

Deep inelastic scattering (DIS) experiments have led to the experimental confirmation of the existence of quarks, and to precision tests of QCD. However, there has never been an experimental facility with sufficient luminosity to measure DIS cross sections in the kinematic regime where the three basic ("valence") quarks of the proton and neutron dominate the wave function, i.e. for values of Bjorken-$x \to 1.0$. The 12 GeV upgrade will allow us to map out the quark distribution functions in this region with high precision.

Figure 2 shows one example of a measurement that is particularly well suited for the upgrade. The neutron polarization asymmetry A_1^n is sensitive to the spin wave function of the quarks. Most dynamical models predict that in the limit where a single quark carries all of the nucleon's momentum, it will also carry all of the spin polarization ($A_1^n \to 1$ as $x \to 1$). Existing data on A_1^n end before reaching the region of valence quark dominance, and show no sign of making the predicted transition $A_1^n \to 1$.

Even in unpolarized DIS, where the available data are best, there are unresolved issues. To extract the ratio of the probability of finding a d quark vs. a u quark at high x requires combining proton and neutron measurements. However, high-x neutron information is difficult to disentangle from nuclear binding corrections. To overcome this problem a planned experiment will exploit the mirror symmetry of $A = 3$ nuclei through simultaneous measurements of the inclusive structure functions for ^{3}H and ^{3}He. In this case the differences in the nuclear corrections is estimated to be quite small, permitting the neutron-to-proton ratio (and thus the d/u ratio) to be extracted with precision.

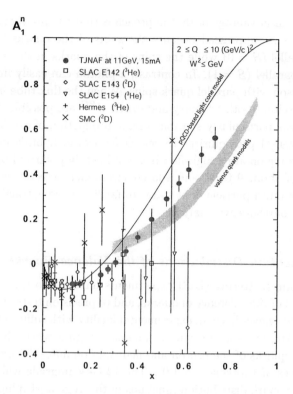

Figure 2. A projected 12 GeV measurement of the neutron polarization asymmetry A_1^n. The shaded band represents the range of predictions of valence quark models; the solid line is the prediction of a pQCD light-cone quark model.

3.3. Quark-Quark Correlations from Deep Exclusive Scattering

The information that can be obtained in DIS is limited to the square of the quark momentum wave function along the direction of the virtual photon (i.e. averaged over momenta transverse to the virtual photon direction). The new insights provided by the discovery of the Generalized Parton Distributions (GPD's) has made it possible, in principle, to map out the complete wave functions [11,12]. The GPD's are sensitive to the wave function at the amplitude level, instead of merely at the probability level, and, in particular, allow one to explore quark-quark correlations. The GPD's can be extracted from the cross sections for deep exclusive scattering (DES) processes with either a photon (Deeply Virtual Compton Scattering, DVCS)

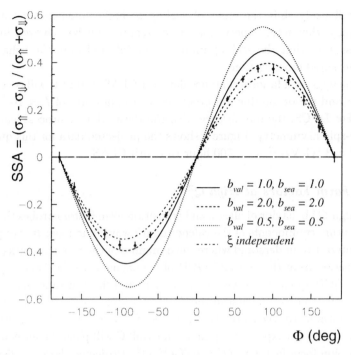

Figure 3. A projected measurement of the single-spin asymmetry for $\vec{e}p \to ep\gamma$ with longitudinally polarized 11 GeV electrons. Bins of $Q^2 = (3 \pm 0.1)$ $(\mathrm{GeV}/c)^2$, $W = (2.8 \pm 0.15)$ GeV, and $-t = (0.3 \pm 0.1)$ $(\mathrm{GeV}/c)^2$ were used. The curves indicate various GPD models, all of which are compatible with the known longitudinal parton momentum distributions.

or a meson in the final state. These processes access a rich body of new information about the full wave function, including non-forward overlaps of their longitudinal parts and their transverse momentum structure.

The necessary condition for the applicability of the GPD formalism is that the underlying processes can be factorized into a "hard" pQCD scattering amplitude and a "soft" amplitude that arises from the quark wave functions. The kinematic range over which measurements must be done to ensure that DES scaling (and thus factorization) holds must be determined experimentally.

The 12 GeV upgrade will allow DES cross sections to be investigated in the relevant kinematic regions for the first time. Scaling is expected to be reached early in the accessible regime in the case of DVCS. However, for deep virtual meson production, an additional hard collision is required;

this will likely shift the onset of scaling to higher momentum transfers. One may, therefore, expect that either scaling will be achieved in these processes, or that the scaling limits can be inferred from the behavior of the measured cross sections.

The upgraded large acceptance detector CLAS in Hall B will cover both DVCS and meson production processes at a luminosity of $L = 10^{35}$ cm^{-2} s^{-1}. For DVCS, the interference with the Bethe-Heitler process leads to a single-spin asymmetry. Figure 3 shows the projected data for this quantity at $Q^2 = 3$ (GeV/c)^2 for a 500 hour run with CLAS.

3.4. *Form Factors at High* Q^2

The high-Q^2 behavior of elastic and transition form factors probes the high-momentum components of the valence quark wave functions. It is of particular interest to understand where the dynamics of the valence quarks makes a transition from the "strong" QCD of confinement to the "weak" perturbative QCD applicable at very high energies and short distance scales. This transition should occur first in the simplest systems; in particular, the pion elastic form factor seems the best prospect for observing this transition experimentally. An experiment planned for Hall C will push the measurement of the pion form factor to $Q^2 = 6$ (GeV/c)^2. Projected data are displayed in Fig. 4.

4. Technical Aspects of the Energy Upgrade

The optimization of the physics reach, in combination with practical considerations, has led to the selection of 12 GeV as the maximum energy for the upgrade proposal [6]. A maximum energy higher than 12 GeV would make it possible to reach higher momentum transfer, Q^2, and thus make covering the deep-inelastic regime easier. However, with increasing energy the requirements on detector resolution and luminosity (to get sufficient count rate at the higher Q^2) will become increasingly difficult and costly to meet, and the cost of the accelerator upgrade will rise sharply above 11 GeV delivery to the present end stations (and above 12 GeV to the proposed new experimental hall).

The 12 GeV upgrade will require increasing the energy gain available from a single traversal of the linear accelerator pair from the present 1.2 GeV to 2.2 GeV; this will be accomplished by installing additional accelerating structures. All of the recirculating arcs will have to be upgraded to higher integral magnetic field. The highest energy of 12 GeV will be avail-

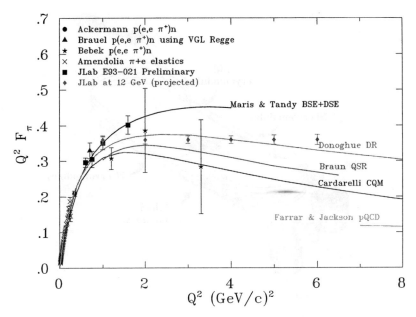

Figure 4. Projected data points (diamonds) for the pion form factor extracted from the measurement of the $ep \rightarrow e\pi^+(n)$ reaction with an 11 GeV electron beam in Hall C.

able in a new experimental area at the end of the north linac, Hall D. This area will be devoted to the photoproduction of exotic mesons using a 9 GeV linearly polarized photon beam (produced using 12 GeV electrons and the coherent bremsstrahlung technique) and a large acceptance solenoidal detector. The three existing areas will be able to use any multiple of 2.2 GeV (or a lower energy), up to 11 GeV. The layout of the facility is shown in Fig. 5.

The equipment in the present halls will be upgraded to take advantage of the higher energy. In Hall A, a broad-range spectrometer (Medium Acceptance Detector, MAD) will be added. The CLAS in Hall B will be upgraded to higher luminosity ($L = 10^{35}$ cm^{-2}s^{-1}) and improved forward coverage. In Hall C, the SOS will be replaced by a new Super High Momentum Spectrometer, SHMS, to cover small angles and high momenta.

5. Long-Range Possibilities

Looking to possibilities beyond the 12 GeV upgrade, the JLab accelerator team has started to investigate the beam dynamics and technical issues

42

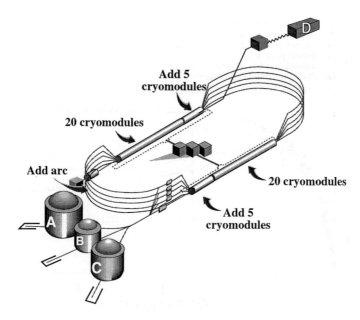

Figure 5.   Layout of the 12 GeV upgrade of CEBAF

associated with an electron-light ion collider. The energy range that has been considered is (3-5) GeV electrons colliding with (30-50) GeV light ions. Such a facility, with adequate luminosity, would permit extending DES studies of the nucleons to much lower values of Bjorken-x, permitting the study of the sea quarks as well as the valence quarks accessible via the 12 GeV upgrade.

Preliminary studies show a maximum luminosity of 6×10^{34} cm^{-2}s^{-1} can be reached by colliding an electron beam out of a linear accelerator with light ions circulating in a ring (linac-ring collider). This requires recovering the energy contained in the electron beam, a technique which has been pioneered at high beam intensities by the JLab infra-red Free Electron Laser. Electron beam cooling will be necessary to reduce the emittance of the light ion beam. It should be pointed out that a major R&D effort will be required to solve the technical difficulties associated with the challenging beam parameters for the collider. An interesting side benefit of this approach to a collider is that a high intensity, 25 GeV external electron beam could be generated by recirculating the beam five times through the 5 GeV linac pair (as in the present accelerator), permitting high-luminosity ($L \approx 10^{38}$ cm^{-2}s^{-1}) fixed-target experiments.

6. Summary

The physics motivation for the proposed 12 GeV upgrade of the Continuous Electron Beam Accelerator at JLab has been discussed. The unique combination of high energy, high duty-cycle, high intensity and small phase space beams from CEBAF's superconducting accelerator, in combination with complementary equipment in four experimental halls, will make it possible to perform experiments that will determine the internal structure of mesons, nucleons, and nuclei with unprecedented precision.

References

1. M. K. Jones *et al.*, Phys. Rev. Lett. **84**, 1398 (2000);
   O. Gayou *et al.*, Phys. Rev. Lett. **88**, 092301 (2002).
2. K. A. Aniol *et al.*, Phys. Rev. Lett. **82**, 1096 (1999).
3. D. Abbott *et al.*, Phys. Rev. Lett. **84**, 5053 (2000).
4. K. Joo *et al.*, Phys. Rev. Lett. **88**, 122001 (2002).
5. V. V. Frolov *et al.*, Phys. Rev. Lett. **82**, 45 (1999).
6. L. S. Cardman *et al.*, White Paper:
   'The Science Driving the 12 GeV Upgrade of CEBAF', 2001.
7. G. S. Bali *et al.*, Phys. Rev. D **62**, 054503 (2000).
8. N. Isgur, and J. Paton, Phys. Rev. D **31**, 2910 (1985);
   N. Isgur, R. Kokoski and J. Paton, Phys. Rev. Lett. **54**, 869 (1985).
9. T. Barnes, Ph.D. thesis, California Institute of Technology, 1977 (unpublished).
10. F. E. Close and P. Page, Nucl. Phys. **B443**, 233 (1995).
11. X. Ji, Phys. Rev. Lett. **78**, 610 (1997); Phys. Rev. D **55**, 7114 (1997).
12. A. V. Radyushkin, Phys. Lett. **B380**, 417 (1996);
    Phys. Lett. **B385**, 333 (1996); Phys. Rev. D **56**, 5524 (1997).

HIGH-T MESON PHOTO- AND ELECTROPRODUCTION: A WINDOW ON THE PARTONIC STRUCTURE OF HADRONS

J.-M. LAGET

CEA-Saclay, DAPNIA/SPhN
F91191 Gif-Sur-Yvette, France
E-mail: jlaget@cea.fr

A consistent description of exclusive photoproduction of mesons at high momentum transfer t relies on a few effective degrees of freedom which can be checked against Lattice calculation or more effective models of QCD.

1. Introduction

Exclusive photoproduction of mesons at high momentum transfer t offers us a fantastic tool to investigate the partonic structure of hadrons. The high momentum transfer t implies that the impact parameter is small enough to allow a short distance interaction between, at least, one constituent of the probe and one constituent of the target. In addition, the exclusive nature of the reaction implies that all the constituents of the probe and the target be in the small interaction volume, in order to be able to recombine into the well defined particles emitted in the final state.

This enables us to identify, to determine the role and to access the interactions between each constituent of hadrons.

The relevant constituents depend on the scale of observation. At low momentum transfer t, a comprehensive description [1,2] of available data is achieved by the exchange of a few Regge trajectories, between the probe and the target. At very high (asymptotic) momentum transfer t, the interaction should reduce to the exchange of the minimal number of gluons, in order to share the momentum transfer between all the current quarks which recombine into the initial and final hadrons [3]. Here, dimensional counting rules lead to the famous power law behavior of the various cross sections, but it is unlikely that the present generation of high luminosity facilities gives access to this regime.

To date, CEBAF at Jefferson Laboratory is the only facility which allows to reach momentum transfers t up to 6 GeV^2 in *exclusive* reactions, at reasonably high energy (s up to 12 GeV^2). This range will be considerably enlarged when the 12 GeV CEBAF upgrade becomes a reality.

The range of momentum transfers accessible at JLab corresponds to a resolving power of the order of 0.1 to 0.2 fm. It is significantly smaller than the size of a nucleon, but comparable to the correlation lengths of partons (the distance beyond which a quark or a gluon cannot propagate and hadronizes). At this scale, the relevant degrees of freedom are the *constituent partons*, whose lifetime is short enough to prevent them to interact and form the mesons which may be exchanged in the t-channel, but is too long to allow to treat them as current quarks or gluons. This is the *non perturbative partonic regime*, where the amplitude can be computed as a set of few dominant Feynman diagrams which involve dressed quarks and gluons, effective coupling constants and quark distributions [7]. This resembles the treatment of meson exchange mechanisms at low energies [4].

The first results [5,6], recently released by JLab, support such a picture and provide a link with Lattice QCD predictions on the gluon propagator [8], modelisations of quark wave functions in hadrons [9] and Regge saturating trajectories [1].

2. Gluon exchange

Measurement of phi photoproduction selects gluon exchanges. Since the ϕ meson is predominantly made of a strange $s\bar{s}$ quark pair, and to the extent that the strangeness content of the nucleon is small, quark interchange mechanisms are suppressed.

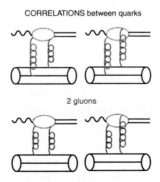

Figure 1.   The four diagrams depicting two-gluon exchange mechanisms.

The destructive interference between the two graphs in the bottom of

Fig. 1 (where each gluon couples to the same quark in the nucleon, but may couple to a different quark in the vector meson) leads to a node in the cross-section (dashed line) depicted in Fig. 2. This node is filled when the two gluons are allowed to couple also to two different quarks in the proton (faint solid line), giving access to their correlation. At the largest momentum transfer t, nucleon exchange in the u-channel takes over, but the corresponding peak moves towards higher t when the incoming photon energy increases (solid curves, marked 3.5 and 4.5 GeV), leaving more room to access the two gluon exchange contribution.

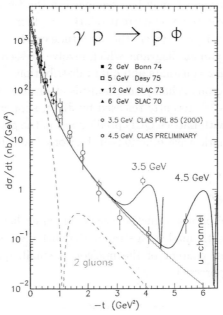

Figure 2. The ϕ meson photoproduction cross section at $< E_\gamma >= 3.5$ and 4.5 GeV.

The combined use of the high luminosity of CEBAF and of the large acceptance of the CLAS set up at JLab made possible the measurement of cross sections, more than two orders of magnitude below previous ones, in a virgin domain up to $t= 5.5$ GeV^2. The data at $E_\gamma= 3.5$ GeV have been published [5], while the data at $E_\gamma= 4.5$ GeV are still preliminary and their errors will soon decrease. This is however good enough to exhibit the trend of the cross section, to confirm the move of the u-channel peak toward higher momentum transfers and to establish the relevance of the two gluon exchange description.

The key to the success of such a good agreement is the use [7] of the correlated quark wave function of Ref. [9] and the Lattice gluon propagator of Ref. [8]. For instance, if a perturbative gluon propagator had been used, one would have obtained a much more steep t dependence of the cross-section (see Ref. [5] for more details). In the gluon loop (Fig. 1), the virtuality of each gluon is on average $t/4$, $i.e.$ about 1 GeV^2 at $t = 4$ GeV^2 where, according to Fig. 2 in Ref. [8], the lattice gluon propagator exhibits strong non perturbative corrections. On the contrary, it reaches its asymptotic behavior around a virtuality of 4 GeV^2: this requires a momentum transfer t of about 16 GeV^2, which will be achievable when CEBAF is upgraded to 12 GeV.

I refer to the talk of K. McCormick [10] for the analysis of the tensor polarisation of the ϕ which confirms the dominance of the u-channel contribution at the highest momentum transfer.

3. Quark exchange

Quark exchange mechanisms are not supressed in the photoproduction of ρ and ω mesons, which are mostly made of light quarks.

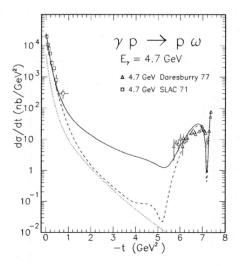

Figure 3.   The ω meson photoproduction cross section at $E_\gamma = 4.7$ GeV.

The most stricking example is the photoproduction of ω meson (Fig. 3). At low momentum transfer, π exchange (dashed line) takes over two gluon exchange (dash-dotted line). At the largest momentum transfer, u-channel proton exchange dominates and accounts for the backward angle cross section. Here the experimental node is well reproduced by the use of a non-degenerate Regge trajectory for the nucleon. At intermediate transfer, the use of a linear Regge trajectory for the π leads to a vanishing cross section, while the use of the saturated Regge trajectory (full line), which already led to a good accounting of the cross section of the $\gamma p \to \pi^+ n$ reaction [1], enhances the cross section by two orders of magnitude.

This is really a parameter free prediction, which has been beautifully confirmed by the recent CLAS data: I refer to the talk of M. Battaglieri [6], for a more detailed discussion in the ω sector as well as the ρ sector.

4. Compton scattering

A few GeV real photon has a significant hadronic component. Due to the uncertainty principle, it fluctuates into vector mesons (or quark anti-quark pairs of various flavors) over a distance which exceeds the nucleon size. For instance, a 4 GeV real photon fluctuates into a ρ meson over about 2.7 fm. The real Compton scattering cross section is therefore related to the ρ meson photoproduction cross section by a simple multiplicative factor: $4\pi\alpha_{em}/f_V^2$, where f_V is the radiative decay constant of the vector meson. As shown in Fig. 4, the comparison with this model [7] and the old Cornell data confirm this conjecture.

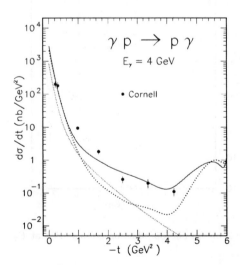

Figure 4. The Real Compton Scattering cross section at $E_\gamma = 4.0$ GeV. Dash-dotted line: two gluon exchange. Full (dotted) line: saturated (linear) Regge trajectories.

More interesting, the model predicts also spin observables. As an example, Fig. 5 compares the predicted longitudinal spin (helicity) tranfer (solid line in the upper part) to the hard scattering models [12] (solid line in the bottom part) and the soft "handbag" model [13] (dashed line). The strong variation at backward angles comes from the u-channel baryon exchange. The preliminary data from Hall A at JLab confirms this prediction. I refer to the talk of A. Nathan [14], for a more detailed account and discussion of future experimental prospects.

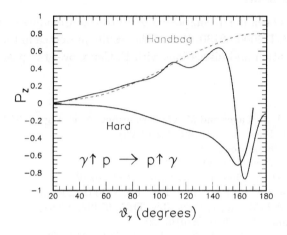

Figure 5.   The longitudinal spin transfer coefficient in Real Compton Scattering at $E_\gamma =$ 4.0 GeV.

5. Conclusion

A consistent picture is emerging from the study of exclusive photoproduction of vector mesons and Compton scattering. At low momentum transfer, it relies on diffractive scattering of the hadronic contents of the photon, in a wide energy range from threshold up to the HERA energy domain. At higher momentum transfer, it relies on a partonic description of hard scattering mechanisms which provides us with a bridge with Lattice Gauge calculations. The dressed gluon and quark propagators have already been estimated on lattice. One may expect that correlated constituent quark wave functions (at least their first moments) will soon be available from lattice. An estimate, on lattice, of the saturating part of the Regge trajectories is definitely called for.

Today, JLab is the only laboratory which allows to explore this regime, thanks to its high luminosity. Its current operation, at 4–6 GeV, has already revealed a few jewels. Its energy upgrade to 12 GeV will expand to higher values the accessible range in momentum transfer and allow to extend these studies to virtual photon induced reactions. It will permit, among others, a more comprehensive study of the onset of asymptotic hard scattering regime and of the role of correlations between quarks.

Acknowledgments

This work has been partly supported by European Commission under Contract HPRN-CT-2000-00130. Part of the results presented in this note arose from collaboration and discussions with F. Cano, over the past two years.

References

1. M. Guidal, J.-M. Laget and M. Vanderhaeghen, Nucl. Phys.**A627**, 645 (1997).
2. J.-M. Laget, Phys. Lett. **B489**, 313 (2000).
3. S. Brodsky, *these proceedings*.
4. J.-M. Laget, Phys. Rep. **69**, 1 (1981).
5. E. Anciant *et al.*, Phys. Rev. Lett. **85**, 4682 (2000).
6. M. Battaglieri *et al.*, Phys. Rev. Lett. **87**, 172002 (2001).
7. F. Cano and J.-M. Laget, Phys. Rev. **D65**, 074022 (2002).
8. A.G. Williams *et al.*, hep-ph/0107029.
9. H.W. Huang and P. Kroll, *Eur. Phys. J.* **C17**, 423 (2000).
10. K. McCormick, *these proceedings*.
11. M. Battaglieri, *these proceedings*.
12. M. Vanderhaeghen *et al.*, Nucl. Phys. **A622**, 144c (1997).
13. P. Kroll, *these proceedings*.
14. A. Nathan, *these proceedings*.

CONNECTION OF GENERALIZED PARTON DISTRIBUTIONS TO HIGH-T PROCESSES

M. VANDERHAEGHEN

Institut für Kernphysik,
Johannes Gutenberg-Universität,
D-55099 Mainz, Germany
E-mail: marcvdh@kph.uni-mainz.de

We discuss the connection of generalized parton distributions (GPDs) to processes at large momentum transfer (high-t), and discuss in particular the links between GPDs and elastic nucleon form factors. These links, in the form of sum rules, represent powerful constraints on parametrizations of GPDs. A Regge parametrization for the GPDs at small-t is extended to the large-t region and it is found to catch the basic features of proton and neutron electromagnetic form factor data.

1. Introduction

Generalized parton distributions (GPDs), are universal non-perturbative objects entering the description of hard exclusive electroproduction processes (see Refs. [1,2,3] for reviews and references). In leading twist there are four GPDs for the nucleon, i.e. H, E, $\tilde{H}$ and $\tilde{E}$, which are defined for each quark flavor (u, d, s). These GPDs depend upon the different longitudinal momentum fractions of the initial (final) quark and upon the overall momentum transfer t to the nucleon.

As the momentum fractions of initial and final quarks are different, one accesses quark momentum correlations in the nucleon. Furthermore, if one of the quark momentum fractions is negative, it represents an antiquark and consequently one may investigate $q\bar{q}$ configurations in the nucleon. Therefore, these functions contain a wealth of new nucleon structure information, generalizing the information obtained in inclusive deep inelastic scattering.

To access this information, a general parametrization for all four GPDs has been given in Ref. [3]. In this paper, based on the work of Ref. [7], we focuss on the t-dependence of the GPDs, which have attracted a considerable interest recently. In particular, it has been shown that by a Fourier transform of the t-dependence of GPDs, it is conceivable to access the distributions of parton in the transverse plane, see Refs. [4,5].

2. Nucleon electromagnetic form factors at low t

The t-dependence of the GPDs is directly related to nucleon elastic form factors (FFs) through sum rules. In particular, the nucleon Dirac and Pauli form factors $F_1(t)$ and $F_2(t)$ can be calculated from the GPDs H and E through the following sum rules for each quark flavor ($q = u, d$)

$$F_1^q(t) = \int_{-1}^{+1} dx\, H^q(x, \xi, t), \qquad F_2^q(t) = \int_{-1}^{+1} dx\, E^q(x, \xi, t) . \quad (1)$$

We can choose $\xi = 0$ in the previous equations, and model $H(x, 0, t)$ and $E(x, 0, t)$ subsequently. For the GPD $H(x, 0, t)$, a plausible ansatz at low $-t$ is a Regge form as discussed in Ref. [3]. This leads to the following integrals to calculate the Dirac FFs for u- and d-quark flavors :

$$F_1^u(t) = \int_0^{+1} dx\, u_v(x)\, \frac{1}{x^{\alpha_1' t}}, \qquad F_1^d(t) = \int_0^{+1} dx\, d_v(x)\, \frac{1}{x^{\alpha_1' t}}, \quad (2)$$

where $u_v(x)$ and $d_v(x)$ are the u- and d-quark valence distributions, and where α_1' is the slope of the leading Regge trajectory. The proton and neutron Dirac FFs then follow from

$$F_1^p(t) = e_u F_1^u(t) + e_d F_1^d(t), \qquad F_1^n(t) = e_u F_1^d(t) + e_d F_1^u(t), \quad (3)$$

where by construction $F_1^p(0) = 1$, and $F_1^n(0) = 0$.

Using the above ansatz, the Dirac mean squared radii of proton and neutron can be calculated as :

$$r_{1,p}^2 = -6\alpha_1' \int_0^{+1} dx\, \left\{ e_u\, u_v(x) + e_d\, d_v(x) \right\} \ln x , \quad (4)$$

$$r_{1,n}^2 = -6\alpha_1' \int_0^{+1} dx\, \left\{ e_u\, d_v(x) + e_d\, u_v(x) \right\} \ln x , \quad (5)$$

which yields for the electric mean squared radii of proton and neutron :

$$r_{E,p}^2 = r_{1,p}^2 + \frac{3}{2} \frac{\kappa_p}{m_N^2} , \qquad r_{E,n}^2 = r_{1,n}^2 + \frac{3}{2} \frac{\kappa_n}{m_N^2} , \quad (6)$$

where κ_p (κ_n) are the proton (neutron) anomalous magnetic moments.

In Fig. 1, we show the proton and neutron rms radii as function of the Regge slope α_1', which is the only free parameter in the ansatz of Eq. (2). One notes that the neutron rms radius is dominated by the Foldy term, which gives $r_{E,n}^2 = -0.126$ fm^2. Therefore, a relatively wide range of values α_1' are compatible with the neutron data. However for the proton, a rather narrow range of values around $\alpha_1' = 1.0 - 1.1$ GeV^{-2} are favored. Such value is close to the expectation from Regge slopes for meson trajectories, therefore supporting the ansatz of Eq. (2).

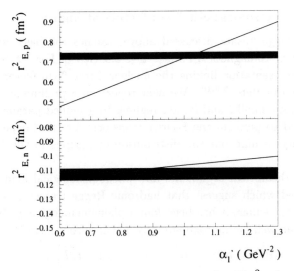

Figure 1. Proton and neutron electric mean squared radii $r_{E,p}^2$ (upper panel) and $r_{E,n}^2$ (lower panel), Eq. (6). The calculations show the dependence of the Regge ansatz according to Eqs. (4,5) on the Regge slope α_1'. For the quark distributions, the MRST01 NNLO parametrization [6] at scale $\mu^2 = 1$ GeV^2 was used in the calculations. The shaded bands correspond to the experimental values.

To calculate the electric and magnetic FFs of the nucleon, one also needs to calculate the Pauli FF F_2, besides F_1. For F_2, we use an ansatz which is based on a valence quark distribution for the valence part of $E(x, 0, t)$ entering in Eq. (1) as :

$$F_2^u(t) = \int_0^{+1} dx\, \kappa_u \frac{1}{2}\, u_v(x)\, \frac{1}{x^{\alpha_2' t}}\,, \qquad F_2^d(t) = \int_0^{+1} dx\, \kappa_d\, d_v(x)\, \frac{1}{x^{\alpha_2' t}}\,, \quad (7)$$

where κ_u and κ_d are given by $\kappa_u = 2\,\kappa_p + \kappa_n$, and $\kappa_d = \kappa_p + 2\,\kappa_n$.

In Fig. 2, we show the predictions of the above discussed Regge ansatz for the proton and neutron FFs. For both proton and neutron magnetic FFs, one sees that the Regge forms reproduce the experimentally observed dipole behavior up to about $-t = 0.5$ GeV^2. Such behavior follows in the present ansatz from the behavior of valence quark distributions at small/intermediate values of x. At larger values of $-t$, the Regge form expectedly falls short of the data as one expects a transition to the perturbative behavior ($\sim 1/t^2$) of the magnetic FFs. For G_E/G_M of the proton, one interestingly sees that the Regge form leads to a decreasing ratio with $-t$, in qualitative agreement with the data. For the neutron electric FF, one obtains a remarkable good description up to $-t \simeq 1$ GeV^2.

3. Nucleon electromagnetic form factors at high t

The simple Regge ansatz discussed above, catches the basic features of the nucleon electromagnetic FFs at $-t < 0.5$ GeV^2. For $-t > 1$ GeV^2, an overlap representation linking the nucleon Dirac form factor to GPDs has been given in Refs. [8,9,10]. We next report on a preliminary study to incorporate both small-t and large-t regimes in a unified parametrization. This is needed to perform the Fourier transform for the t-dependence of GPDs in order to map out the distribution of partons in the transverse plane.

Recently, theoretical arguments and phenomenological evidence have been presented which suggest that hadronic Regge trajectories are non-linear [11]. In particular, it has been found phenomenologically that Regge trajectories can well be approximated by a square-root form [11] :

$$\alpha(t) = \alpha(0) + \alpha' \, 2T \left[1 - \sqrt{1 - t/T} \right] , \tag{8}$$

where α' is the slope of the Regge trajectory, and where the parameter T describes the non-linearity of the trajectory. In a model calculation for a heavy quark-antiquark system through a potential [11], it was shown that the corresponding real parts of the Regge trajectories are non-linear and terminate as a consequence of the flux tube breaking due to pair creation.

It is therefore interesting to see how the original Regge ansatz of Eq. (2) with a linear Regge trajectory changes when allowing for the non-linear form. The generalization of Eq. (2) for a non-linear Regge trajectory yields :

$$F_1^u(t) = \int_0^{+1} dx u_v(x) \frac{1}{x^{\alpha_1(t) - \alpha_1(0)}} , F_1^d(t) = \int_0^{+1} dx d_v(x) \frac{1}{x^{\alpha_1(t) - \alpha_1(0)}} . \tag{9}$$

In following calculations, Eq.(9) will be evaluated with $\alpha_1(t) - \alpha_1(0)$ given by the square-root form of Eq. (8) with parameters α_1' and T_1.

At small values of $-t$, the ansatz of Eqs. (9) reduces to the linear Regge trajectory ansatz of Eq. (2). Therefore, the Dirac mean squared radii of proton and neutron for the ansatz of Eqs. (9) are exactly the same as given by Eqs. (4,5). Furthermore, one can verify that when $q_v(x) \sim (1 - x)^3$ for $x \to 1$, the square-root form of the Regge trajectory leads to a leading power behavior of $F_1(t) \sim 1/t^2$ at large $-t$, and therefore also satisfies the Drell-Yan-West relation.

To calculate F_2, we modify the Regge ansatz of Eq. (7) for a general Regge trajectory and include an additional factor $(1-x)^\eta$ in the ansatz for $E(x, 0, t)$. This leads to the following integrals for the Pauli form factors:

$$F_2^u(t) = \int_0^{+1} dx \, \kappa_u \frac{1}{N_u} (1 - x)^{\eta_u} u_v(x) \frac{1}{x^{\alpha_2(t) - \alpha_2(0)}} , \tag{10}$$

$$F_2^d(t) = \int_0^{+1} dx \, \kappa_d \, \frac{1}{N_d} \, (1-x)^{\eta_d} \, d_v(x) \frac{1}{x^{\alpha_2(t) \, - \, \alpha_2(0)}} \,, \tag{11}$$

where $\alpha_2(t) - \alpha_2(0)$ is given by the square-root form of Eq. (8) with parameters α_2' and T_2. The normalization factors N_u and N_d are fixed so as to yield as first moments κ_u and κ_d respectively. In Eqs. (10,11), the powers η_u and η_d are treated as free parameters, and determine the high-t behavior of $F_2(t)/F_1(t)$. Note that for $e(x) \sim (1-x)^5$ in the limit $x \to 1$, the square-root Regge trajectory form leads to a leading power behavior of $F_2(t) \sim 1/t^3$ at large $-t$.

Summarizing, this preliminary study shows that the phenomenological Regge parametrization (with 4 free parameters) is able to provide a quantitative description of the nucleon form factors from low-t to high-t.

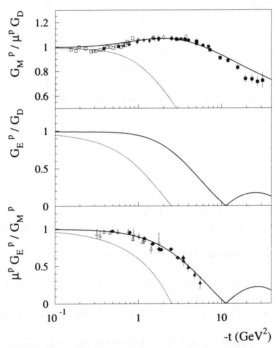

Figure 2. Proton magnetic (upper panel) and electric (middle panel) form factors compared to the dipole form $G_D(t) = 1/(1-t/0.71)^2$, as well as the ratio of both form factors (lower panel). The dotted curves correspond to the Regge ansatz of Eqs.(2) and (7) , with $\alpha_1' = 1.1$ GeV^{-2}, $\alpha_2' = 1.1$ GeV^{-2}. The solid curves correspond to the non-linear Regge ansatz, with $\alpha_1' = 0.99$ GeV^{-2}, $\alpha_2' = 0.95$ GeV^{-2}, $T_1 = 1.06$ GeV^2, $T_2 = 0.68$ GeV^2, $\eta_u = 5.39$ and $\eta_d = 2.44$. The recent data for the ratio G_E^p/G_M^p are from Refs. [12] (solid circles), [13] (open triangles), and [14] (solid triangles).

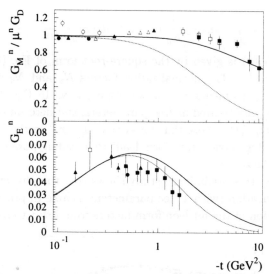

Figure 3. Neutron magnetic form factor compared to the dipole form (upper panel), and neutron electric form factor (lower panel), with curve conventions as in Fig. 2. The data for the neutron electric form factor G_E^n are from different double polarization experiments at MAMI (triangles), NIKHEF (open diamond), and JLab (solid circle).

References

1. X. Ji, J. Phys. G **24**, 1181 (1998).
2. A.V. Radyushkin, in the Boris Ioffe Festschrift 'At the Frontier of Particle Physics / Handbook of QCD', edited by M. Shifman (World Scientific, Singapore, 2001).
3. K. Goeke, M.V. Polyakov, M. Vanderhaeghen, Prog. Part. Nucl. Phys. **47**, 401 (2001).
4. M. Burkardt, Phys. Rev. D **62**, 071503 (R) (2000); hep-ph/0207047.
5. M. Diehl, hep-ph/0205208.
6. A.D. Martin, R.G. Roberts, W.J. Stirling, R.S. Thorne, Phys. Lett. B **531**, 216 (2002).
7. M. Guidal, M.V. Polyakov, and M. Vanderhaeghen, forthcoming.
8. A.V. Radyushkin, Phys. Rev. D **58**, 114008 (1998).
9. M. Diehl, T. Feldmann, R. Jakob, and P. Kroll, Eur. Phys. J. **C8**, 409 (1999).
10. P. Stoler, these proceedings.
11. M.M. Bridsudová, L. Burakovsky, and T. Goldman, Phys. Rev. D **61**, 054013 (2000).
12. M.K. Jones et al., Phys. Rev. Lett. **84**, 1398 (2000).
13. O. Gayou et al., Phys. Rev. C **64**, 038202 (2001).
14. O. Gayou et al., Phys. Rev. Lett. **88**, 092301 (2002).

EXCLUSIVE VECTOR MESON PRODUCTION AT HIGH MOMENTUM TRANSFER

M. BATTAGLIERI

(FOR THE CLAS COLLABORATION)

Istituto Nazionale di Fisica Nucleare,
Via Dodecaneso 33, 16146 Genova, Italy
E-mail: battaglieri@ge.infn.it

Exclusive reactions at large momentum transfer ($t \sim$ 3-6 GeV^2) and moderate c.m. energy ($s \sim$ 5-10 GeV^2) provide a unique tool to investigate the transition between soft and hard physics. Recent JLab data on vector meson photoproduction indicates that this regime can be successfully and consistently described by using effective partonic degrees of freedom.

1. Introduction

One of the key issues in nuclear physics is how to link the low energy and low momentum transfer regime to the high energy and large momentum transfer domain. While the former can be described by constituent quarks or phenomenological models using effective degrees of freedom, the latter uses true QCD fields in the perturbative approach. The specific kinematic region of large momentum transfer but moderate energy can help to understand the link between the the the 'soft' and the 'hard' physics. Large momentum transfer corresponds to a small impact parameter ($b \sim 1/\sqrt{-t}$): at short distances where QCD asymptotic freedom dominates, one expects the complexity of hadron scattering to be reduced to a simpler picture based on elementary partons. The phenomenology of large transverse momentum reactions in hadron scattering [1] indicates that reactions in the kinematic region of $s \sim$ 5-10 GeV^2 and $t \sim$ 3-6 GeV^2 can be described using dimensional counting rules [a] [2].

In this paper I will mainly focus on recent JLab photoproduction data. The electromagnetic interaction is weak enough to be treated perturba-

---

[a] Counting rules relate in a model independent way the large angle cross section behavior to the number of constituent partons involved in the scattering. In this sense the behavior predicted by counting rules is an indication of 'hard' scattering.

tively and exclusive processes involving one or more photons are, at leading twist, insensitive to long-distance physics revealing a precocious 'hard' behavior [3]. In particular, according to the vector dominance hypothesis, photoproduction shows common features with hadronic scattering: if the photon energy is high enough ($E_\gamma > 2$ GeV) the photon can fluctuate into a ($q\bar{q}$) state that propagates and interacts with the target nucleon. This picture is confirmed by comparing the hadronic elastic total cross sections to the photo-absorption [4]: all of them are simply explained by the exchange of the Pomeron and other few Regge trajectories (f_2 and π). The same 'universal' behavior is also evident when the differential cross sections are compared (e.g., $\pi^+ p \to \pi^+ p$ and $\gamma p \to \pi^+ n$ [5]).

It is well known that Regge theory is an economic and powerful tool to describe both exclusive and inclusive soft hadronic processes but it is a phenomenological approach not dealing with quarks and gluons. In the last years several attempts have been made to evaluate the photo production cross sections at leading twist within pQCD [6] assuming the factorization of the hard scattering amplitudes (pQCD calculable) and the distribution amplitudes (containing the non-perturbative dynamics). Only a qualitative agreement between calculations and experimental data was obtained. Soft physics corrections (e.g quark-diquark models [7]) improve the agreement but still fail to achieve a complete description in this regime within the hard-scattering approach.

An alternative approach [8,9,10], describes the transition domain by using effective partonic degrees of freedom. It uses the Regge phenomenology but interprets the Regge quanta in terms of QCD fields: the Pomeron exchange in terms of two-gluon exchange and the Regge trajectories in terms of quark exchange. Non-perturbative effects are taken into account by using dressed gluons, constituent quark propagators, and parton correlations in the nucleon wave function. It is therefore easy to relate experimental observables (total and differential cross section, polarization etc.) to fundamental and truly non-perturbative quantities like propagators or distribution amplitudes, which can be calculated using lattice-QCD. In this approach the link between the forward (t-exchange) and the backward (u-exchange) peaks is obtained by using the "saturated" Regge trajectories, i.e. t-independent at large momentum transfer [11,12,13]. Saturated trajectories lead to the asymptotic quark counting rules [2]. This approach was successfully adopted to explain the large momentum transfer $hadron - hadron$ interactions as well as several photon induced reactions. The simultaneous description of many channels has been used to fix all free parameters of the model and test the validity of this picture.

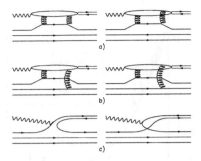

Figure 1. The Feynman diagrams corresponding to a) two-gluon exchange from a single quark, b) two-gluon exchange taking into account quark correlations in the nucleon, and c) quark exchange.

2. Vector meson photoproduction

Vector mesons (ρ, ω and ϕ) have the same quantum numbers as the photon ($J^{PC} = 1^{--}$) but different flavor composition. For all of them, the differential cross section shows two well defined dynamical regimes depending on the momentum transfer $-t$. The low $-t$ region ($-t < 1$ GeV^2) shows a diffractive behavior interpreted in the frame of the Vector Meson Dominance model as the elastic scattering of vector mesons off the proton target. At high $-t$ ($-t > 1$ GeV^2), where the cross section should be sensitive to the microscopic details of the interaction, the underlying physics can be described using parton degrees of freedom. The small impact parameter ($\approx 1/\sqrt{-t}$) resolves the two-gluon (two-quark) exchange mechanisms by preventing the the constituent gluons (quarks) of the exchange to interact and form a Pomeron (Reggeon). Moreover, small transverse sizes select configurations where each gluon couples to different quarks both in the vector meson and the nucleon, giving access to the correlation function in the proton (Fig. 1-b) [9]. Due to the dominant $s\bar{s}$ component of the ϕ, and to the extent that the strangeness of the proton is small, the ϕ photoproduction at large $-t$ is a good tool to resolve the Pomeron into its simplest 2-gluon component. In the ρ and ω cases, their light quark composition also allows valence quarks to be exchanged between the baryon and the meson states (Fig. 1-c) [9]. The comparison between the three channels puts stringent constraints on our understanding of reaction mechanisms.

3. Experimental setup and data analysis

The vector meson production data at high $-t$ were obtained with the CE-BAF Large Acceptance Spectrometer (CLAS) in the Hall B of the Thomas

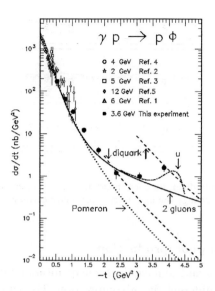

Figure 2. The differential cross section of the reaction $\gamma p \to p\phi$. See text for explanation.

Jefferson National Accelerator Facility. CLAS is a nearly 4π spectrometer based on a toroidal magnetic field generated by six superconducting coils. Three drift chambers regions allowed tracking of charged particles; time of flight scintillators were used for hadron identification. The experiment was performed using a high intensity $(6 \times 10^6 \gamma/s)$ bremsstrahlung photon beam, tagged in the E_γ range between 3 to 4 GeV.

The exclusive production of ϕ, ρ and ω were uniquely identified by direct measurements of the decay products in combination with appropriate missing-mass constraints. The reaction $\gamma p \to p\phi$ was reconstructed by measuring the proton and K^+ and determining the K^- momentum using energy and momentum conservation. The number of ϕ mesons in each t-bin was determined by subtracting the non-resonant contribution under the phi peak in the invariant (K^+K^-) mass spectrum in different ways (fitting the shape of the data and using a simple model). The reaction $\gamma p \to p\rho^0$ with $\rho^0 \to \pi^+\pi^-$ was measured by detecting all possible final-state with at least two hadrons and using the missing mass technique when necessary. For each $(E_\gamma, -t)$ bin, a simultaneous fit of both the invariant mass $(p\pi^+)$ and $(\pi^+\pi^-)$ was performed in order to separate the ρ production from the other channels $(\Delta^0\pi^+, \Delta^{++}\pi^-$, phase space, $f_2(1270))$ contributing to the $p\pi^+\pi^-$ final state. The reaction $\gamma p \to p\omega$ was detected via the main decay channel $\omega \to \pi^+\pi^-\pi^0$ using a combination of the procedures described above.

4. Results

The measured $d\sigma/dt$ for elastic ϕ photoproduction on the proton [14], is shown in Fig. 2: at low momentum transfer the two-gluon exchange model of J.M. Laget and R. Mendez-Galain [9,8] (solid line) coincides with the well known diffractive behavior (dotted line). For $-t > 2$ GeV^2 the data rule out the diffractive Pomeron exchange and strongly favor its two-gluon realization. In the calculation the gluons can couple to any quarks in the baryon (see Fig. 1-a,b) and quark correlations in the proton are taken into account assuming the simplest form of its wave function with three valence quarks equally sharing the proton longitudinal momentum. In the same plot two perturbative QCD calculations are also shown (dashed lines). Neither is able to describe the data demonstrating that the asymptotic regime is not yet reached. The ρ [15] and the preliminary ω differential cross sections are shown in Fig. 3. The parameters of the two-gluon exchange mechanism are fixed from the analysis of the ϕ. The low momentum transfer region is well reproduced when the two Regge trajectories (f_2 and σ for the ρ and f_2 and π for the ω) are included while the model underestimates the large $-t$ region, indicating that other reaction mechanisms are needed. The good agreement with data is obtained (solid line) when quark-exchange processes (see Fig. 1-c) are added trough the "saturated" trajectory technique. The same model using non-saturated trajectories is also shown (dot-dashed line). The power law s^{-C} fit to $d\sigma/dt$ at $90°$ in the center of mass was performed using CLAS and world-data. The fit yields $C = 7.9 \pm 0.3$ ($\chi^2 = 0.6$) for the ρ and $C = 7.2 \pm 0.7$ ($\chi^2 = 0.4$) for the ω channel. It is the first time that such a power law behavior has been observed in the ω channel, although it has been seen in other exclusive reactions. The quark exchange diagrams of Fig. 1-c-left (point-like interaction) and 1-c-right (hadronic component of the photon) have a s^{-7} and s^{-8} power-law behavior by dimensional counting [2]. Note that also the saturated σ and π Regge trajectories behave like s^{-8}. In addition to the differential cross section at fixed energy, the s-dependence is a strong hint of the presence of hard quark interchange mechanisms in addition to the two-gluon exchange.

5. Conclusion

Vector meson photoproduction at large momentum transfer is a unique tool to study and understand the quark-gluon structure of hadronic matter. The Jefferson Lab has completed a set of dedicated experiments with photons tagged between 4 to 5 GeV covering the $-t$ range up to 5 GeV^2. The production of ϕ, ρ and ω mesons were analysed in the same framework

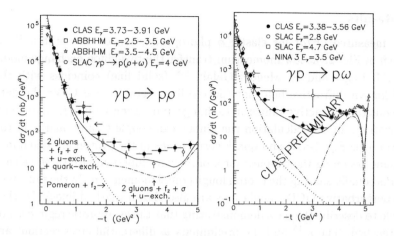

Figure 3. The differential cross section of reactions $\gamma p \to p\rho$ and $\gamma p \to p\omega$. See text for explanation.

simultaneously. At large momentum transfer the ϕ data support a model where the Pomeron is resolved into its simplest component, two gluons, which may couple to any quark in the proton and in the vector meson. The ρ and the ω data indicate that other processes, such as quark interchange, are required to fully describe these channels. In conclusion, a consistent picture is emerging from ϕ, ρ and ω photoproduction showing that this kinematical region can be described as a *non perturbative partonic regime*, where microscopic degrees of freedom (gluons and quarks) are combined with a low-energy picture of hadrons (constituent quarks, dressed gluons).

References

1. C. White *et al.*, Phys. Rev. **D49**, 58 (1994).
2. S.J. Brodsky and G.R. Farrar, Phys. Rev. Lett. **31**, 1153 (1973).
3. G.R. Farrar *et al.*, Phys. Rev. Lett. **62**, 2229 (1989).
4. A. Donnachie , P.V. Landshoff, Phys. Lett.**B296**, 227 (1992).
5. R.L. Anderson *et al.*, Phys. Rev. **D14**, 679 (1976).
6. G.R. Farrar *et al.*, *et al.*, Nucl. Phys. **B349**, 655 (1991).
7. P. Kroll *et al.*, Phys. Rev. **D55**, 4315 (1997).
8. J.M. Laget and R. Mendez-Galain, Nucl. Phys. **A581**, 397 (1995).
9. J.M. Laget, Phys. Lett. **489B**, 313 (2000).
10. F. Cano and J.M. Laget, Phys. Rev. **D65**, 074022 (2002).
11. M.N. Sergeenko, Z. Phys. **C64**, 315 (1994).
12. D.D. Coon *et al.*, Phys. Rev. **D5**, 1451 (1978).
13. M. Brisudova *et al.*, Phys. Rev. **D61**, 054013 (2000).
14. E. Anciant *et al.*, Phys. Rev. Lett. **85**, 4682 (2000).
15. M. Battaglieri *et al.*, Phys. Rev. Lett. **87**, 172002 (2001).

CHARGED PHOTOPION PRODUCTION AT JLAB

H. GAO

for the JLab Hall A Collaboration, E94-104 Collaboration and E02-010
Collaboration
Laboratory for Nuclear Science and Department of Physics,
Massachusetts Institute of Technology,
Cambridge, MA 02139, USA
E-mail: haiyan@mit.edu

The $\gamma n \to \pi^- p$ and $\gamma p \to \pi^+ n$ reactions are essential probes of the transition from meson-nucleon degrees of freedom to quark-gluon degrees of freedom in exclusive processes. The cross sections of these processes are also advantageous, for investigation of the oscillatory behavior around the quark counting prediction, since they decrease relatively slower with energy compared with other photon-induced processes. Moreover, these photoreactions in nuclei can probe the QCD nuclear filtering effects. In this talk, I discuss the preliminary results for the $\gamma n \to \pi^- p$ and $\gamma p \to \pi^+ n$ processes at a center-of-mass angle of 90° from Jefferson Lab experiment E94-104. I also discuss a new proposal in which singles $\gamma p \to \pi^+ n$ measurement from hydrogen, and coincidence $\gamma n \to \pi^- p$ measurements at the quasifree kinematics from deuterium and ^{12}C for photon energies between 2.25 GeV to 5.8 GeV in fine steps at a center-of-mass angle of 90° are planned. The proposed measurement will allow detailed investigation of the oscillatory scaling behavior in the photopion production differential cross-section and the study of the nuclear dependence of rather mysterious oscillations with energy that previous experiments have indicated.

1. Introduction

Exclusive processes are essential to studies of transitions from the non-perturbative to the perturbative regime of QCD. The differential cross-section for many exclusive reactions [1] at high energy and large momentum transfer appear to obey the quark counting rule [2,3]. Despite many successes, a model-independent test of the approach, called the hadron helicity conservation rule, tends not to agree with data in the similar energy and momentum region. In recent years, a renewed trend has been observed in deuteron photo-disintegration experiments at SLAC and JLab [4-6]. Onset of the scaling behavior has been observed [6] in deuteron photodisintegration at a surprisingly low momentum transfer of 1.0 $(GeV/c)^2$ to the nucleon involved. However, a recent polarization measurement on deuteron photo-

64

disintegration [7], carried out in Hall A at JLab, shows disagreement with hadron helicity conservation in the same kinematic region where the quark counting behavior is apparently observed.

Moreover, it is important to look closely at claims of agreement between the differential cross section data and the quark counting prediction. In fact, the re-scaled $90°$ center-of-mass pp elastic scattering data, $s^{10} \frac{d\sigma}{dt}$ show substantial oscillations about the power law behavior. With new high luminosity experimental facilities such as CEBAF, these oscillatory scaling behavior can be investigated with significantly improved precision. This will help identify the exact nature and the underlying mechanism responsible for the scaling behavior. The energy and nuclear dependence of such oscillatory behavior is also crucial in the search for signatures of the nuclear filtering effect.

2. Constituent Counting Rule and Oscillations

The constituent counting rule predicts the energy dependence of the differential cross section at a fixed center-of-mass angle for an exclusive two-body reaction at high energy and large momentum transfer as follows:

$$d\sigma/dt = h(\theta_{cm})/s^{n-2}, \tag{1}$$

where s and t are the Mandelstam variables, The quantity n is the total number of elementary fields in the initial and final states, while $h(\theta_{cm})$ depends on details of the dynamics of the process. In the case of pion photoproduction from a nucleon target, the quark counting rule predicts a $\frac{1}{s^7}$ scaling behavior for $\frac{d\sigma}{dt}$ at a fixed center-of-mass angle. The quark counting rule based on dimensional analysis [2] was also confirmed later within the framework of perturbative QCD analysis [3] up to a logarithmic factor of α_s and is believed to be valid at high energy. Although the quark counting rule agrees with data from a variety of exclusive processes, the other natural consequence of pQCD: the hadron helicity conservation selection rule, tends not to agree with data in the experimentally tested region. The same dimensional analysis which predicts the quark counting rule also predicts hadron helicity conservation for exclusive processes at high energy and large momentum transfers.

Apart from the early onset of scaling and the disagreement with hadron helicity conservation rule, several other striking phenomena have been observed in pp elastic scattering. One such phenomenon is the oscillation of the differential cross-section about the scaling behavior predicted by the quark counting rule (s^{-10} for pp scattering) [9]. Secondly, the spin correlation experiments in pp scattering [10,11] show striking behavior: at the largest momentum transfer $(p_T{}^2 = 5.09 \ (\text{GeV}/c)^2, \ \theta_{c.m.} = 90°)$, it is ~ 4 times more likely for protons to scatter when their spins are both parallel and normal to the scattering plane than when they are anti-parallel. Theoretical interpretation for such an oscillatory behavior $(s^{10}\frac{d\sigma}{dt})$ and the striking spin-correlation in pp scattering was attempted [12] within the framework of quantum chromodynamic quark and gluon interactions, where interference between hard pQCD short-distance and long-distance (Landshoff) [13] amplitudes due to soft gluon radiation was discussed. This effect is believed to be analogous to the coulomb-nuclear interference that is observed in low-energy charged-particle scattering.

Lastly, an anomalous energy dependence of nuclear transparency from quasi-elastic A(p,2p) process has been observed [14,15]: the nuclear transparency first rises followed by a decrease. Ralston and Pire [16] explained the free pp oscillatory behavior in the scaled differential cross section and the A(p,2p) nuclear transparency results using the ideas of interference between the short-distance and long-distance amplitudes and the QCD nuclear filter effect. Such an approach [17] was able to explain the pp spin correlation data.

It was previously thought that the oscillatory $s^{10}\frac{d\sigma}{dt}$ feature is unique to pp scattering or to hadron induced exclusive processes. However, it has been suggested that similar oscillations should occur in deuteron photo-disintegration [18], and photopion productions at large angles [19]. If these predictions are correct, such oscillatory behavior may be a general feature of high energy exclusive photoreactions. Thus it is very important to search experimentally for these oscillations in photoreactions. Farrar, Sterman and Zhang [20] have shown that the Landshoff contributions are suppressed at leading-order in large-angle photoproduction but they can contribute at subleading order in $\frac{1}{Q}$ as pointed out by the same authors. The most recent data on $d(\gamma,p)n$ reaction [6] show that the oscillations, if present, are very weak in this process, and the rapid drop of the cross section $(\frac{d\sigma}{dt} \propto \frac{1}{s^{11}})$ makes it impractical to investigate such oscillatory behavior. Given that the nucleon photopion production has a much larger cross-section at high energies $(\frac{d\sigma}{dt} \propto \frac{1}{s^{7}})$, it is very desirable to use these reactions to verify the existence of such oscillations.

3. Nuclear Filtering

Nuclear filtering refers to the suppression of the long distance amplitude (Landshoff amplitude) in the strongly interacting nuclear environment. This leads to suppression of the oscillation phenomena arising from interference of the long distance amplitude with the short distance amplitude. Nuclear transparency measurements of A(p,2p) process [14,15] have shown a rise in transparency for $Q^2 \approx$ 3 - 8 $(GeV/c)^2$, and a decrease in the transparency at higher momentum transfers. If the oscillatory behavior of the cross-section is suppressed in nuclei one would expect to see oscillations in the transparency, which are 180^0 out of phase with the oscillations in the free pp cross-section. This is because the transparency is formed by dividing the A(p,2p) cross-section by the pp cross-section scaled by the proton number Z of the nuclear target. Brodsky and de Teramond [21] proposed that the structure seen in $s^{10}\frac{d\sigma}{dt}(pp \rightarrow pp)$, the A_{NN} spin correlation at $\sqrt{s} \sim$ 5 GeV (around center-of-mass angle of 90°) [10,11], and the $A(p, 2p)$ transparency result can be attributed to $c\bar{c}uuduud$ resonant states. The opening of this channel gives rise to an amplitude with a phase shift similar to that predicted for gluonic radiative corrections. Thus interpretations of the elastic $pp \rightarrow pp$ cross section, the analyzing power A_{NN} and the transparency data remain controversial.

The nuclear filtering effect is complementary to color transparency (CT) effect, which refers to the suppression of final (and initial) state interactions of hadrons with the nuclear medium in exclusive processes at high momentum transfers. The expansion time relative to the time to traverse the nucleus is an essential factor for the observation of the CT effect, based on the quantum diffusion model [22]. Thus, while one expects to observe the onset of CT effect sooner in light nuclei compared to heavier nuclei [23], the large A limit provides a perturbatively calculable limit for the nuclear filtering effect. This makes ^{12}C a good choice as the target for the search of nuclear filtering and color transparency effects.

4. JLab Experiment E94-104 and Future Extension

Experiment E94-104 was proposed to carry out coincidence measurements of the exclusive $\gamma + n \rightarrow \pi^- + p$ process to investigate the onset of the constituent counting rule behavior in the unexplored region of $\sqrt{s} > 2.0$ GeV. Furthermore, a coincidence measurement of the exclusive reaction $\gamma + n \rightarrow \pi^- + p$ in ^4He target to study the nuclear transparency of this fundamental process was also proposed for the first time.

The experiment was performed in JLab Hall A using unpolarized electrons incident upon a copper radiator, generating bremsstrahlung photons. Final state protons and pions from the photopion process of interest were detected in the two high resolution spectrometers in Hall A, with the left arm configured for optimum π^+/proton separation and the right arm for optimum π^-/electron separation. While the left arm spectrometer was used for the proton detection, the right arm was employed for the π^- detection for the coincidence measurement. The photon energy was reconstructed from the measured momenta and scattering angles of the detected particles after well defined particle identification cuts. This experiment was carried out in the spring of 2001. The preliminary results [24] on the $\gamma n \to \pi^- p$ differential cross-section at the 90° center-of-mass angle suggest a general agreement with the constituent quark counting rule prediction. The data also suggest hints of oscillatory scaling behavior.

Recently a new experiment [25] was proposed to carry out a measurement of the photopion production cross-section for the fundamental $\gamma n \to \pi^- p$ process from a ^{2}H and ^{12}C target and for the $\gamma p \to \pi^+ n$ process from a hydrogen target at a center-of-mass angle of 90°, at $\sqrt{s} \sim 2.25$ GeV to 3.41 GeV in steps of approximately 0.07 GeV. The nuclear transparency for the $\gamma n \to \pi^- p$ process will be formed by taking the ratio of the production cross-section from ^{12}C to that from ^{2}H. The new experiment will make individual cross-section measurements with a 2% statistical uncertainty and point-to-point systematic uncertainties of $< 3\%$, which will allow the test of the oscillatory behavior in the scaled free cross section. The systematic uncertainties for the transparency measurement will be greatly reduced when one takes the ratio of carbon to ^{2}H. Thus, the proposed transparency measurement is expected to have combined statistical and systematic uncertainties of $< 5\%$, which should be sufficient to provide evidence for or against the nuclear filtering effect in nuclear photopion production processes.

In summary, the preliminary E94-104 results in a rather coarse step of $\sqrt{s}$, seem to suggest oscillatory behavior in $s^7 \frac{d\sigma}{dt}$. Thus, it is essential to confirm such oscillatory behavior in finer step of $\sqrt{s}$ in the $\gamma p \to \pi^+ n$ and the $\gamma n \to \pi^- p$ processes. Furthermore, a nuclear transparency measurement of the $\gamma n \to \pi^- p$ process from a ^{12}C target will allow the investigation of the nuclear filtering effect. Such an experiment is currently planned at JLab.

5. Acknowledgement

The author acknowledges stimulating discussions with S. Brodsky, P. Hoyer, P. Jain, G.A. Miller, J.P. Ralston, M. Sargsian, M. Strikman. This work is supported by the U.S. Department of Energy under contract number DE-FC02-94ER40818. The author also acknowledges the support of the Outstanding Junior Faculty Invesitgator Award in Nuclear Physics from the U.S. Department of Energy.

68

References

1. G. White *et al.*, Phys. Rev. **D49**, 58 (1994).
2. S.J. Brodsky and G.R. Farrar, Phys. Rev. Lett.**31**, 1153 (1973); Phys. Rev. D **11**, 1309 (1975); V. Matveev *et al.*, Nuovo Cimento Lett. **7**, 719 (1973).
3. G.P. Lepage, and S.J. Brodsky, Phys. Rev. D **22**, 2157 (1980).
4. J. Napolitano *et al.*, Phys. Rev. Lett. **61**, 2530 (1988); S.J. Freedman *et al.*, Phys. Rev. C **48**, 1864 (1993); J.E. Belz *et al.*, Phys. Rev. Lett. **74**, 646 (1995).
5. C. Bochna *et al.*, Phys. Rev. Lett. **81**, 4576 (1998).
6. E.C. Schulte, *et al.*, Phys. Rev. Lett. **87**, 102302 (2001).
7. K. Wijesooriya, *et al.*, JournalPhys. Rev. Lett.86, 2975 (2001).
8. Jefferson Lab Experiment E94-104, Spokespersons: H. Gao, R.J. Holt.
9. A.W. Hendry, Phys. Rev. D **10**, 2300 (1974).
10. D.G. Crabb *et al.*, Phys. Rev. Lett. **41**, 1257 (1978).
11. G.R. Court *et al.*, Phys. Rev. Lett. **57**, 507 (1986); T.S. Bhatia *et al.*, Phys. Rev. Lett. **49**, 1135 (1982); E.A. Crosbie *et al.*, Phys. Rev. D **23**, 600 (1981).
12. S.J. Brodsky, C.E. Carlson, and H. Lipkin, Phys. Rev. D **20**, 2278 (1979).
13. P. V. Landshoff, Phys. Rev. D **10**, 1024 (1974).
14. A.S. Carroll *et al.*, Phys. Rev. Lett. **61**, 1698 (1988).
15. Y. Mardor *et al.*, Phys. Rev. Lett. **81**, 5085 (1998); A. Leksanov *et al.*, Phys. Rev. Lett. **87**, 212301-1 (2001).
16. J.P. Ralston and B. Pire, Phys. Rev. Lett. **61**, 1823 (1988); J.P. Ralston and B. Pire, Phys. Rev. Lett. **65**, 2343 (1990).
17. C.E. Carlson, M. Chachkhunashvili, and F. Myhrer, Phys. Rev. D **46**, 2891 (1992).
18. L.L. Frankfurt, G.A. Miller, M.M. Sargsian, and M.I. Strikman, Phys. Rev. Lett. **84**, 3045 (2000); M.M. Sargsian, private communication.
19. P. Jain, B. Kundu, and J. Ralston, hep-ph/0005126; B. Kundu, J. Samuelsson, P. Jain and J.P. Ralston, Phys. Rev. D **62**, 113009 (2000).
20. G.R. Farrar, G. Sterman, and H. Zhang, Phys. Rev. Lett. **62**, 2229 (1989).
21. S. J. Brodsky, and G. F. de Teramond, Phys. Rev. Lett. **60**, 1924 (1988).
22. G.R. Farrar, H. Liu, L.L. Frankfurt, and M.I. Strikman, Phys. Rev. Lett. **61**, 686 (1988).
23. H. Gao, R.J. Holt, V. Pandharipande, Phys. Rev. C **54**, 2779 (1996).
24. L.Y. Zhu, private communications.
25. Jefferson Lab Proposal PR02-010, Spokespersons: D. Dutta, H. Gao and R.J. Holt.

KAON ELECTROPRODUCTION AT LARGE MOMENTUM TRANSFER

P. MARKOWITZ

For the Jefferson Lab Hall A and E98-108 Collaborations
Physics Department
Florida International University
Miami, FL USA
E-mail: markowit@fiu.edu

Exclusive H(e,e'K)Y data were taken in January, March and April of 2001 at the Jefferson Lab Hall A. The electrons and kaons were detected in coincidence in the hall's two High Resolution Spectrometers (HRS). The kaon arm of the pair had been specially outfitted with two aerogel Čerenkov threshold detectors, designed to separately provide pion and proton particle identification thus allowing kaon identification. The data show the cross section's dependence on the invariant mass, W, and 4-momentum transfer, Q^2, along with results of systematic studies. Ultimately the data will be used to perform a Rosenbluth Separation as well, separating the longitudinal from the transverse response functions. Preliminary data on this L/T ratio are presented.

1. Introduction

Exploring the electromagnetic structure of the hadronic spectrum is part of Jefferson Lab's primary mission of basic research into the nuclear building blocks. The kinematical region accessible with beam energies upto 6 GeV allows investigation of both nucleon and light meson structure.

The electromagnetic production of kaons allows measurement of the structure of mesons containing a strange quark. Jefferson Lab operates in an ideal kinematical range for such studies; Jefferson Lab is able to measure strange quark electroproduction from threshold through the deep-inelastic scattering (DIS) region.

Experiment E98-108[1] was approved to measure kaon electroproduction over a broad kinematical range. The experiment separates the longitudinal, transverse, and longitudinal-transverse interference responses to the unpolarized cross section. One goal is to obtain a data set allowing the extrapolation (in the Mandelstam variable t) of the isolated longitudinal response to the kaon mass pole. The reaction at the kaon mass pole would correspond to scattering an electron off of a free kaon and is correspondingly sensitive to the internal elecromagnetic structure of the kaon. By extrapolating the data taken to the mass pole, the goal is to constrain that kaon electromagnetic form factor.

A second goal is to examine the behavior of transverse response in this kinematical region, which overlaps with both the resonance region and extends DIS region. The kaon production reaction been calculated in terms of the hard scattering model of Brodsky and Lepage with three different baryon Distribution Amplitudes[2]. A quark calculation to leading twist using the Born approximation is also available[3]. Both calculations are for photoproduction; similar electroproduction calculations for the transverse response are not available.

A third goal is to use the longitudinal-transverse interference response function to constrain which reaction models contribute to the measurement. Once these reaction models are understood, the longitudinal response will provide sensitivity to the kaon electromagnetic form factor.

2. Present Status

Measurements of the $\gamma + p \to K^+ + Y$ and $e + p \to e' + K^+ + Y$ ($Y = \Lambda, \Sigma^0$) reactions are limited by short lifetimes ($c \cdot t_K = 370$ cm, $c \cdot t_\Lambda = 8$ cm), small production rates (an order of magnitude smaller than for pions) and high thresholds [$E_{th}(K\Lambda)= 911$ MeV, $E_{th}(K\Sigma^0)= 1.05$ GeV]. The unseparated cross sections are known with an accuracy of about 10%[11,9,12] and provide an empirical fit to the unseparated cross section based on phase space, a simple monopole Q^2-form factor, and an exponential drop with t. The past ten years have seen the first new data on electromagnetic kaon production.[4,5]

Recent data on kaon photoproduction shows a more complicated relation between the coupling of resonances and the kaon-hyperon final state system. The single polarization asymmetries are available for Λ production with errors of 25% to 50%. The photon energy range in the published data is limited to $0.9 \le E_\gamma \le 1.4$ GeV (a few additional points were measured[6,4] at a fixed momentum transfer $t = -0.147$ GeV^2 in the energy range $E_\gamma = 1.05 - 2.2$ GeV). On the theoretical side, new calculations based on Regge models now provide fits comparable to the hadrodynamic models which include higher spin resonances to fit the photoproduction data.[7,8]

However the current situation for kaon electroproduction remains less satisfactory, both from the experimental and theoretical point of view. Jefferson Lab experiment E93-108[5] was the first actual Rosenbluth separation, and demonstrated that the longitudinal response is large (approximately three-quarters the size of the transverse response) at $t = t_{min}$ or $\theta_{cm} = 0$. That experiment also demonstrated that Jefferson Lab is well suited for such precision separations, with the small emittance of the beam, the excellent particle identification of the detectors, and the accuracy of the spectrometers for cross section measurements.

In kaon electroproduction, the mass pole is in the unphysical region where $t > 0$. Instead, the electroproduction process attempts to isolate the t-channel by extrapolating the longitudinal response in t to the unphysical kaon mass pole. Such an extrapolation is model dependent meaning that it is desirable to check the extrapolation against available data on the kaon form factor at lower Q^2.

The only unambigous data on the kaon form factor is from high-energy scattering of kaon beams from atomic electrons.[13] Due to the inverse kinematics, the measurements were limited to low 4-momentum transfers $(.02 < Q^2 < .12$ (Gev/c)$^2)$. The measurements were able to determine the kaon charge radius by looking at the slope of the cross section with respect to Q^2. [Interestingly, within the next year the addition of the septum magnets to Hall A would let Hall A do a comparison of electroproduction with actual meson-electron elastic scattering in about 3 days for the kaon, plus about the same for the pion.]

3. Experimental Setup

The experiment took place in Hall A at the Thomas Jefferson National Accelerator Facility's CEBAF accelerator. The standard equipment in Hall A has been described elsewhere.[14] Electron beams of energies upto 5.7 GeV were incident on liquid hydrogen targets of nominal 4 cm and 15 cm lengths. Electrons and kaons were detected in coincidence in the two magnetically symmetric high resolution spectrometers (HRS).

The HRS spectrometers were outfitted with special particle identification: on the electron side there were a gas Čerenkov counter and a lead glass calorimeter to veto π^- mesons, while on the hadron side two different aerogel Čerenkov counters were used to separately veto pions and protons.

The pions were vetoed in the A1 aerogel which has a refractive index of 1.015 by requiring that the detector A1 not fire. Protons were vetoed in the A2 aerogel which has a refractive index of 1.055 by requiring that the detector did fire.

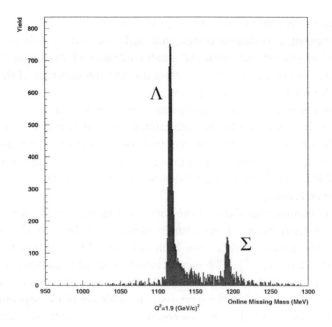

Figure 1. The Λ-hyperon reconstructed from the (H(e,e′ K)Y missing mass.

4. Preliminary Results

Figure 1 shows the reconstructed mass of the unobserved baryon in the H(e,e′ K) reaction, in this case either a Λ or Σ hyperon. The ratio of these two production cross sections is observed to drop rapidly with Q^2, however because the phase space acceptance for the two reactions also changes rapidly, final results are dependent on the ongoing acceptance studies. The quality of the particle identification can be judged by the lack of background in the plot.

The range in W covered by the experiment does not show any striking behavior; the unseparated cross sections smoothly follows the nearly flat reaction phase space as shown by the photoproduction data. Similarly, the unseparated cross sections drop off approximately as $1/(Q^2 + 2.67)^2$ as given by the E93-108 data. Here minor discrepancies to the global fit should be expected since the data were all taken at different ε-values (the virtual photon polarization).

Figure 2 shows the very preliminary separations for the data at $Q^2 = 2.4$ $(GeV/c)^2$ and $W = 1.8, 1.9$ GeV plotted versus ε, the polarization of the

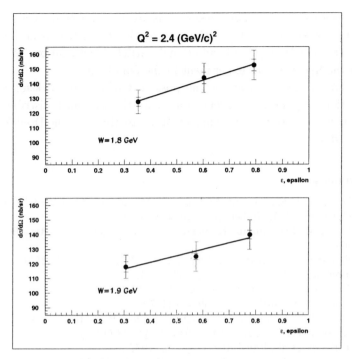

Figure 2. The cross sections at $Q^2 = 2.4$ (GeV/c)^2 and $W = 1.8, 1.9$ GeV plotted versus ε.

virtual photon. A straight line fit to those data will provide the separated longitudinal and transverse contributions. The inner errror bars are the statistical errors while the outer error bars reflect a conservative estimate of what the systematic uncertainties are at this point in the analysis. The ultimate error bars will be dominated by the statistics. As can be seen from the plot, at the larger momentum transfers of this experiment (as compared to E93-018), the longitudinal cross section is decreasing relative to the transverse but is still appreciable.

5. Outlook

The data analysis is continuing. The preliminary results indicate that the quality of the data is excellent and show that the longitudinal response decreases faster than the transverse with increasing Q^2. Additonal results are expected in the coming months.

The collaboration is finishing construction of two septum magnets which will be used with nuclear targets for hypernuclear production experiments. The resolution is critical for hypernuclear measurements making the Hall

A HRS a good match. The backgrounds are also expected to rise in these experiments, and the collaboration, led by the Roma'/INFN group is constructing a RICH to provide additional particle identification. We expect to run the hypernuclear experiment in the fall of 2002.

At higher energies there is the possiblity of looking at semi-inclusive production off polarized targets to measure the sea quark distributions as well; such data would be complementary to the existing Drell-Yan and Hermes data.

References

1. *Jefferson Lab Experiment E98-108*, P. Markowitz, M. Iodice, S. Frullani, C. C. Chang, O. K. Baker spokespersons (1998).
2. P. Kroll, M. Schurmann, K. Passek, W. Schweiger, Phys. Rev. D 55 4315 (1997).
3. G. R. Farrar, K. Huleihel, H. Zhang, Nucl. Phys. B349, 655 (1991).
4. M. Bockhurst *etal.*, Z. Phys. C 63 37 (1994).
5. G. Niculescu, *etal.*, Phys. Rev. Lett. 81, 1805 (1998).
6. P. Feller, *et.al.*, Nucl. Phys. B39 413 (1979).
7. J.C. David, C. Fayard, G.H. Lamot, B. Saghai, Phys. Rev. C 53 2613 (1996).
8. R. A. Adelseck, B. Saghai, Phys. Rev. C 42 108 (1990).
9. C. J. Bebek *et al.*, Phys. Rev. D 15, 47 (1977).
10. T. Azemoon *et al.*, Advance in Nucl. Phys. 17, 47 (1987).
11. P. Brauel *et al.*, Z. Physik, 3, 101 (1979).
12. A. Bodek *et al.*, Phys. Lett. 51B, 417 (1974).
13. S. R. Amendolia *et.al.*, Phys. Lett. B 178, 435 (1986).
14. The Jefferson Lab Hall A Collaboration, Paper in progress, to be submitted to Nuclear Methods and Instrumentation.

TENSOR POLARIZATION OF THE PHI
PHOTOPRODUCED AT HIGH-T

K. MCCORMICK

Rutgers, The State University of New Jersey
Piscataway, NJ 08855, USA
E-mail: mccormic@jlab.org

for The CLAS Collaboration

As part of a measurement [1] of the cross section of phi photoproduction to high momentum transfer, we measured the polar angular decay distribution of the outgoing K^+ in the channel $\phi \to K^+K^-$ in the phi center of mass frame (the helicity frame). If the phi is produced with the same polarization as the real photon, then this distribution should have a $\sin^2 \theta_{CM}$ behavior, which indicates that S-channel helicity conservation (SCHC) holds. We find that SCHC is respected up to $-t \sim 2.7$ GeV^2. Above this momentum threshold, u-channel production of a phi meson dominates and induces a violation of SCHC.

1. Introduction

The decay distribution of the vector meson will be presented in its helicity system: the z-direction is chosen opposite to the direction of the outgoing nucleon in the vector meson rest frame, as illustrated in Fig. 1. The decay angles θ and ϕ are the polar and azimuthal angles, respectively, of one of the decay particles, in our case the K^+.

The decay angular distribution is given by [2]

$$\frac{dN}{d\cos\theta d\phi} = W^0(\cos\theta, \phi) + \sum_{\alpha=1}^{3} P_\gamma^\alpha W^\alpha(\cos\theta, \phi). \tag{1}$$

For unpolarized photons, only the first term survives. Assuming pure phi production and integrating over the azimuthal angle phi, the decay angular distribution can then be written in terms of three variables:

$$\frac{dN}{d\theta} = \frac{3}{4}\left[(1-\rho_{00}^0)\sin^3\theta + 2\rho_{00}^0\cos^2\theta\sin\theta\right] + \alpha\cos\theta\sin\theta + \frac{1}{2}\kappa\sin\theta. \tag{2}$$

The first is the spin density matrix element ρ_{00}^0, which describes the probability that a longitudinally polarized phi is produced by a transversely

76

polarized photon. If SCHC holds then this term is zero and there is no contribution from longitudinally polarized phi's. The second, α, describes the interference between any longitudinal phi's and the s-wave K^+K^- continuum and the third, κ, describes the s-wave continuum:

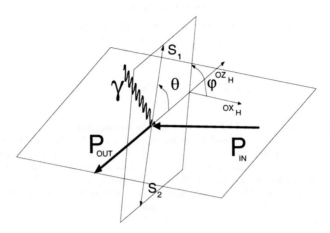

Figure 1. Schematic diagram of the kinematics of the $\gamma p \rightarrow p\phi$ reaction in the helicity frame (the center of mass frame of the phi). The angles θ and ϕ are respectively the polar and azimuthal decay angles of the K^+ from the decay of the ϕ into K^+K^-.

As described in the following sections, the angular decay distribution of the K^+ in the phi center of mass frame was measured for eight bins in t. The resulting distributions were fit with Eq. 2 to extract values of ρ^0_{00} and α (κ is fixed from the data, as described in Sec. 3).

2. Experimental Setup

Measurements of phi photoproduction to $-t = 4$ GeV^2 were made possible by the high duty factor of the Jefferson Lab electron beam and the large acceptance of the CLAS spectrometer. For this measurement, a 4.1 GeV electron beam was incident on a gold radiator of 10^{-4} radiation lengths, producing a Bremstrahlung photon beam which was tagged in the energy range of 3.3 – 3.9 GeV. A photon tagging system [3] detected the scattered electrons, with a resolution of 0.1% of the incident beam energy. The photon beam was then incident on a liquid hydrogen target, a mylar cylinder 6 cm in diameter and 18 cm long, which was maintained at 20.4 K. The photon flux was determined with a pair spectrometer located downstream of the target, which was calibrated via comparison to a total absorption counter.

The hadrons were detected in CLAS, the CEBAF large acceptance spectrometer [4]. The toroidal field of CLAS is generated with a six-coil super-conducting magnet, effectively leading to six independent spectrometers capable of measuring particles with polar angles between 10° and 140° in a large percentage of the phi acceptance. Particle momentum is determined via the path through the drift chambers, plus each sector is completely covered by scintillators, allowing particle separation via time-of-flight techniques. The phi decay into a K^+K^- pair was identified through the missing mass in the reaction $\gamma p \to pK^+(X)$. This technique is preferable to measuring the K^- directly, since because of the magnetic field configuration, negative particles are deflected into the inert forward region of CLAS where they are lost.

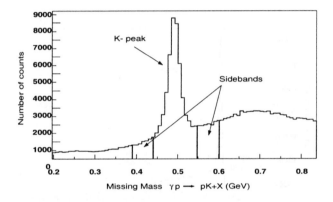

Figure 2.   Missing mass in the reaction $\gamma p \to pK^+(X)$.

Figure 2 shows the missing mass in the reaction $\gamma p \to pK^+(X)$. A well-defined K^- peak can be seen above a background which corresponds to a combination of misidentified channels, the contributions of multiparticle channels and accidentals between CLAS and the tagger. The multiparticle background is thought to come mostly from multipion channels such as $\gamma p \to p\pi^+(\pi^-\pi^0)$. For these types of channels, if the π^+ is misidentified as a K^+, then the missing mass of the remaining two pion system is close to that of a K^-, leading to a background event.

For each event, the invariant mass is then calculated (see Fig. 3). For the events whose invariant masses fall within the phi mass peak (1–1.050 GeV), the angle of the K^+ in the phi center of mass frame is calculated. As noted above, this angle provides information on SCHC. The events that come from the background under the K^- peak must still be subtracted from the angular decay distribution. This is done by calculating the invariant mass for each of the events in the sidebands (upper and lower, respectively), assuming that the mass of the missing particle is the average mass of the sideband under consideration. The thresholds for the different reactions being considered, $\gamma p \to pK^+K^-$ and $\gamma p \to pK^+X$, vary according to the mass, as can be seen in Fig. 3 from the invariant mass distributions of the two sidebands. To take the varying kinematics into account, the contributions of the sidebands are taken at the same distance from the threshold. For these events the angle of the K^+ in the phi center of mass is calculated. Then the angular distributions for the left and right sidebands are subtracted separately from the distribution resulting from the phi events. Some sample distributions are shown in Fig. 4 for four different bins in t. All together eight bins in t were measured, spanning $0.4 \leq -t \leq 5.0$ GeV^2.

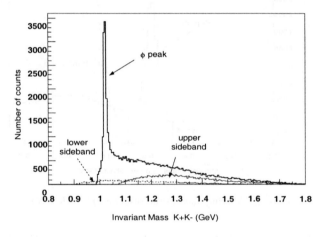

Figure 3. Invariant Mass of $K^+(X)$. The invariant mass of the events which fall into the lower sideband of the K^- distribution are shown as the dotted curve, the ones of the upper sideband as the solid curve.

3. ρ^0_{00} extraction

The extraction of ρ^0_{00} was done by fitting the angular decay distribution with the left and right sideband contributions subtracted seperately. The difference between the values for these different fits gives an evaluation of the systematic errors bars due to the continuum subtraction. As explained above, three components are taken into account in the fitting procedure: the resonant ϕ contribution (ρ^0_{00}), the (κ) and the interference term (α). Two different hypotheses were made for the continuum (see ref. [1]: either

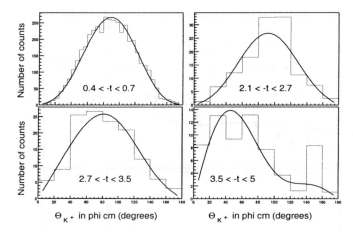

Figure 4.   The theta angle of the K^+ in the phi center of mass system for four different bins in t. The solid lines are the fits of Eq. 2.

a flat contribution or a phase space distribution plus a contribution from the $f_0(980)$). The ratio of the phi to the continuum, κ, was taken from the data and imposed in the fitting procedure. This ratio varied as a function of t, generally increasing as t increased, except in the high t bins where statistics were low.

The results for our extraction of ρ^0_{00} and α are shown in Fig. 5. The error bars show the spread in the values taking different hypotheses for the background subtraction, as described above. Also shown in the figure are the calculations of Ref. [5]. The dotted curve models pomeron exchange as the exchange of two non-perturbative gluons. The solid curve is the same model but with the addition of u-channel phi production contributions. The distributions show that there is essentially no violation of s-channel helicity conservation at lower momentum transfers. Above $-t \sim 2.7$ GeV^2, the u-channel contribution to phi production begins to dominate and a large violation of SCHC is observed. The values for α are shown in the lower part of the figure. They indicate the interference between the helicity zero phi and the s-wave K^+K^- continuum, and are strongly correlated with the SCHC violation seen in the ρ^0_{00} distribution.

Acknowledgments

I would like to thank Jean-Marc Laget, Gerard Audit, Eric Anciant and Claude Marchand for many valuable discussions on the subject of phi photoduction. The Southeastern Universities Research Association operates

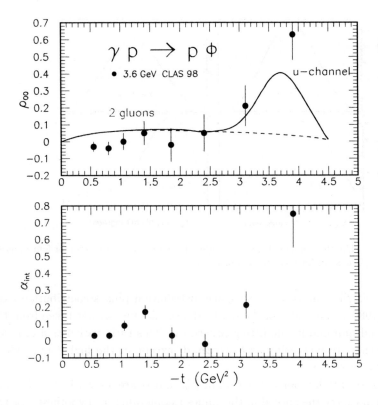

Figure 5. The values for ρ_{00}^0 and α extracted from the decay angular distribution of the K^+ in the phi center of mass system. As described in the text, the error bars show the spread in the values taking different hypotheses for the background subtraction. Also shown are the calculations of Ref. [5]. The dotted curve models pomeron exchange as the exchange of two non-perturbative gluons. The solid curve is the same model but with the addition of u-channel phi production contributions.

the Thomas Jefferson National Accelerator Facility under Department of Energy contract DE-AC05-84ER40150.

References

1. E. Anciant et al., Phys. Rev. Lett. **85**, 4682 (2000).
2. K. Schilling, P. Seyboth and G. Wolf, Nucl. Phys. **B15**, 397 (1970).
3. D.I. Sober et al., Nucl. Instrum. Methods Phys. Res., Sect. A **440**, 263 (2000).
4. W. Brooks et al., Nucl. Phys. **A663/A664**, 1077 (2000).
5. J.-M. Laget, Phys. Lett. **B489**, 313 (2000)

PROTON POLARIZATION IN NEUTRAL PION PHOTO-PRODUCTION

K. WIJESOORIYA

Argonne National Laboratory, Argonne, IL, 60439, USA
E-mail: krishniw@jlab.org

FOR THE JEFFERSON LAB HALL A COLLABORATION

We present measurements of recoil proton polarization for $^1H(\vec{\gamma}, \vec{p})\pi^0$ in and above the resonance region. These are the first data in this reaction for polarization transfer with circularly polarized photons. The results are compared to phase shift analyses and quark model calculations.

1. INTRODUCTION

At photon energies below about 2 GeV, corresponding to $W = \sqrt{s}$ below 2.15 GeV, π^0 photo-production is dominated by the production and decay of baryon resonances, indicated by structure in the cross section [1] and polarization observables. There are extensive data for only a few spin observables, and only for $E_\gamma \leq 1.5$ GeV. We present recoil proton polarization measurements which extend across the resonance region, up to $W \gg$ 2 GeV, at large scattering angles and four momentum transfers.

Above the resonance region, the cross section follows the constituent counting rules [2], which can be derived from perturbative QCD (pQCD), as is the case for a number of exclusive photo-reactions [3,4]. Hadron helicity conservation [5] (HHC) is also taken to be a consequence of pQCD, although this has come under question due to the lack of HHC in hadronic reactions [6,7,8]. In hadronic photo-reactions, HHC may hold if a single photon interacts only with a single quark in the target [9], and if orbital angular momenta may be neglected. A recent paper by Miller and Frank [10] suggests that helicity conservation is not satisfied for exclusive processes involving protons. HHC predicts that the induced polarization p_y and the transferred polarization $C_{x'\ c.m.}$ vanish. In the one photo-reaction tested, deuteron photodisintegration [11], helicity conservation was not valid.

The experiment ran in Hall A of the Thomas Jefferson National Accelerator Facility (JLab). Space limitations preclude discussion of the details, which may be found in Ref. [12].

2. MODELS

Predictions for $H(\vec{\gamma},\vec{p})\pi^0$ come from the phase shift analysis codes SAID [13] and MAID [14], and from quark model [15], and pQCD [16] calculations. In SAID, both an energy-dependent and a set of single energy partial wave analyses of single-pion photo-production data are performed. These analyses extend from threshold to 2.0 GeV in laboratory photon energy. Photo decay amplitudes are extracted from Breit-Wigner fits for the baryon resonances within this energy range. The MAID model contains Born terms, vector mesons and nucleon resonances up to the third resonance region ($P_{33}(1232)$, $P_{11}(1440)$, $D_{13}(1520)$, $S_{11}(1535)$, $F_{15}(1680)$ and $D_{33}(1700)$). This model is fitted to data up to $E_\gamma = 1.25$ GeV. The resonance contributions are included taking into account unitarity to provide the correct phases of the pion photo-production multipoles.

Afanasev et al. [15] use a pQCD approach for large transverse momenta, p_T, where mesons are directly produced by short range processes – see also Ref. [17]. This approach is similar to the factorization approach [17] used to describe Compton scattering from the proton. The calculation assumes helicity conservation, which leads to the vanishing of p_y and $C_{x'c.m.}$. In the lab, $C_{x'}$ does not generally vanish as it has contributions from both $C_{x'c.m.}$ and $C_{z'c.m.}$. This gives a simple result for exclusive photo-production of neutral pions:

$$p_y = C_{x'c.m.} = 0 \tag{1}$$

$$C_{z'c.m.} = \frac{s^2-u^2}{s^2+u^2}. \tag{2}$$

This model assumes that the polarization of the struck quark is the same as the polarization of the outgoing proton, but wave function effects can reduce the polarization. Farrar et al. [16] use pQCD to calculate all lowest-order (α_s^3) Feynman diagrams. They assume HHC, so $p_y = C_{x'c.m.} = 0$; $C_{z'c.m.}$ is constant at fixed $\theta_{c.m.}$. The exact magnitude of $C_{z'c.m.}$ can be large, but it depends on the hadronic distribution amplitudes. The calculations shown here used the asymptotic distribution amplitudes for both the proton and the pion. These pQCD approaches underpredict the cross section, and are not expected to work until $-t$ is very large.

3. RESULTS

Figure 1 compares our data for p_y to previous measurements – see Ref. [12] for a complete list of references – and theoretical predictions [13,14]. Our

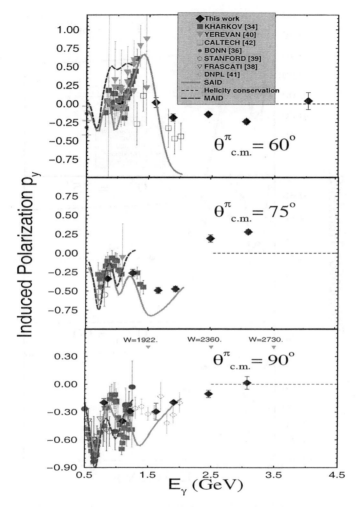

Figure 1. Induced polarization p_y at three c.m. angles. Only statistical uncertainties are shown. The curves are described in the text.

low-energy data agree well with the world data. At higher energy, the data follow the trend predicted by SAID, but not by MAID. There is no general indication of an approach to helicity conservation, $p_y \to 0$; similarly, we find that $C_{x'c.m.}$ does not tend to vanishing.

The longitudinal in-plane polarization transfer $C_{z'}$, shown in Fig. 2, does not show large polarizations as predicted in Ref. [15], nor does it appear to reach a constant value at each angle, as predicted by Ref. [16]. The data agree reasonably well with the phase shift analysis curves, SAID and MAID, at the lowest energies, but the agreement tends to increasingly deteriorate

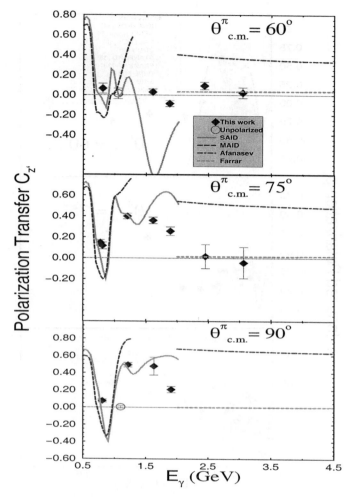

Figure 2. Polarization transfer $C_{z'}$ at three c.m. angles. Only statistical uncertainties are shown. The curves are described in the text.

with energy. There is insufficient space to show that similar observations can be made for $C_{x'}$. Data with unpolarized beam at $E_\gamma = 1.1$ GeV, are consistent with zero, as expected.

Figure 3 shows some angular distributions for p_y. The data at 2.5 GeV suggests a strong oscillatory behavior, and the $\sin(12\theta)$ curve indicates high partial waves in either the background or resonances. More finely binned angular distributions are evidently needed. The strong angular dependence appears to persist up to at least 3.1 GeV, suggesting that resonances are needed to explain the data, and quark models which sum over the reso-

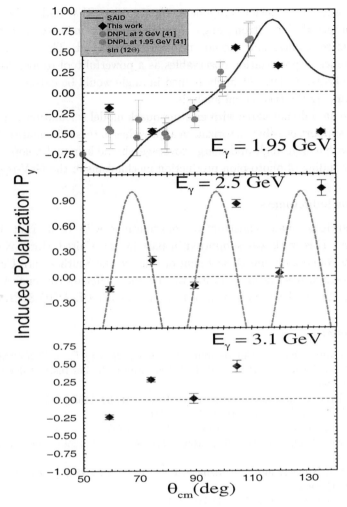

Figure 3. Angular distributions of induced polarization p_y at three beam energies. The curves are described in the text.

nances will be unable to reproduce the polarizations.

4. CONCLUSIONS

Our induced polarizations extend the world data for π^0 photo-production from the proton to much higher photon energies. Our data agree reasonably well with previous data and with the SAID and MAID phase shift analyses. We present the first data set for polarization transfer observables $C_{x'}$ and $C_{z'}$ for the $^1H(\vec{\gamma}, \vec{p})\pi^0$ reaction. The nonzero nature of p_y and the polarization transfer component, $C_{x'c.m.}$, show that hadron helicity is not conserved. The strong angle dependence of p_y indicates interference between resonances and the non-resonant background. This confirms the importance of polarization observables as a powerful tool to look for resonance effects. A data set finely binned in angle would be extremely useful in advancing the theoretical analysis.

The data do not agree with existing quark model calculations, and thus do not support possible factorization approaches in these kinematics. These data, with a large angle and energy coverage, should help to develop a better understanding of photo-pion production mechanisms in the GeV-region.

Acknowledgements

I acknowledge many constructive conversations with R. J. Holt and R. Gilman. This work was supported in part by the United States National Science Foundation and Department of Energy; the Southeastern Universities Research Association operates the Thomas Jefferson National Accelerator Facility under Department of Energy contract DE-AC05-84ER40150.

References

1. Data and references can be found in *Compilation of Pion Photoproduction Data*, D. Menze, W. Pfeil, and R. Wilcke, Physikalisches Institut der Universitat Bonn (1977).
2. S.J. Brodsky and G.R. Farrar, Phys. Rev. Lett. **31** 1153 (1973); V. Matveev *et al.*, Nuovo Cimento Lett. **7**, 719 (1973).
3. R.L. Anderson *et al.*, Phys. Rev. D **14** 679 (1976).
4. See E.C. Schulte *et al.*, Phys. Rev. Lett. **87** 102302 (2001), and references therein.
5. See S.J. Brodsky and G.P. Lepage, Phys. Rev. D **24** 2848 (1981), and references therein.
6. P.V. Landshoff, Phys. Rev. D **10** 1024 (1974).
7. T. Gousset, B. Pire, and J.P. Ralston, Phys. Rev. D **53** 1202 (1996).
8. C. Carlson and M. Chachkhunashvili, Phys. Rev. D **45** 2555 (1992).
9. A. Afanasev, C. Carlson, and C. Wahlquist, Phys. Rev. D **61** 034014 (2000).
10. G.A. Miller, M.R. Frank, arXiv:nucl-th/0201021.
11. K. Wijesooriya *et al.*, Phys. Rev. Lett. **86** 2975 (2001).
12. K. Wijesooriya *et al.*, accepted by Phys. Rev. C.
13. R.A. Arndt *et al.*, Phys. Rev. C **53** 430 (1996); R.A. Arndt*et al.*, Phys. Rev. C **56** 577 (1997).
14. D. Drechsel *et al.*, Nucl. Phys. A **645** 145 (1999); Version MAID 2000.
15. A. Afanasev, C. Carlson, and C. Wahlquist, Phys. Lett. B **398** 393 (1997).
16. G.R. Farrar, K. Huleihel, and H. Zhang, Nucl. Phys. B **349** 655 (1991).
17. A.V. Radyushkin, Phys. Rev. D **58** 11408 (1998); M. Diehl, Eur. Phys. J. C **8** 409 (1999); H.W. Huang and P. Kroll, Eur. Phys. J. C **17** 423 (2000).

NUCLEON HOLOGRAM WITH EXCLUSIVE LEPTOPRODUCTION

A.V. BELITSKY

Department of Physics, University of Maryland, College Park, MD 20742-4111, USA

D. MÜLLER

Fachbereich Physik, Universität Wuppertal, D-42097 Wuppertal, Germany

Hard exclusive leptoproductions of real photons, lepton pairs and mesons are the most promising tools to unravel the three-dimensional picture of the nucleon, which cannot be deduced from conventional inclusive processes like deeply inelastic scattering.

1. From macro to micro

Why do we see the world around us the way it is? Human eyes can detect electromagnetic waves in a very narrow range of wavelength, $\lambda_\gamma \sim 0.4 - 0.7\mu$m, which we call visible light. The light from a source, say the sun, is reflected from the surface of macro-objects and is absorbed by the eye's retina which transforms it into a neural signal going to the brain which forms the picture. The same principle is used in radars which detect reflected electromagnetic waves of a meter wavelength. The only requirement to "see" an object is that the length of resolving waves must be comparable to or smaller than its size. The same conditions have to be obeyed in case one wants to study the microworld, e.g., the structure of macromolecules (DNA, RNA) or assemblies (viruses, ribosomes). Obviously, when one puts a chunk of material in front of a source of visible light, see Fig. 1, the object merely leaves a shadow on a screen behind it and one does not see its elementary building blocks, i.e., atoms. Obviously, visible light is not capable to resolve the internal lattice structure of a crystal since the size of an individual atom, say hydrogen, is of order $r_{\rm atom} \sim (\alpha_{\rm em} m_e)^{-1} \sim (10 \text{ KeV})^{-1}$ and the light does not diffract from it. Therefore, to "see" atoms in crystals one has to have photons with the wavelength $\lambda_\gamma \leq r_{\rm atom}$, or equivalently, of

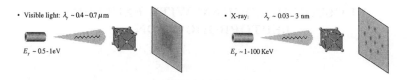

Figure 1. Left: A beam of visible light does not resolve the crystal's structure. Right: An X-ray beam does and creates a diffraction pattern on the photo-plate.

the energy $E_\gamma \geq r_{\text{atom}}^{-1}$. To do this kind of "nano-photography" one needs a beam of X-rays which after passing through the crystal creates fringes on a photo-plate, see Fig. 1. Does one get a three-dimensional picture from such a measurement? Unfortunately, no. In order to reconstruct atomic positions in the crystal's lattice one has to perform an inverse Fourier transform. This requires knowledge of both the magnitude and the phase of diffracted waves. However, what is measured experimentally is essentially a count of number of X-ray photons in each spot of the photo-plate. The number of photons gives the intensity, which is the square of the amplitude of diffracted waves. There is no practical way of measuring the relative phase angles for different diffracted spots experimentally. Therefore, one cannot unambiguously reconstruct the crystal's lattice. This is termed as "The Phase Problem". None of techniques called to tackle the problem provides a parameter-free answer.

When we study hadronic matter at the fundamental level we attempt to perform the "femto-photography" of the interior constituents (quarks and gluons) of strongly interacting "elementary" particles such as the nucleon. Quantum $\chi\rho\omega\mu\alpha$ dynamics, the theory of strong interaction, is not handy at present to solve the quark bound state problem. Therefore, phenomenological approaches, based on accurate analyses of high-energy scattering experimental data and making use of rigorous perturbative QCD predictions, are indispensable for a meticulous understanding of the nucleon's structure. As we discuss below most of high-energy processes resolving the nucleon content, such as described in terms of form factors and inclusive parton densities, suffer from the same "Phase Problem" and therefore they lack the opportunity to visualize its three-dimensional structure. A panacea is found in newborn generalized parton distributions [1], which are measurable in exclusive leptoproduction experiments.

2. Form factors

Nucleon form factors are measured in the elastic process $\ell N \to \ell' N'$. Its amplitude is given by the lepton current $L_\mu(\Delta) \equiv \bar{u}_\ell(k - \Delta)\gamma_\mu u_\ell(k)$ interacting via photon exchange with the nucleon matrix element of the quark electromagnetic current $j_\mu(x) = \sum_q e_q \bar{q}(x)\gamma_\mu q(x)$:

$$A_{NN'} = \frac{1}{\Delta^2}L_\mu(\Delta)\langle p_2|j_\mu(0)|p_1\rangle \equiv \frac{1}{\Delta^2}L_\mu(\Delta)\left\{h_\mu F_1(\Delta^2) + e_\mu F_2(\Delta^2)\right\}. \tag{1}$$

Here the matrix element of the quark current is decomposed in terms of Dirac and Pauli form factors ($\Delta \equiv p_2 - p_1$), accompanied by the Dirac bilinears $h_\mu \equiv \bar{u}_N(p_2)\gamma_\mu u_N(p_1)$ and $e_\mu \equiv \bar{u}_N(p_2)i\sigma_{\mu\nu}\Delta_\nu u_N(p_1)/(2M_N)$. In the Breit frame $\vec{p}_2 = -\vec{p}_1 = \vec{\Delta}/2$ there is no energy exchange $E_1 = E_2 = E$ and thus relativistic effects are absent. The momentum transfer is three-dimensional $\Delta^2 = -\vec{\Delta}^2$, so that

$$\langle p_2|j_0(0)|p_1\rangle = \tilde{\varphi}_2^*\tilde{\varphi}_1 G_E(-\vec{\Delta}^2),$$
$$\langle p_2|\vec{j}(0)|p_1\rangle = -\frac{i}{2M_N}\tilde{\varphi}_2^*[\vec{\Delta} \times \vec{\sigma}]\tilde{\varphi}_1 G_M(-\vec{\Delta}^2), \tag{2}$$

are expressed in terms of Sachs electric $G_E(\Delta^2) \equiv F_1(\Delta^2) + \Delta^2/(4M_N^2)F_2(\Delta^2)$ and magnetic $G_M(\Delta^2) \equiv F_1(\Delta^2) + F_2(\Delta^2)$ form factors. Introducing the charge $q \equiv \frac{1}{V}\int d^3\vec{x}\, j_0(\vec{x})$ and magnetic moment $\vec{\mu} \equiv \frac{1}{V}\int d^3\vec{x}\,[\vec{x} \times \vec{j}](x)$ operators, one finds the normalization

$$\langle p|q|p\rangle = \tilde{\varphi}_2^*\tilde{\varphi}_1 G_E(0), \quad \langle p|\vec{\mu}|p\rangle = \frac{\tilde{\varphi}_2^*\vec{\sigma}\tilde{\varphi}_1}{2M_N}G_M(0). \tag{3}$$

The interpretation of Sachs form factors as Fourier transforms of charge and magnetization densities in the nucleon requires to introduce localized nucleon states in the position space $|\vec{x}\rangle$ as opposed to the plane-wave states used above $|p\rangle$,

$$|\vec{x}\rangle = \sum_{\vec{p}} \frac{e^{i\vec{p}\cdot\vec{x}}}{\sqrt{V}}\Psi(\vec{p})|\vec{p}\rangle, \quad \text{with} \quad \sum_{\vec{p}}|\Psi(\vec{p})|^2 = 1. \tag{4}$$

Here a very broad wave packet $\Psi(\vec{p}) \approx$ const is assumed in the momentum space. Then the charge density $\rho(\vec{x})$ of the nucleon, localized at $\vec{x} = 0$, is

$$\langle \vec{x} = 0|j_0(\vec{x})|\vec{x} = 0\rangle \equiv \tilde{\varphi}_2^*\tilde{\varphi}_1\rho(\vec{x}) = \tilde{\varphi}_2^*\tilde{\varphi}_1\int \frac{d^3\vec{\Delta}}{(2\pi)^3}e^{-i\vec{\Delta}\cdot\vec{x}}G_E(-\vec{\Delta}^2), \tag{5}$$

and similar for the magnetic form factor. The famous Hofstadter's experiments established that the proton is not a point-like particle $\rho_{\text{point}}(\vec{x}) =$

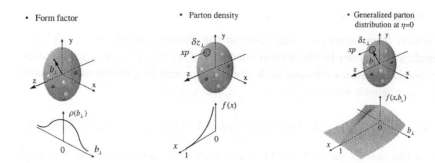

• Form factor   • Parton density   • Generalized parton
                                                                       distribution at $\eta=0$

Figure 2. Probabilistic interpretation of form factors, parton densities and generalized
parton distributions at $\eta = 0$ in the infinite momentum frame $p_z \to \infty$.

$\delta^3(\vec{x})$ which would have $G_E^{\text{point}} = \text{const}$, but rather $G_E(-\vec{\Delta}^2) \approx (1 + \vec{\Delta}^2 r_N^2/6)^{-2}$ with the mean square radius $r_N \approx 0.7\,\text{fm}$.

The Breit frame is not particularly instructive for an interpretation of high-energy scattering. Here an infinite momentum frame (IMF) is more useful, see discussion below. In this frame, obtained by a z-boost, the nucleon momentum is $p_z = (p_1 + p_2)_z \to \infty$. In the IMF one builds a nucleon state localized in the transverse plane at $\boldsymbol{b}_\perp = (x, y)$

$$|p_z, \boldsymbol{b}_\perp\rangle = \sum_{\boldsymbol{p}_\perp} \frac{e^{i\boldsymbol{p}_\perp \cdot \boldsymbol{b}_\perp}}{\sqrt{V_\perp}} \Psi(\boldsymbol{p}_\perp)|p_z, \boldsymbol{p}_\perp\rangle. \tag{6}$$

Then one finds that the transverse charge distribution of the nucleon wave packet, see Fig. 2, is given by the two-dimensional Fourier transform of form factors

$$\langle p_z, \boldsymbol{b}_\perp = 0|j_+(\boldsymbol{b}_\perp)|p_z, \boldsymbol{b}_\perp = 0\rangle = h_+ \int \frac{d^2\boldsymbol{\Delta}_\perp}{(2\pi)^2} e^{-i\boldsymbol{\Delta}_\perp \cdot \boldsymbol{b}_\perp} F_1\left(-\boldsymbol{\Delta}_\perp^2\right) + \dots. \tag{7}$$

As previously one assumes a rather delocalized transverse momentum wave function $\sum_{\boldsymbol{p}_\perp} \Psi^*(\boldsymbol{p}_\perp + \boldsymbol{\Delta}_\perp/2)\Psi(\boldsymbol{p}_\perp - \boldsymbol{\Delta}_\perp/2) \approx 1$. Thus, we can interpret form factors as describing the transverse localization of partons in a fast moving nucleon, irrespective of their longitudinal momenta and independent on the resolution scale.

3. Parton densities

The deeply inelastic lepton-nucleon scattering $\ell N \to \ell' X$ probes, via the amplitude

$$A_{NX} = \frac{1}{Q^2} L_\mu(q)\langle p_X|j_\mu(0)|p\rangle, \tag{8}$$

the nucleon with the resolution $\hbar/Q \approx (0.2\,\text{fm})/(Q\,\text{in GeV})$, set by the photon virtuality $q^2 \equiv -Q^2$. Recalling that the nucleon's size is $r_N \sim 1\,\text{fm}$, one concludes that for Q^2 of order of a few GeV, the photon penetrates the nucleon interior and interacts with its constituents. The cross section of the deeply inelastic scattering is related, by the optical theorem, to the imaginary part of the forward Compton scattering amplitude

$$d\sigma_{\text{DIS}}\left(x_{\text{B}}, Q^2\right) \sim \sum_X |A_{NX}|^2\, \delta^4(p + q - p_X)$$

$$\sim \frac{1}{\pi}\Im\, i\int d^4z\, e^{iq\cdot z}\langle p|T\left\{j_\mu^\dagger(z)j_\mu(0)\right\}|p\rangle\,. \qquad (9)$$

The very intuitive parton interpretation has its clear-cut meaning in the IMF. A typical interaction time of partons is inversely proportional to the energy deficit of a given fluctuation of a particle with the energy E_0 and three-momentum $p_0 = (p_{\perp 0}, x_0 p_z)$ into two partons with energies $E_{1,2}$ and three-momenta $p_{1,2} = (p_{\perp 1,2}, x_{1,2} p_z)$. It scales, for $p_z \to \infty$, as

$$\Delta t \sim \frac{1}{\Delta E} = \frac{1}{E_0 - E_1 - E_2} \sim \frac{p_z}{p_{\perp 0}^2/x_0 - p_{\perp 1}^2/x_1 - p_{\perp 2}^2/x_2} \to \infty\,. \qquad (10)$$

Therefore, one can treat partons as almost free in the IMF due to the time dilation. The virtual photon "sees" nucleon's constituents in a frozen state during the time of transiting the target which is, thus, describable by an instantaneous distribution of partons. Here again the analogy with X-ray crystallography is quite instructive: Recall that an X-ray, scattered off atoms, reveals crystal's structure since rapid oscillations of atoms in the lattice sites can be neglected. Atoms can be considered being at rest during the time X-rays cross the crystal. The transverse distance probed by the virtual photon in a Lorentz contracted hadron, is of order $\delta z_\perp \sim 1/Q$, see Fig. 2. One can conclude therefore that simultaneous scattering off an n-parton cascade is suppressed by an extra power of $\left(1/Q^2\right)^{n-1}$. The leading contribution to $d\sigma_{\text{DIS}}$ is thus given by a handbag diagram, i.e., the photon–single-quark Compton amplitude. The character of relevant distances in the Compton amplitude (9) is a consequence of the Bjorken limit which implies large Q^2 (small distances) and energies $\nu \equiv p \cdot q$ (small times) at fixed $x_{\text{B}} \equiv Q^2/(2\nu)$. By going to the target rest frame one immediately finds that at large Q^2 the dominant contribution comes from the light-cone distances $z^2 \approx O\left(1/Q^2\right)$ between the points of absorption and emission of the virtual photon in (9) because $z_- \sim 1/(Mx_{\text{B}})$, $z_+ \sim Mx_{\text{B}}/Q^2$.

Since the hard quark-photon subprocess occupies a very small space-time volume but the scales involved in the formation of the nucleon are much larger, hence, they are uncorrelated and will not interfere. The quantum mechanical incoherence of physics at different scales results into the factorization property of the cross section (9),

$$d\sigma_{\mathrm{DIS}}\left(x_{\mathrm{B}}, Q^2\right) \sim \sum_q e_q^2 \int_0^1 dx\, \delta\left(x - x_{\mathrm{B}}\right) f_q\left(x; Q^2\right),\qquad(11)$$

where f_q is a parton distribution, — the density of probability to find partons of a given longitudinal momentum fraction x of the parent nucleon with transverse resolution $1/Q$,

$$\langle p|\bar{q}(0)\gamma_+ q(z_- n)|p\rangle = 2p_+ \int_0^1 dx\, \{f_q(x)e^{-ixz_- p_+} - \bar{f}_q(x)e^{ixz_- p_+}\}.\qquad(12)$$

No information on the transverse position of partons is accessible here, Fig. 2.

4. Generalized parton distributions

Both observables addressed in the previous two sections give only one-dimensional slices of the nucleon since only the magnitude of scattering amplitudes is accessed in the processes but its phase is lost. These orthogonal spaces are probed simultaneously in generalized parton distributions (GPDs), which arise in the description of deeply virtual Compton scattering (DVCS) $\ell N \to \ell' \gamma^* N \to \ell' N' \gamma$ in the Bjorken limit. In the same spirit as in deeply inelastic scattering, the latter consists of sending $q^2 \equiv (q_1 + q_2)^2/4 \to -\infty$ to the deep Euclidean domain while keeping $\Delta^2 \equiv (p_2 - p_1)^2 \ll -q^2$ small and $\xi \equiv -q^2/p \cdot q$ fixed, $p \equiv p_1 + p_2$. By the reasoning along the same line as in the previous section one finds that the Compton amplitude factorizes into GPDs parametrizing the twist-two light-ray operator matrix element

$$\langle p_2|\bar{q}(-z_- n)\gamma_+ q(z_- n)|p_1\rangle$$
$$= \int_{-1}^1 dx e^{-ixz_- p_+} \{h_+ H_q(x, \eta, \Delta^2) + e_+ E_q(x, \eta, \Delta^2)\},\qquad(13)$$

and a handbag coefficient function, so that one gets

$$A_{\mathrm{DVCS}} = \varepsilon_\mu^*(q_2)L_\nu(q_1) \int d^4 z\, e^{iq\cdot z} \langle p_2|T\{j_\mu^\dagger(z/2)j_\nu(-z/2)\}|p_1\rangle$$
$$\sim \sum_q e_q^2 \int_{-1}^1 dx\, \frac{F_q(x, \eta, \Delta^2)}{\xi - x - i0},\qquad(14)$$

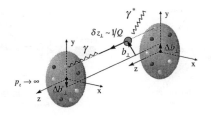

Figure 3. Geometric picture of deeply virtual Compton scattering.

where $F_q = H_q, E_q$ and the contribution from a crossed diagram is omit-
ted. GPDs depend on the s-channel momentum fraction x, measured with
respect to the momentum p, and t-channel fraction $\eta \equiv q \cdot \Delta / q \cdot p$, which
is the longitudinal component of the momentum transfer $\Delta \approx \eta p + \Delta_\perp$,
as well as its square $\Delta^2 \approx - \left(\Delta_\perp^2 + 4M_N^2\eta^2\right) / \left(1 - \eta^2\right)$. Due to the reality
of the final state photon $\eta \approx -\xi$. A geometric picture underlying DVCS
is as follows, see Fig. 3. The electric field of lepton's virtual fluctuation
$\ell \to \ell'\gamma^*$ accelerates a quark localized in the transverse area $(\delta z_\perp)^2 \sim 1/Q^2$
at the impact parameter $b_\perp$ and carrying a certain momentum fraction of
the parent nucleon. The accelerated parton tends to emit the energy via
electromagnetic radiation and fall back into the nucleon, see Fig. 3. The
incoming-outgoing nucleon system is localized at the center of coordinates
$b_\perp = 0$, however, due to non-zero longitudinal momentum exchange in the
t-channel the individual transverse localizations of incoming and outgoing
nucleons are shifted in the transverse plane by amounts[a] $\Delta b_\perp \sim \eta(1+\eta)b_\perp$
and $\Delta b'_\perp \sim \eta(1 - \eta)b_\perp$, respectively [2].

Generally, GPDs are not probabilities rather they are the interference
of amplitudes of removing a parton with one momentum and inserting it
back with another. In the limit $\Delta = 0$ they reduce to inclusive parton
densities and acquire the probabilistic interpretation. This is exhibited
in a most straightforward way in the light-cone formalism [3], where one
easily identifies the regions $-1 < x < -\eta$ and $\eta < x < 1$ with parton
densities while $-\eta < x < \eta$ with distribution amplitudes. This latter
domain precludes the density interpretation for $\eta \neq 0$.

The first moment of GPDs turns into form factors (1). The second
moment of Eq. (13) gives form factors of the quark energy-momentum

---

[a]Note the difference in the definition of our impact parameter space GPDs as compared
to Ref. [2]. In our frame $p_{\perp 2} = -p_{\perp 1} = \Delta_\perp/2$. We define the Fourier transform with
respect to $\Delta_\perp$ which has its advantages that $b_\perp$ does not depend implicitly on η.

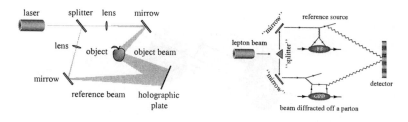

Figure 4. Left: Conventional setup for taking the holographic picture. Right: Nucleon hologram with leptoproduction of a photon: interference of the Bethe-Heitler (reference) and DVCS (sample) amplitudes.

tensor, $\Theta^q_{\mu\nu} = \bar{q}\gamma_{\{\mu}D_{\nu\}}q$. Since gravity couples to matter via $\Theta_{\mu\nu} = \frac{\delta}{\delta g_{\mu\nu}}\int d^4x\sqrt{-\det g_{\mu\nu}(x)}\,L(x)$, these form factors are the ones of the nucleon scattering in a weak gravitational field [4]

$$\langle p_2|\Theta_{\mu\nu}|p_1\rangle = A(\Delta^2)h_{\{\mu}p_{\nu\}} + B(\Delta^2)e_{\{\mu}p_{\nu\}} + C(\Delta^2)\Delta_\mu\Delta_\nu. \qquad (15)$$

Analogously to the previously discussed electromagnetic case, the combination $A(\Delta^2) + \Delta^2/(4M_N^2)B(\Delta^2)$ arising in the Θ_{00} component measures the mass distribution inside the nucleon [5]. It is different from the charge distribution due to presence of neutral constituents inside hadrons not accounted in electromegnetic form factors. The gravitomagnetic form factor $A(\Delta^2) + B(\Delta^2)$ at zero recoil encodes information on the parton angular momentum [4] $\vec{J} = \frac{1}{V}\int d^3\vec{x}\,[\vec{x}\times\vec{\Theta}](x)$ expressed in terms of the momentum flow operator $\Theta_{0i} \equiv \Theta_i$ in the nucleon and gives its distribution when Fourier transformed to the coordinate space. These form factor are accessible once GPDs are measured:

$$\int_{-1}^{1} dx\,x\,H(x,\eta,\Delta^2) = A(\Delta^2) + \eta^2 C(\Delta^2),$$

$$\int_{-1}^{1} dx\,x\,E(x,\eta,\Delta^2) = B(\Delta^2) - \eta^2 C(\Delta^2). \qquad (16)$$

GPDs regain a probabilistic interpretation once one sets $\eta = 0$ but $\Delta_\perp \neq 0$ [6,7]. When Fourier transformed to the impact parameter space they give a very intuitive picture of measuring partons of momentum fraction x at the impact parameter $b_\perp$ with the resolution of order $1/Q$ set by the photon virtuality in the localized nucleon state (6),

$$f(x,b_\perp) = \int \frac{d^2\Delta_\perp}{(2\pi)^2}e^{-i\Delta_\perp\cdot b_\perp} H\left(x,0,-\Delta_\perp^2\right). \qquad (17)$$

To visualize it, see Fig. 2, one can stick to the Regge-motivated ansatz $H(x, 0, -\mathbf{\Delta}_\perp^2) \sim x^{-\alpha_R(-\mathbf{\Delta}_\perp^2)}(1 - x)^3$ with a linear trajectory $\alpha_R(-\mathbf{\Delta}_\perp^2) = \alpha_R(0) - \alpha'_R \mathbf{\Delta}_\perp^2$ where $\alpha_R(0) \approx 0.5$ and $\alpha' \approx 1 \, \mathrm{GeV}^2$.

5. Hard leptoproduction of real photon and lepton pair

The light-cone dominance in DVCS is a consequence of the external kine-matical conditions on the process in the same way as in deeply inelastic scattering. Therefore, one can expect precocious scaling starting as early as at $-q^2 \sim 1 \, \mathrm{GeV}^2$. It is not the case for hard exclusive meson produc-tion, giving access to GPDs as well, where it is the dynamical behavior of the short-distance parton amplitude confined to a small transverse volume near the light cone that drives the perturbative approach to the process. Here the reliability of perturbative QCD predictions is postponed to larger momentum transfer.

Although GPDs carry information on both longitudinal and transverse degrees of freedom, their three-dimensional experimental exploration re-quires a complete determination of the DVCS amplitude, i.e., its magni-tude and phase. One way to measure the phase at a given spot is known as holography, for visible light. This technology allows to make three-dimensional photographs of objects, see Fig. 4: The laser beam splits into two rays. One of them serves as a reference source and the other reflects from the object's surface. The reflected beam, which was in phase with the reference beam before hitting the "target", interferes with the refer-ence beam and forms fringes on the plate with varying intensity depending on the phase difference of both. (Unfortunately, the same method cannot be used for X-ray holography of crystals and scattering experiments due to the absence of practical "splitters".) For the exclusive leptoproduction of a photon, however, there are two contributions to the amplitude: the DVCS one A_{DVCS}, we are interested in, and A_{BH} from the 'contaminating' Bethe-Heitler (BH) process, in which the real photon spills off the scattered lepton rather than the quark, see Fig. 4. The BH amplitude is completely known since the only long-distance input turns out to be nucleon form factors measured elsewhere. The relative phase of the amplitudes can be measured by the interference of DVCS and BH amplitudes in the cross sec-tion $d\sigma_{\ell N \to \ell' N' \gamma} \sim |A_{\mathrm{DVCS}} + A_{\mathrm{BH}}|^2$ and, thus, the nucleon hologram can be taken. The most straightforward extraction of the interference term is achieved by making use of the opposite lepton charge conjugation prop-erties of DVCS and BH amplitudes. The former is odd while the latter

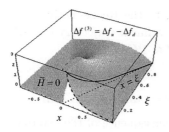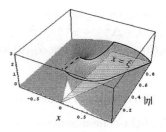

Figure 5. What is extractable from DVCS (left) and DVCS lepton pair production (right).

is even under change of the lepton charge. The unpolarized beam charge asymmetry gives

$$d\sigma_{\ell N \to \ell' N' \gamma}(+e_\ell) - d\sigma_{\ell N \to \ell' N' \gamma}(-e_\ell) = (A_{\mathrm{DVCS}} + A_{\mathrm{DVCS}}^*) A_{\mathrm{BH}}$$

$$\sim \Re e \int \frac{F_q(x, \xi, \Delta^2)}{\xi - x - i0} \cos \phi'_\gamma$$

and measures the real part of the DVCS amplitude modulated by the harmonics of the azimuthal angle between the lepton and photon scattering planes ϕ'_γ [8,9]. If on top of the charge asymmetry one further forms either beam or target polarization differences, this procedure would allow to cleanly extract the imaginary part of the DVCS amplitude where GPDs enter in diverse combinations. These rather involved measurements have not yet been done. Luckily, since the ratio of BH to DVCS amplitude scales like $[\Delta^2/q_1^2(1-y)]^{1/2}/y$, for large y or small $-\Delta^2$, it is safe to neglect $|A_{\mathrm{DVCS}}|^2$ as compared to other terms. Thus, in such kinematical settings one has access to the interference in single spin asymmetries,

$$d\sigma_{\ell N \to \ell' N' \gamma}(+\lambda_\ell) - d\sigma_{\ell N \to \ell' N' \gamma}(-\lambda_\ell) \approx (A_{\mathrm{DVCS}} - A_{\mathrm{DVCS}}^*) A_{\mathrm{BH}}$$

$$\sim \Im m \int \frac{F_q(x, \xi, \Delta^2)}{\xi - x - i0} \sin \phi'_\gamma$$

which measure GPDs directly on the line $x = \xi$ as shown in Fig. 5. Experimental measurements of these asymmetries were done by HERMES [11,13] and CLAS [12] collaborations. The comparison to current GPD models is demonstrated in Fig. 6.

In order to go off the diagonal $x = \xi$ one has to relax the reality constraint on the outgoing γ-quantum, i.e., it has to be virtual and fragment into a lepton pair $\bar{L}L$ with invariant mass $q_2^2 > 0$. Thus, one has to study the process $\ell N \to \ell' \bar{L}LN'$. In these circumstances, the skewedness parameter η independently varies for fixed Bjorken variable since $\xi \approx -\eta(|q_1^2| - q_2^2)/(|q_1^2| + q_2^2)$, and one is able to scan the three-

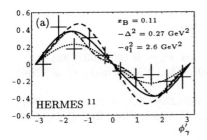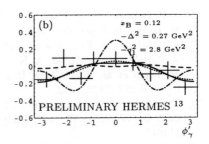

Figure 6. Beam spin asymmetry (a) in $e^+p \to e^+p\gamma$ and unpolarized charge asymmetry (b) from HERMES with $E = 27.6$ GeV are predicted making use of the complete twist-three analysis for input GPDs from Ref. [8]: model A without the D-term (solid) and C with the D-term (dashed) in the Wandzura-Wilczek approximation [10] as well as the model B with the D-term (dash-dotted) and included quark-gluon correlations. The dotted lines on the left and right panels show $0.23 \sin \phi'_\gamma$ and $-0.05+0.11 \cos \phi'_\gamma$ HERMES fits, respectively. Note that a toy model for quark-gluon correlations while only slightly changing the beam asymmetry, however, strongly alter the charge asymmetry.

dimensional shape of GPDs, see Fig. 5. Unfortunately, the cross section for DVCS lepton pair production is suppressed by α_{em}^2 as compared to DVCS and also suffers from resonance backgrounds, see, e.g., [14].

Finally, perturbative next-to-leading (NLO) and higher-twist effects are shortly discussed. Estimates of the former are, in general, model dependent. NLO contributions to the hard-scattering amplitude [15] of a given quark species are rather moderate, i.e., of the relative size of 20%, however, the net result in the DVCS amplitude can be accidentally large [8,16]. This can be caused by a partial cancellation that occurs in tree amplitudes. Evolution effects [17] in the flavor non-singlet sector are rather small. In the case of gluonic GPD models we observed rather large NLO corrections to the DVCS amplitude for the naive scale setting $\mu_F^2 = -q_1^2$ [8]. For such models one also has rather strong evolution effects, which severely affect LO analysis. However, one can tune the factorization scale μ_F so that to get rid of these effects. The renormalon-motivated twist-four [18] and target mass corrections [19] await their quantitative exploration.

References

1. D. Müller et al., Fortschr. Phys. 42 (1994) 101; X. Ji, Phys. Rev. D 55 (1997) 7114; A.V. Radyushkin, Phys. Rev. D 56 (1997) 5524.
2. M. Diehl, hep-ph/0205208;
3. M. Diehl et al., Nucl. Phys. B 596 (2001) 33; S.J. Brodsky, M. Diehl, D.S. Hwang, Nucl. Phys. B 596 (2001) 99.
4. X. Ji, Phys. Rev. Lett. 78 (1997) 610.

5. A.V. Belitsky, X. Ji, Phys. Lett. B 538 (2002) 289.
6. M. Burkardt, Phys. Rev. D 62 (2000) 071503.
7. J.P. Ralston, B. Pire, hep-ph/0110075.
8. A.V. Belitsky, D. Müller, A. Kirchner, Nucl. Phys. B 629 (2002) 323.
9. M. Diehl et al., Phys. Lett. B 411 (1997) 193; A.V. Belitsky et al., Nucl. Phys. B 593 (2001) 289.
10. A.V. Belitsky, D. Müller, Nucl. Phys. B 589 (2000) 611; N. Kivel et al., Phys. Lett. B 497 (2001) 73; A.V. Radyushkin, C. Weiss, Phys. Rev. D 63 (2001) 114012.
11. A. Airapetian et al. (HERMES Coll.), Phys. Rev. Lett. 87 (2001) 182001.
12. S. Stepanyan et al. (CLAS Coll.), Phys. Rev. Lett. 87 (2001) 182002.
13. F. Ellinghaus, these proceedings.
14. E.R. Berger, M. Diehl, B. Pire, Eur. Phys. J. C 23 (2002) 675.
15. A.V. Belitsky, D. Müller, Phys. Lett. B 417 (1997) 129; L. Mankiewicz et al., Phys. Lett. B 425 (1998) 186; X. Ji, J. Osborne, Phys. Rev. D 58 (1998) 094018.
16. A. Freund, M. McDermott, Phys. Rev. D 65 (2002) 074008.
17. D. Müller, Phys. Rev. D 49 (1994) 2525; A.V. Belitsky, D. Müller, Nucl. Phys. B 537 (1999) 397; A.V. Belitsky, A. Freund, D. Müller, Nucl. Phys. B 574 (2000) 347.
18. A.V. Belitsky, A. Schäfer, Nucl. Phys. B 527 (1998) 235.
19. A.V. Belitsky, D. Müller, Phys. Lett. B 507 (2001) 173.

POSITION SPACE INTERPRETATION FOR GENERALIZED PARTON DISTRIBUTIONS

M. BURKARDT

Dept. of Physics, New Mexico State University, Las Cruces, NM 88003, USA
*E-mail: burkardt@nmsu.edu**

For an unpolarized target, the generalized parton distribution $H_q(x, 0, t)$ is related to the distribution of partons in impact parameter space. The transverse distortion of this distribution for a transversely polarized target is described by $E_q(x, 0, t)$.

1. Introduction

Generalized parton distributions (GPDs) [1] have attracted significant interest since it has been recognized that they can not only be probed in deeply virtual Compton scattering experiments but can also be related to the orbital angular momentum carried by quarks in the nucleon [2]. However, remarkably little is still known about the physical interpretation of GPDs, and one may ask the question: *suppose, about 10-15 years from now, after a combined effort from experiment, simulation and theory, we know how GPDs look like for the nucleon. What will we have learned about the structure of the nucleon?* Of course, we will have learned something about the orbital angular momentum carried by the quarks [2], but is that all there is? In these notes, I will discuss another interesting piece of information that can be extracted from GPDs, namely *how partons are distributed in the transverse plane.*

In nonrelativistic quantum mechanics, the physics of form factors is illucidated by transforming to the center of mass frame and by interpreting the Fourier transform of form factors as charge distributions in that frame.

GPDs [1] are the form factors of the same operators [light cone correlators $\hat{O}_q(x, \mathbf{0}_\perp)$] whose forward matrix elements also yield the usual (forward) parton distribution functions (PDFs). For example, the unpolarized PDF $q(x)$ can be

---

*this work was supported by the doe (de-fg03-95er40965)

expressed in the form[a]

$$q(x) = \langle p, \lambda | \hat{O}_q(x, 0_\perp) | p, \lambda \rangle, \tag{1}$$

while the GPDs $H_q(x, \xi, t)$ and $E_q(x, \xi, t)$ are defined as

$$\langle p', \lambda' | \hat{O}_q(x, 0_\perp) | p, \lambda \rangle = \frac{1}{2\bar{p}^+} \bar{u}' \left(\gamma^+ H_q(x, \xi, t) + i \frac{\sigma^{+\nu} \Delta_\nu}{2M} E_q(x, \xi, t) \right) u, \tag{2}$$

where $\Delta = p' - p$, $2\bar{p} = p + p'$, $t = \Delta^2$, $2\bar{p}^+ \xi = \Delta^+$, and

$$\hat{O}_q(x, \mathbf{b}_\perp) = \int \frac{dx^-}{4\pi} \bar{q} \left(-\frac{x^-}{2}, \mathbf{b}_\perp \right) \gamma^+ q \left(\frac{x^-}{2}, \mathbf{b}_\perp \right) e^{ixp^+ x^-}. \tag{3}$$

In the case of form factors, non-forward matrix elements provide information about how the charge (i.e. the forward matrix element) is distributed in position space. By analogy with form factors, one would therefore expect that the additional information (compared to PDFs) contained in GPDs helps to understand how the usual PDFs are distributed in position space [3]. Of course, since the operator $\hat{O}_q(x, 0_\perp)$ already 'filters out' quarks with a definite momentum fraction x, Heisenberg's uncertainty principle does not allow a simultaneous determination of the partons' longitudinal position, but determining the distributions of partons in impact parameter space is conceiveable. Making these intuitive expectation more precise (e.g. what is the 'reference point', 'are there relativistic corrections', 'how does polarization enter', 'is there a strict probability interpretation') will be the main purpose of these notes.

2. Impact parameter dependent PDFs

In nonrelativistic quantum mechanics, the Fourier transform of the form factor yields the charge distribution in the center of mass frame. In general, the concept of a center of mass has no analog in relativistic theories, and therefore the position space interpretation of form factors is frame dependent.

The infinite momentum frame (IMF) plays a distinguished role in the physical interpretation of regular PDFs as momentum distributions in the IMF. It is therefore natural to attempt to interpret GPDs in the IMF. This task is facilitated by the fact that there a is Galilean subgroup of transverse boosts in the IMF, whose generators are defined as

$$B_x \equiv M^{+x} = (K_x + J_y)/\sqrt{2} \qquad B_y \equiv M^{+y} = (K_y - J_x)/\sqrt{2}, \tag{4}$$

---

[a] We will suppress the scale (i.e. Q^2) dependence of these matrix elements for notational convenience. In the end, the $\perp$ 'resolution' will be limited by $1/Q$.

where $M_{ij} = \varepsilon_{ijk}J_k$, $M_{i0} = K_i$, and $M^{\mu\nu}$ is the familiar generator of Lorentz transformations. The commutation relations between $\mathbf{B}_\perp$ and other Poincaré generators

$$[J_3, B_k] = i\varepsilon_{kl}B_l \qquad\qquad [P_k, B_l] = -i\delta_{kl}P^+$$
$$[P^-, B_k] = -iP_k \qquad\qquad [P^+, B_k] = 0 \qquad\qquad (5)$$

where $k, l \in \{x, y\}$, $\varepsilon_{xy} = -\varepsilon_{yx} = 1$, and $\varepsilon_{xx} = \varepsilon_{yy} = 0$, are formally identical to the commutation relations among boosts/translations for a nonrelativistic system in the plane provided we make the identification [4]

$$\begin{array}{ll}
\mathbf{P}_\perp \longrightarrow \text{momentum in the plane} & P^+ \longrightarrow \text{mass} \\
L_z \longrightarrow \text{rotations around } z\text{-axis} & P^- \longrightarrow \text{Hamiltonian} \qquad (6) \\
\mathbf{B}_\perp \longrightarrow \text{generator of boosts in the } \perp \text{ plane.}
\end{array}$$

Because of this isomorphism it is possible to transfer a number of results and concepts from nonrelativistic quantum mechanics to the infinite momentum frame. For example, for an eigenstate of P^+, one can define a (transverse) *center of momentum* (CM)

$$\mathbf{R}_\perp \equiv -\frac{\mathbf{B}_\perp}{p^+} = \frac{1}{p^+}\int dx^- \int d^2\mathbf{x}_\perp T^{++}\mathbf{x}_\perp, \qquad (7)$$

where $T^{\mu\nu}$ is the energy momentum tensor. Like its nonrelativistic counterpart, it satisfies $[J_3, R_k] = i\varepsilon_{kl}R_l$ and $[P_k, R_l] = -i\delta_{kl}$. These simple commutation relations enable us to form simultaneous eigenstates of $\mathbf{R}_\perp$ (with eigenvalue $\mathbf{0}_\perp$), P^+ and J_3

$$|p^+, \mathbf{R}_\perp = \mathbf{0}_\perp, \lambda\rangle \equiv N \int d^2\mathbf{p}_\perp |p^+, \mathbf{p}_\perp, \lambda\rangle, \qquad (8)$$

where N is some normalization constant, and λ corresponds to the helicity when viewed from a frame with infinite momentum. For details on how these IMF helicity states are defined, as well as for their relation to usual rest frame states, see Ref. [5].

In the following we will use the eigenstates of the $\perp$ center of momentum operator[b] (8) to define the concept of a parton distributions in impact parameter space [c]

$$q(x, \mathbf{b}_\perp) \equiv \langle p^+, \mathbf{R}_\perp = \mathbf{0}_\perp, \lambda | \hat{O}_q(x, \mathbf{b}_\perp) | p^+, \mathbf{R}_\perp = \mathbf{0}_\perp, \lambda\rangle. \qquad (9)$$

---

[b]Note that the Galilei invariance in the IMF is crucial for being able to construct a useful CM concept.

[c]In Ref. [6], wave packets were used in order to avoid states that are normalized to δ functions. The final results are unchanged. This was also verified in Ref. [8].

The impact parameter dependent PDFs defined above (9) are the Fourier transform of H_q [6,7,8] (without relativistic corrections!) [d]

$$q(x, \mathbf{b}_\perp) = \frac{|N|^2}{(2\pi)^2} \int d^2\mathbf{p}_\perp \int d^2\mathbf{p}'_\perp \left\langle p^+, \mathbf{0}_\perp, \lambda \left| \hat{O}_q(x, \mathbf{b}_\perp) \right| p^+, \mathbf{0}_\perp, \lambda \right\rangle \quad (10)$$

$$= \frac{|N|^2}{(2\pi)^2} \int d^2\mathbf{p}_\perp \int d^2\mathbf{p}'_\perp H_q(x, -(\mathbf{p}'_\perp - \mathbf{p}_\perp)^2) e^{i\mathbf{b}_\perp \cdot (\mathbf{p}'_\perp - \mathbf{p}_\perp)}$$

$$= \int \frac{d^2\mathbf{\Delta}_\perp}{(2\pi)^2} H_q(x, -\mathbf{\Delta}_\perp^2) e^{i\mathbf{b}_\perp \cdot \mathbf{\Delta}_\perp}$$

and its normalization is $\int d^2\mathbf{b}_\perp q(x, \mathbf{b}_\perp) = q(x)$. Furthermore, $q(x, \mathbf{b}_\perp)$ has a probabilistic interpretation. Denoting $\tilde{b}_s(k^+, \mathbf{b}_\perp)$ $[\tilde{d}_s(k^+, \mathbf{b}_\perp)]$ the canonical destruction operator for a quark [antiquark] with longitudinal momentum k^+ and $\perp$ position $\mathbf{b}_\perp$, one finds [7]

$$q(x, \mathbf{b}_\perp) = \begin{cases} \sum_s \left| \tilde{b}_s(xp^+, \mathbf{b}_\perp) \, |p^+, \mathbf{0}_\perp, \lambda\rangle \right|^2 \geq 0 & \text{for} \quad x > 0 \\ -\sum_s \left| \tilde{d}_s^\dagger(xp^+, \mathbf{b}_\perp) \, |p^+, \mathbf{0}_\perp, \lambda\rangle \right|^2 \leq 0 & \text{for} \quad x < 0 \end{cases} \quad (11)$$

For large x, one expects $q(x, \mathbf{b}_\perp)$ to be not only small in magnitude (since $q(x)$ is small for large x) but also very narrow (localized valence core!). In particular, the $\perp$ width should vanish as $x \to 1$, since $q(x, \mathbf{b}_\perp)$ is defined with the $\perp$ CM as a reference point. A parton representation for $\mathbf{R}_\perp$ (7) is given by $\mathbf{R}_\perp = \sum_{i \in q, g} x_i \mathbf{r}_{i,\perp}$, where x_i ($\mathbf{r}_{i,\perp}$) is the momentum fraction ($\perp$ position) of the ith parton, and for $x = 1$ the position of active quark coincides with the $\perp$ CM.

In order to gain some intuition for the kind of results that one might expect for impact parameter dependent PDFs, we consider a simple model

$$H_q(x, 0, -\mathbf{\Delta}_\perp^2) = q(x) e^{-a\mathbf{\Delta}_\perp^2 (1-x) \ln \frac{1}{x}}. \quad (12)$$

The precise functional form in this ansatz should not be taken too seriously, and the model should only be considered a simple parameterization which is consistent with both Regge behavior at small x and Drell-Yan-West duality at large x. A straightforward Fourier transform yields (Fig. 1)

$$q(x, \mathbf{b}_\perp) = q(x) \frac{1}{4\pi(1-x) \ln \frac{1}{x}} e^{-\frac{\mathbf{b}_\perp^2}{4a(1-x) \ln \frac{1}{x}}}. \quad (13)$$

---

[d] A similar interpretation exists for $\tilde{H}_q(x, 0, t)$ in terms of impact parameter dependent polarized quark distributions $\Delta q(x, \mathbf{b}_\perp) = \int \frac{d^2\mathbf{\Delta}_\perp}{(2\pi)^2} \tilde{H}_q(x, 0, -\mathbf{\Delta}_\perp^2) e^{i\mathbf{\Delta}_\perp \cdot \mathbf{b}_\perp}$.

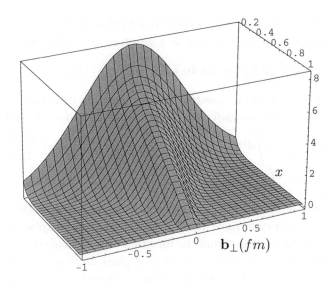

Figure 1. Impact parameter dependent parton distribution $q(x, \mathbf{b}_\perp)$ for the model (13).

3. Position Space Interpretation for $E(x, 0, -\boldsymbol{\Delta}_\perp^2)$

While both $H(x, 0, t)$ and $\tilde{H}(x, 0, t)$ are diagonal in helicity, $E(x, 0, t)$ contributes only to helicity flip matrix elements. In fact for $p^+ = p^{+\prime}$ [9] [10]

$$\int \frac{dx^-}{4\pi} e^{ip^+x^-x} \left\langle P{+}\Delta, \uparrow \left| \bar{q}(0)\, \gamma^+ q\left(x^-\right) \right| P, \uparrow \right\rangle = H(x, 0, -\boldsymbol{\Delta}_\perp^2) \tag{14}$$

$$\int \frac{dx^-}{4\pi} e^{ip^+x^-x} \left\langle P{+}\Delta, \uparrow \left| \bar{q}(0)\, \gamma^+ q\left(x^-\right) \right| P, \downarrow \right\rangle = -\frac{\Delta_x - i\Delta_y}{2M} E(x, 0, -\boldsymbol{\Delta}_\perp^2).$$

Since E is off diagonal in helicity, it will therefore only have a nonzero expectation value in states that are *not* eigenstates of helicity, i.e. if we search for a probabilistic interpretation for $E(x, 0, t)$ we need to look for it in states that are superpositions of helicity eigenstates. For this purpose, we consider the state $|X\rangle \equiv |p^+, \mathbf{R}_\perp = 0, X\rangle \equiv (|p^+, \mathbf{R}_\perp = 0, \uparrow\rangle + |p^+, \mathbf{R}_\perp = 0, \downarrow\rangle)/\sqrt{2}$. In this state, we find for the (unpolarized) impact parameter dependent PDF

$$q_X(x, \mathbf{b}_\perp) \equiv \langle X \,| O_q(x, \mathbf{b}_\perp)| \, X\rangle \tag{15}$$

$$= \int \frac{d^2\boldsymbol{\Delta}_\perp}{(2\pi)^2} \left[H_q(x, 0, -\boldsymbol{\Delta}_\perp^2) + \frac{i\Delta_y}{2M} E_q(x, 0, -\boldsymbol{\Delta}_\perp^2) \right] e^{-i\mathbf{b}_\perp \cdot \boldsymbol{\Delta}_\perp}$$

Upon introducing the Fourier transform of E_q

$$e_q(x, \mathbf{b}_\perp) \equiv \int \frac{d^2\boldsymbol{\Delta}_\perp}{(2\pi)^2} E_q(x, 0, -\boldsymbol{\Delta}_\perp^2) e^{-i\mathbf{b}_\perp \cdot \boldsymbol{\Delta}_\perp} \tag{16}$$

we find that E_q describes how the unpolarized PDF in the $\perp$ plane gets distorted when the nucleon target is polarized in a direction other than the z direction

$$qx(x,\mathbf{b}_\perp) = q(x,\mathbf{b}_\perp) + \frac{1}{2M}\frac{\partial}{\partial b_y}e_q(x,\mathbf{b}_\perp). \tag{17}$$

Since $q_X(x,\mathbf{b}_\perp)$ must still be positive further positivity constraints [7,11] follow. Above distortion (17) also shifts the $\perp$CM of the partons in the y-direction

$$\langle x_q b_y^q \rangle \equiv \int dx \int d^2\mathbf{b}_\perp x b_y^q qx(x,\mathbf{b}_\perp) = -\int dx \int d^2\mathbf{b}_\perp x \frac{e_q(x,\mathbf{b}_\perp)}{2M}$$

$$= -\int dx\, x \frac{E_q(x,0,0)}{2M}, \tag{18}$$

i.e. the second moment of E_q describes the displacement of the $\perp$CM in a state with $\perp$ polarization (note that the direction of the displacement is perpendicular to both the z axis and the direction of the polarization. Of course, the net (summed over all quark flavors plus glue) displacement of the $\perp$CM vanishes $\sum_{i \in q,g}\langle x_i b_y^i \rangle = 0$ [12]. The resulting $\perp$ dipole moment is given by

$$d_q^y \equiv \int dx \int d^2\mathbf{b}_\perp b_y qx(x,\mathbf{b}_\perp) = -\int dx \int d^2\mathbf{b}_\perp \frac{e_q(x,\mathbf{b}_\perp)}{2M} = -\int dx \frac{E_q(x,0,0)}{2M}$$

$$= -\frac{\kappa_q(0)}{2M}. \tag{19}$$

κ_q is the contribution from flavor q to the anomalous Dirac moment $F_2(0)$. In order to get some feeling for the order of magnitude, we consider a very simple model where only $q = u, d$ contribute to $F_2(0)$, one finds for example $\kappa_d \approx -2$ and therefore a mean displacement of d quarks of by about $0.2 fm$. For u quarks the effect is about half as large and in the opposite direction.

References

1. X. Ji, J. Phys. G24 (1998) 1181; A.V. Radyushkin, Phys. Rev. D56 (1997) 5524; K. Goeke et al., Prog. Part. Nucl. Phys. 47 (2001) 401.
2. X. Ji, Phys. Rev. Lett. **78** (1997) 610.
3. J.P. Ralston and B. Pire, hep-ph/0110075; J.P. Ralston, B. Pire and R.V. Buniy, hep-ph/0206074.
4. J. Kogut and D.E. Soper, Phys. Rev. D **1** (1970) 2901.
5. D.E. Soper, Phys. Rev. D **5** (1972) 1956.
6. M. Burkardt, Phys. Rev. D **62** (2000) 071503.
7. M. Burkardt, proceedings of the workshop on *Lepton Scattering, Hadrons and QCD*, Eds. W.Melnitchouk et al., Adelaide, March 2001; hep-ph/0105324.
8. M. Diehl, hep-ph/0205208.
9. M. Diehl, Eur. Phys. J. C19, 485 (2001); P.V. Pobylitsa, hep-ph/0201030.
10. M. Diehl *et al.*, Nucl. Phys. B **596**, 33 (2001).
11. P.V. Pobylitsa, hep-ph/0204337.
12. S.J. Brodsky et al., Nucl. Phys. B **593**, 311 (2001).

RESOLVING THE MICROSCOPIC LANDSCAPE OF THE PROTON

JOHN P. RALSTON

University of Kansas, Lawrence, KS 66045, USA
E-mail: ralston@ku.edu

PANKAJ JAIN

Physics Department, I.I.T. Kanpur, India 208016
E-mail: pkjain@iitk.ac.in

Progress in understanding exclusive reactions points to a broad kinematic region of *large* Q^2 and *small* $|t|$. Measurements in this region will explore a new landscape of great detail that can reveal surprising features. Topics of the physical size of the proton, the role of quark orbital angular momentum, and using amplitudes measured by interference to make *images* of hadron microstructure are highlighted.

A New Landscape of Exclusive Reactions

Hadron exclusive reactions have a new geography. It is seen in the plane of two momentum transfers, Q^2 and t. The Q^2 variable is a *virtuality* of a virtual photon; $t = (p - p')^2$ is Mandlestam's variable for initial and final target momenta p, p'.

Virtual photon microscopy[1] can be done with DVCS and related experiments in the region of $Q^2 > GeV^2$, $|t| < 1/size$ (Fig. 1). We are interested to know just what is the hadron size, and to explore the hadron's internal structure.[1,2] Hadron sizes were thought to be known very accurately, from the second derivative of elastic form factors. However for small Q^2 the impulse approximation is not motivated. As discussed elsewhere[2], even the basis in history of the historical association between a charge density and the Fourier transform of the form factor is rather questionable. The existence of an independent probe of the "quark radius" using DVCS is new. Very basic concepts of quantum mechanics are involved. Even the uncertainty principle can contain surprises.

The 21st Century Heisenburg Microscope: Textbooks describe a thought-experiment attributed to Heisenberg. Take an elementary particle, say a proton, and look at it with a microscope. Scatter gamma ray

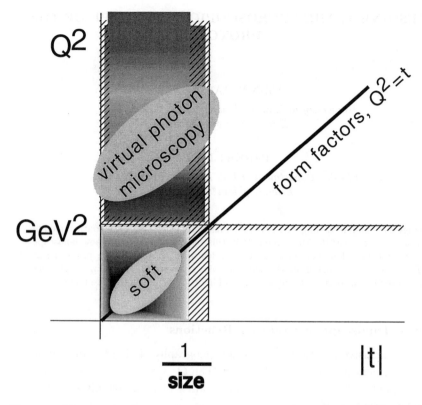

Figure 1. The new landscape of exclusive reactions. The region of $Q^2 > GeV^2$, $|t| < 1/size$ is the area of *virtual photon microscopy*. Amplitudes measured in DVCS can make well-resolved images of the target in this region. The line of $Q^2 = |t|$ is the kinematic constraint of *elastic form factors*.

photons of wavelength λ, from the target. If the microscope has aperture R, the angular resolution $\Delta\theta$ is of order λ/R, and the position resolution $\Delta x > \frac{\lambda}{\sin\Theta}$, where Θ is the angle subtended by the aperture. Meanwhile the gamma ray delivers momentum $\Delta p_x \sim \frac{\hbar \sin\Theta}{\lambda}$. The factors obey the uncertainty principle $\Delta p_x \Delta x > \hbar$, with the conventional conclusion that we cannot make a microscope to see an elementary particle in any meaningful way.

The uncertainty principle is right, but the conclusion is wrong.[3] First arrange an *exclusive* reaction where the momentum transfer t is measured, rather than being uncertain. Next use a point-like *virtual photon* γ^* from a relativistic electron beam. Finally, radiate away *almost all* of the virtual photon momentum into a second, real photon with momentum k^μ. This is the process called *deeply virtual Compton scattering* (DVCS), which has generated so much interest in recent years.

The perturbative QCD analysis of DVCS via the "handbag diagram" is very interesting. One half of the reaction, where the γ^* strikes a quark, is just like deeply inelastic scattering (DIS). The other half emits a real photon of $k^2 = 0$. Since the intermediate quark in the handbag diagram can be far off-shell, there is no time for it to do anything but emit the real photon, shedding the big-momentum $Q - k$ *quickly*. The handbag approach satisfies the impulse approximation (aka "factorization") in pQCD.

From the viewpoint of optics[1], the process of *image formation* starts with frequency ω emitted from a source at $S(\vec{x}')$. The struck quark is the source. The photon amplitude of optics propagates out to an observation point $\vec{x}$ via *just the same on-shell photon kernel* used in particle and nuclear physics, whose 4-dimensional Fourier transform is $\delta(k^\mu k_\mu)\theta(k_0)$. The calculation is Lorentz covariant. The scattering amplitude $M(\vec{x})$ at infinity is given by

$$4\pi M(\vec{x})\frac{e^{i\omega R}}{R} = \int d^3x'\, S(\vec{x}')\frac{e^{i\omega|\vec{x}-\vec{x}'|}}{|\vec{x}, \vec{x}'|} \sim \frac{e^{i\omega R}}{R} \int d^3x'\, S(\vec{x}')e^{-i\vec{k}\cdot\vec{x}'}. \quad (1)$$

The factors of $\frac{e^{i\omega R}}{R}$ are removed by definitions in scattering theory.

The source $S(\vec{x})$ is reconstructed as the *inverse Fourier transform* of the momentum-space amplitude. The *image* is the coordinate-space intensity map, namely the square of the coordinate-space amplitude. The data should extend over a large enough range to interpolate a smooth function reliably. A useful image also needs *optical spatial resolution* small compared to the *target size*. The real photon originates within a spatial resolution $\sim 1/Q$. The transverse location $\vec{b}_T$ is conjugate to $\Delta_T = (p - p')_T$, with $\Delta_t^2 \sim |t|$ at high energies. Meanwhile we need $Q^2 > GeV^2$ for the reaction time scale to beat the internal time scales of hadrons. The optically well-resolved region is the virtual photon microscopy region shown in Fig. 1. It is the same region where the handbag should be used[4], as found on more pedestrian grounds!

It's What's Observable

To carry out this lovely program we need the scattering amplitude for the process $\gamma^* + p \to p' + k$. It is hard enough to get a cross section, and asking for an *amplitude* might seem impossible.

By the best of luck the DVCS process *allows the direct measurement of the amplitude*. Interference with the Bethe-Heitler (BH) emissions of the photon from the electron beam allows extraction of the DVCS amplitude model-independently. Analysis of the handbag predictions for real (charge) and imaginary (spin) azimuthal angular asymmetries was given previously[4]. There are tests of the handbag mechanism itself, independent of models of generalized parton distributions[9] (GPD). Scaling relations of $sin\phi$, $sin2\phi$, $sin3\phi$, . . . $cos3\phi$ are consequences of helicity-conservation of the quarks (not hadrons!!). The $sin\phi$, $cos\phi$ azimuthal spin asymmetries were predicted to dominate[4]. The predictions are robust yet very detailed and leave little wiggle-room. The beautiful HERMES and CEBAF azimuthal asymmetry data[14] so far confirms the predictions spectacularly. This discovery is on just the same footing as the DIS rule $\sigma_L/\sigma_T \sim, 0$, the so-called *Callan Gross* relation. I call the absence of $n\phi$, $n > 1$ the *Diehl-Gousset* relations after the first authors of the paper.[4] *Probably the data is seeing quarks!*

In physics we must work with what is observable, and it is *the virtual Compton amplitude* that is observable. The handbag DVCS amplitude is a convolution of a quark kernel and some GPD's, and the GPDs are *not* observable. Fortunately the quark kernel is "pointlike" up to $1/Q$ spatial resolution in the transverse direction, so that the images are good in the transverse plane. Worries about non-perturbative complications of the kernel[15] do not affect the observability of the amplitude, or the process of image formation! It is certainly permissible to integrate over the longitudinal variables to improve statistics[1,2]. Ideal data might eventually take "slices" through the proton like a hospital CAT-scan.

A Remark: A comment on the probability interpretation of GPD's is useful. Any set of matrix elements can be diagonalized by taking appropriate linear combinations, i. e. by transforming coordinates. In the new coordinates diagonal matrix elements then look like "expectations" of beginning quantum mechanics. This is the approach to the *transverse spatial quark distribution* by diagonalization, used in the paper of Buniy *et al.*[7]. Simultaneously with Buniy, Burkardt[13] also discovered the diagonalization of GPD's in the transverse spatial coordinate. Both groups had simply rediscovered the good basis for GPD's published by Soper[10] in 1977! It is such a powerful thing that it will be rediscovered again and again. But there is no contradiction with diagonal DVCS amplitudes themselves remaining *amplitudes*. The matrix elements of DVCS are of the form $< out| \ldots |in >$, and only related to diagonal expectations by identifying $|in > = |out >$.

Sub-Summary: The new way of looking at DVCS, with emphasis on what is observable, represents an evolution of emphasis in the field. The tables are turned, and experiments will soon be ahead of theory. If you have an amplitude, you cannot beat what will be learn from making the *image* and *looking at it.*

What Will Images Show: and How Are They Related to Form Factors? We do not know what images of the proton will show because we have never seen the proton and we do not know what it is.

Some of us are convinced that the historical interpretation in the low-Q form factors has been misleading. Vector meson dominance teaches us that the strong meson probe strongly polarizes the proton, and reports back a grossly distorted measure of "size". There is every reason to predict[2] that the "quark-size" of DVCS will be substantially smaller than the so-called charge radius. *Both* the DVCS-type measures and the low Q form factors become more interesting. Form factors will remain superb observables, even more valuable because there will be a "control" for comparisons.[16]

Perhaps the most interesting current controversy is $F_2(Q^2)$, the proton's Pauli electromagnetic form factor. Several theory groups are interested in this[12]. Our interest comes from the rather general connection with quark orbital angular momentum (OAM)[a] and hadron helicity non-conservation in pQCD.[5] On the basis of OAM, angular momentum conservation, and quark helicity conservation, we predicted[7] that $QF_2/F_1 \sim const.$ at large Q^2. The astounding lower Q data of Jones *et al*[8] was helpful to find our mistakes[6]. The data of Gayou *et al*[8] agrees with $QF_2/F_1 \sim const.$ up to $Q^2 = 5.8\,GeV^2$. Moments of the transverse quark location b confirm that there should be no suppression of OAM at CEBAF or even at foreseeable future facilities: see the plots elsewhere[2]. A prediction yielding $QF_2/F_1 \sim const.$ in a constituent quark model was given earlier by Miller and collaborators: it has been updated recently.[11] There is a remarkable coincidence of interpretation. *Both* approaches find that the light-cone quarks carry substantial OAM.

If the current observations of F_2 are indeed due to quark OAM, then the proton is not the small round dot of tradition. Instead it is oblong, oriented by the transverse spin which couples to F_2, and the lack of symmetry should show up in the proton's femto-photographs. Quark orbital angular momentum, as well as flavor, spin, momentum, and every other attribute, can be *localized* in the transverse plane. Navigating by the "map of the future" geography seen in Fig. 1, the proton's basic size, shape and observablity are being challenged.

---

[a]Our quark OAM is defined via a light-cone $SO(2)$ group of rotations about the hadron momentum, not the $SO(3)$ of non-relativistic physics.

110

Acknowledgments

Supported in part under DOE grant.

References

1. J. P. Ralston and B. Pire, hep-ph/0110075.
2. J. P. Ralston and P. Jain, in *Proceedings of Conference on Testing QCD with Spin Observables and Nuclear Targets*(Charlottesville, VA April, 2002); hep-ph/0207129.
3. JPR directed X. Liu, M.S. Thesis 1988, University of Kansas, in an early exploration of the mathematical reconstruction of images of elementary particles. We used real photons in- and out- and a Higgs particle to minimize recoil effects.
4. M. Diehl, T. Gousset, B. Pire and J. P. Ralston, Phys. Lett. **B411** 193, (1997). Early emphasis on BH interference for the charge asymmetry is given in S. J. Brodsky, F. E. Close and J. F. Gunion, Phys. Rev. D **5**, 1384 (1972).
5. T. Gousset, B. Pire and J. P. Ralston, Phys. Rev. **D53** 1202 (1996).
6. P. Jain and J. P. Ralston, in *Future Directions in Particle and Nuclear Physics at Multi-GeV Hadron Beam Facilities* (Proceedings of the Workshop held at BNL, 4-6 March, 1993), hep-ph/9305250.
7. R. Buniy, J. P. Ralston, and P. Jain, in *VII International Conference on the Intersections of Particle and Nuclear Physics*(Quebec City, 2000) edited by Z. Parseh and W. Marciano (AIP, NY 2000); hep-ph/0206074. See also J. P. Ralston, R. V. Buniy and P. Jain *Proceedings of DIS 2001, 9th International Workshop on Deep Inelastic Scattering*, Bologna, 27 April - 1 May, 2001; hep-ph/0206063
8. M.K. Jones *et al*, Phys. Rev. Lett. **84**, 1398 (2000); O. Gayou *et al*, Phys. Rev. Lett. **88** 092301, (2002).
9. See Soper, Ref. 9, for what appears to be the earliest GPD reference. We were not aware of this previously. See also: J. Bartles, Zeit. Phys. C **12** (1982) 263; B. Geyer et al, Zeit. Phys. C **26** (1985) 591; T. Braunschweig et al, Zeit. Phys. C **33** (1986) 275; F. Dittes et al, Phys. Lett. B **209** (1988) 325; I. Balitsky and V. Braun, Nucl. Phys. B **311** (1989) 1541; P. Jain, J. P. Ralston and B. Pire, Proceedings of the DPF92 meeting, Fermilab, November 10-14 (1992), hep-ph/9212243, and Ref. 5; X. Ji, Phys. Rev. D **55** (1997) 7114; A. Radyushkin, Phys. Lett. B **380** (1996) 417; Phys. Rev. **D56** (1997) 5524.
10. D. E. Soper. Phys. Rev. **D** 15, 1141, 1977.
11. G.A. Miller and M. R. Frank, nucl-th/0201021; M. R. Frank, B. K. Jennings and G.A. Miller, Phys. Rev. **C54** 920 (1996). See also F. Schlumpf, Phys. Rev. **D47**, 4114 (1993); Erratum-ibid. **D49** 6246 (1994).
12. See, e.g. P. Stoler, these proceedings, and Phys. Rev. **D65**, 053013 (2002); M. Diehl, Th. Feldmann, R. Jacob and P. Kroll, Nucl. Phys. **B596**, 33 (2001); A. Afanasev, hep-ph/9910565, *Proceedings of the JLAB-INT Workshop on Exclusive and Semi-Exclusive Processes at High Momentum Transfer*, C. Carlson and A. Radyushkin, eds., World Scientific (2000), May 1999; W. R. B. de Araujo, E. F. Suisso, M. Beyer and H.J. Weber, Phys.Lett. **B478**, 86 (2000), and Virginia preprint (2002); F. Wang and X.-S. Chen, hep-ph/9802346; Di.

Qing, X-S Chen, and F. Wang, Phys. Rev. **D58**(1998) 114032; F. Gross, D. S. Hwang, comments at the meeting.

13. M. Burkardt, Phys. Rev. **D62** 071503, 2000; hep-ph/0010082; hep-ph/0008051; hep-ph/0007036; M. Diehl, hep-ph/0205208.
14. A. Airapetian *et al*, Phys. Rev. Lett. **87**, 182001 (2001); S. Stepanyan *et al*, Phys. Rev. Lett. **87**, 182002 (2001).
15. S. J. Brodsky, D. S. Hwang and I. Schmidt, Phys. Lett. **B530**, 99 (2002).
16. The stunning recent data on the neutron form factor for instance, will finally have a baseline, of better than circular logic, for comparison. See A. Semenov, Invited talk in *Proceedings of DNP 2002* (Albuquerque, April 2002) for preliminary data..

GENERALIZED PARTON DISTRIBUTIONS IN TERMS OF LIGHT-CONE WAVE FUNCTIONS

DAE SUNG HWANG

Depatment of Physics, Sejong University, Seoul 143-747, Korea
Stanford Linear Accelerator Center, Stanford Univ., Stanford, CA 94309, USA

The matrix elements of local operators such as electromagnetic current, energy momentum tensor, angular momentum, and generalized parton distributions have exact representations in terms of light-cone Fock state wavefunctions of bound states such as hadrons. We present formulae which express the form factors and the generalized parton distributions in terms of the light-cone wavefunctions.

1. Introduction

The light-cone expansion is constructed by quantizing QCD at fixed light-cone time $\tau = t + z$ and forming the invariant light-cone Hamiltonian: $H_{LC}^{QCD} = P^+ P^- - \vec{P}_\perp^2$ where $P^\pm = P^0 \pm P^z$. The proton state, for example, satisfies: $H_{LC}^{QCD} |\psi_p\rangle = M_p^2 |\psi_p\rangle$. The expansion of the proton eigensolution $|\psi_p\rangle$ on the color-singlet $B = 1$, $Q = 1$ eigenstates $\{|n\rangle\}$ of the free Hamiltonian $H_{LC}^{QCD}(g = 0)$ gives the light-cone Fock expansion [1,2]:

$$\left| \psi_p(P^+, \vec{P}_\perp) \right\rangle = \sum_n \prod_{i=1}^{n} \frac{dx_i \, d^2 \vec{k}_{\perp i}}{\sqrt{x_i} \, 16\pi^3} \, 16\pi^3 \delta \left(1 - \sum_{i=1}^{n} x_i \right) \delta^{(2)} \left(\sum_{i=1}^{n} \vec{k}_{\perp i} \right)$$

$$\times \psi_n(x_i, \vec{k}_{\perp i}, \lambda_i) \left| n; \, x_i P^+, x_i \vec{P}_\perp + \vec{k}_{\perp i}, \lambda_i \right\rangle . \quad (1)$$

The light-cone momentum fractions $x_i = k_i^+ / P^+$ and $\vec{k}_{\perp i}$ represent the relative momentum coordinates of the QCD constituents. The physical transverse momenta are $\vec{p}_{\perp i} = x_i \vec{P}_\perp + \vec{k}_{\perp i}$. The λ_i label the light-cone spin projections s^z of the quarks and gluons along the quantization direction z. The physical gluon polarization vectors $\epsilon^\mu(k, \lambda = \pm 1)$ are specified in light-cone gauge by the conditions $k \cdot \epsilon = 0$, $\eta \cdot \epsilon = \epsilon^+ = 0$. The n-particle states are normalized as

$$\langle n; \, p_i'^+, \vec{p}_{\perp i}', \lambda_i' \, | n; \, p_i^+, \vec{p}_{\perp i}, \lambda_i \rangle$$

$$= \prod_{i=1}^{n} 16\pi^3 p_i^+ \delta(p_i'^+ - p_i^+) \, \delta^{(2)}(\vec{p}_{\perp i}' - \vec{p}_{\perp i}) \, \delta_{\lambda_i' \lambda_i} . \quad (2)$$

The light-cone wavefunctions $\psi_n(x_i, \vec{k}_{\perp i}, \lambda_i)$ are universal, process independent, and thus control all hadronic reactions. Given the light-cone wavefunctions, one can compute the moments of the helicity distributions measurable in polarized deep inelastic experiments [2]. Exclusive semi-leptonic B-decay amplitudes involving timelike currents [3], electromagnetic and gravitational form factors of hadrons [4], and generalized parton distributions appearing in deeply virtual Compton scattering can also be evaluated exactly in terms of light-cone wavefunctions [5].

2. An Example of Light-Cone Fock State Decomposition and Wavefunction

In the language of light-cone quantization, the electron anomalous magnetic moment $a_e = \alpha/2\pi$ is due to the one-fermion one-gauge boson Fock state component of the physical electron. We employ light-cone gauge $A^+ = 0$ so that the gauge boson polarizations are physical. The light-cone fractions $x_i = k_i^+/P^+$ are positive: $0 < x_i \leq 1$, $\sum_i x_i = 1$. We adopt a non-zero boson mass λ for the sake of generality.

The two-particle Fock state for an electron with $J^z = +\frac{1}{2}$ has four possible spin combinations:

$$\left| \Psi_{\text{two particle}}^{\uparrow}(P^+, \vec{P}_\perp = \vec{0}_\perp) \right\rangle \tag{3}$$

$$= \int \frac{d^2\vec{k}_\perp dx}{\sqrt{x(1-x)}16\pi^3} \left[\ \psi_{+\frac{1}{2}\,+1}^{\uparrow}(x, \vec{k}_\perp) \left| +\frac{1}{2}\,+1\,;\ xP^+,\ \vec{k}_\perp \right\rangle \right.$$

$$+ \psi_{+\frac{1}{2}\,-1}^{\uparrow}(x, \vec{k}_\perp) \left| +\frac{1}{2}\,-1\,;\ xP^+,\ \vec{k}_\perp \right\rangle$$

$$+ \psi_{-\frac{1}{2}\,+1}^{\uparrow}(x, \vec{k}_\perp) \left| -\frac{1}{2}\,+1\,;\ xP^+,\ \vec{k}_\perp \right\rangle$$

$$\left. + \psi_{-\frac{1}{2}\,-1}^{\uparrow}(x, \vec{k}_\perp) \left| -\frac{1}{2}\,-1\,;\ xP^+,\ \vec{k}_\perp \right\rangle \right] ,$$

where the two-particle states $|s_f^z, s_b^z;\ xP^+, \vec{k}_\perp\rangle$ are normalized as in (2). Here s_f^z and s_b^z denote the z-component of the spins of the constituent fermion and boson, respectively. The wavefunctions can be evaluated explicitly in QED perturbation theory using the rules given in Refs. [2,4,6]:

$$\begin{cases} \psi_{+\frac{1}{2}\,+1}^{\uparrow}(x, \vec{k}_\perp) = -\sqrt{2}\frac{(-k^1+ik^2)}{x(1-x)}\,\varphi\ , \\[2mm] \psi_{+\frac{1}{2}\,-1}^{\uparrow}(x, \vec{k}_\perp) = -\sqrt{2}\frac{(+k^1+ik^2)}{1-x}\,\varphi\ , \\[2mm] \psi_{-\frac{1}{2}\,+1}^{\uparrow}(x, \vec{k}_\perp) = -\sqrt{2}(M - \frac{m}{x})\,\varphi\ , \\[2mm] \psi_{-\frac{1}{2}\,-1}^{\uparrow}(x, \vec{k}_\perp) = 0\ , \end{cases} \tag{4}$$

where

$$\varphi = \varphi(x, \vec{k}_\perp) = \frac{\frac{e}{\sqrt{1-x}}}{M^2 - \frac{\vec{k}_\perp^2 + m^2}{x} - \frac{\vec{k}_\perp^2 + \lambda^2}{1-x}} \cdot \tag{5}$$

Similarly, the wavefunctions for an electron with $J^z = -\frac{1}{2}$ are given by

$$\begin{cases} \psi^\downarrow_{+\frac{1}{2}+1}(x, \vec{k}_\perp) = 0 \,, \\ \psi^\downarrow_{+\frac{1}{2}-1}(x, \vec{k}_\perp) = -\sqrt{2} \left(M - \frac{m}{x} \right) \varphi \,, \\ \psi^\downarrow_{-\frac{1}{2}+1}(x, \vec{k}_\perp) = -\sqrt{2} \frac{-k^1 + ik^2}{1-x} \varphi \,, \\ \psi^\downarrow_{-\frac{1}{2}-1}(x, \vec{k}_\perp) = -\sqrt{2} \frac{k^1 + ik^2}{x(1-x)} \varphi \,. \end{cases} \tag{6}$$

The coefficients of φ in (4) and (6) are the matrix elements of $\frac{\bar{u}(k^+, k^-, \vec{k}_\perp)}{\sqrt{k^+}} \gamma \cdot$
$\epsilon^* \frac{u(P^+, P^-, \vec{P}_\perp)}{\sqrt{P^+}}$ which are the numerators of the wavefunctions corresponding to each constituent spin s^z configuration. The two boson polarization vectors in light-cone gauge are $\epsilon^\mu = (\epsilon^+ = 0 \,, \epsilon^- = \frac{2\vec{\epsilon}_\perp \cdot \vec{k}_\perp}{k^+}, \vec{\epsilon}_\perp)$ where $\vec{\epsilon} = \vec{\epsilon}_\perp^{\uparrow, \downarrow} = \mp(1/\sqrt{2})(\hat{x} \pm i\hat{y})$. The polarizations also satisfy the Lorentz condition $k \cdot \epsilon = 0$. In the above we have generalized the framework of QED by assigning a mass M to the external electrons, but a different mass m to the internal electron lines and a mass λ to the internal photon line [6]. The idea behind this is to model the structure of a composite fermion state with mass M by a fermion and a vector constituent with respective masses m and λ. In next sections we will express the form factors in terms of the example light-cone wavefunctions presented in this section in order to show the meaning of the formulae clearly. These formulae can be generalized to those expressed in terms of general light-cone wavefunctions [4,5].

3. Electromagnetic Form Factors

For a spin-$\frac{1}{2}$ composite system, the Dirac and Pauli form factors $F_1(q^2)$ and $F_2(q^2)$ are defined by

$$\langle P' | J^\mu(0) | P \rangle = \bar{u}(P') \left[F_1(q^2) \gamma^\mu + F_2(q^2) \frac{i}{2M} \sigma^{\mu\alpha} q_\alpha \right] u(P) \,, \tag{7}$$

where $q^\mu = (P' - P)^\mu$ and $u(P)$ is the bound state spinor. In the light-cone formalism the Dirac and Pauli form factors are conveniently identified from the spin-conserving and spin-flip vector current matrix elements of the J^+ current [6]:

$$\left\langle P + q, \uparrow \left| \frac{J^+(0)}{2P^+} \right| P, \uparrow \right\rangle = F_1(q^2) \,, \tag{8}$$

$$\left\langle P+q, \uparrow \left| \frac{J^+(0)}{2P^+} \right| P, \downarrow \right\rangle = -(q^1 - iq^2)\frac{F_2(q^2)}{2M} . \qquad (9)$$

We use the light-cone frame: $q = (q^+, \vec{q}_\perp, q^-) = \left(0, \vec{q}_\perp, \frac{-q^2}{P^+}\right)$, $P = (P^+, \vec{P}_\perp, P^-) = \left(P^+, \vec{0}_\perp, \frac{M^2}{P^+}\right)$, where $q^2 = -2P \cdot q = -\vec{q}_\perp^2$. We use the component notation $a = (a^+, \vec{a}_\perp, a^-)$ and our metric defined by $a^\pm = a^0 \pm a^3$ and $a \cdot b = \frac{1}{2}(a^+ b^- + a^- b^+) - \vec{a}_\perp \cdot \vec{b}_\perp$.

We can express the Dirac and Pauli form factors $F_1(q^2)$ and $F_2(q^2)$ using the light-cone wavefunction. From (4) and (8) we have

$$F_1(q^2) = \left\langle \Psi^\uparrow(P^+, \vec{P}_\perp = \vec{q}_\perp)) | \Psi^\uparrow(P^+, \vec{P}_\perp = \vec{0}_\perp) \right\rangle \qquad (10)$$

$$= \int \frac{d^2\vec{k}_\perp dx}{16\pi^3} \left[\psi^{\uparrow \, *}_{+\frac{1}{2}+1}(x, \vec{k}'_\perp)\psi^\uparrow_{+\frac{1}{2}+1}(x, \vec{k}_\perp) + \psi^{\uparrow \, *}_{+\frac{1}{2}-1}(x, \vec{k}'_\perp)\psi^\uparrow_{+\frac{1}{2}-1}(x, \vec{k}_\perp) \right.$$

$$\left. + \psi^{\uparrow \, *}_{-\frac{1}{2}+1}(x, \vec{k}'_\perp)\psi^\uparrow_{-\frac{1}{2}+1}(x, \vec{k}_\perp) \right] ,$$

where

$$\vec{k}'_\perp = \vec{k}_\perp + (1 - x)\vec{q}_\perp . \qquad (11)$$

The Pauli form factor is obtained from the spin-flip matrix element of the J^+ current. From (4), (6) and (9) we have

$$-\frac{(q^1 - iq^2)}{2M} F_2(q^2) = \left\langle \Psi^\uparrow(P^+, \vec{P}_\perp = \vec{q}_\perp)) | \Psi^\downarrow(P^+, \vec{P}_\perp = \vec{0}_\perp) \right\rangle \qquad (12)$$

$$= \int \frac{d^2\vec{k}_\perp dx}{16\pi^3} \left[\psi^{\uparrow \, *}_{+\frac{1}{2}-1}(x, \vec{k}'_\perp)\psi^\downarrow_{+\frac{1}{2}-1}(x, \vec{k}_\perp) + \psi^{\uparrow \, *}_{-\frac{1}{2}+1}(x, \vec{k}'_\perp)\psi^\downarrow_{-\frac{1}{2}+1}(x, \vec{k}_\perp) \right].$$

4. Generalized Parton Distributions

We begin with the kinematics of virtual Compton scattering

$$\gamma^*(q) + p(P) \to \gamma(q') + p(P') . \qquad (13)$$

We specify the frame by choosing a convenient parametrization of the light-cone coordinates for the initial and final proton: $P = \left(P^+ , \vec{0}_\perp , \frac{M^2}{P^+} \right)$, $P' = \left((1 - \zeta)P^+ , -\vec{\Delta}_\perp , \frac{M^2 + \vec{\Delta}_\perp^2}{(1-\zeta)P^+} \right)$, where M is the proton mass. The four-momentum transfer from the target is $\Delta = P - P' = \left(\zeta P^+ , \vec{\Delta}_\perp , \frac{t + \vec{\Delta}_\perp^2}{\zeta P^+} \right)$, where $t = \Delta^2$. In addition, overall energy-momentum conservation requires $\Delta^- = P^- - P'^-$, which connects $\vec{\Delta}_\perp^2$, ζ, and t according to $t = 2P \cdot \Delta = -\frac{\zeta^2 M^2 + \vec{\Delta}_\perp^2}{1 - \zeta}$.

In the limit $Q^2 = -q^2 \to \infty$ at fixed ζ and t the Compton amplitude is given by [7,8]

$$M^{IJ}(\vec{q}_\perp, \vec{\Delta}_\perp, \zeta) = \epsilon_\mu^I \epsilon_\nu^{*J} M^{\mu\nu}(\vec{q}_\perp, \vec{\Delta}_\perp, \zeta) = -e_q^2 \frac{1}{2\overline{P}^+} \int_{\zeta-1}^1 dx \quad (14)$$

$$\times \left\{ t^{IJ}(x,\zeta)\, \overline{U}(P') \left[H(x,\zeta,t)\, \gamma^+ + E(x,\zeta,t)\, \frac{i}{2M}\, \sigma^{+\alpha}(-\Delta_\alpha) \right] U(P) \right.$$

$$\left. + s^{IJ}(x,\zeta)\, \overline{U}(P') \left[\widetilde{H}(x,\zeta,t)\, \gamma^+\gamma_5 + \widetilde{E}(x,\zeta,t)\, \frac{1}{2M}\, \gamma_5(-\Delta^+) \right] U(P) \right\},$$

where $\overline{P} = \frac{1}{2}(P' + P)$. For circularly polarized initial and final photons (I, J are $\uparrow$ or $\downarrow$)) presented below (6), we have $t^{\uparrow\uparrow}(x,\zeta) = t^{\downarrow\downarrow}(x,\zeta) = \frac{1}{x-i\epsilon} + \frac{1}{x-\zeta+i\epsilon}$, $s^{\uparrow\uparrow}(x,\zeta) = -s^{\downarrow\downarrow}(x,\zeta) = \frac{1}{x-i\epsilon} - \frac{1}{x-\zeta+i\epsilon}$, and $t^{\uparrow\downarrow}$, $t^{\downarrow\uparrow}$, $s^{\uparrow\downarrow}$ and $s^{\downarrow\uparrow}$ are zero.

Since the coupling of the electromagnetic current $e_q J^+(0)$ on the quark line is identical to the Compton amplitude with $e_q^2 t^{IJ}$ replaced simply by the quark charge e_q, one finds

$$\int_{\zeta-1}^1 \frac{dx}{1 - \frac{\zeta}{2}}\, H(x,\zeta,t) = F_1(t)\,, \qquad \int_{\zeta-1}^1 \frac{dx}{1 - \frac{\zeta}{2}}\, E(x,\zeta,t) = F_2(t)\,. \quad (15)$$

Analogous sum rules relate $\widetilde{H}$ and $\widetilde{E}$ with the form factors of the axial vector current $J_5^\mu(y) = \overline{\psi}(y)\gamma^\mu\gamma_5\psi(y)$. The factors $(1-\zeta/2)$ in (15) appear because we use the normalization convention for the Compton form factors which involves $\overline{P}^+$ on the right-hand side of (14), and at the same time parametrize light-cone momentum fractions with respect to $P^+ = (1 - \zeta/2)\overline{P}^+$.

In the domain $\zeta < x < 1$, for a general value of ζ between 0 and 1, we can express the generalized parton distributions as the overlap of the light-cone wavefunctions [5]:

$$\frac{\sqrt{1-\zeta}}{1-\frac{\zeta}{2}}\, H_{(2\to2)}(x,\zeta,t) - \frac{\zeta^2}{4(1-\frac{\zeta}{2})\sqrt{1-\zeta}}\, E_{(2\to2)}(x,\zeta,t) \quad (16)$$

$$= \int \frac{d^2\vec{k}_\perp}{16\pi^3} \left[\psi_{+\frac{1}{2}+1}^{\uparrow\,*}(x',\vec{k}_\perp')\psi_{+\frac{1}{2}+1}^{\uparrow}(x,\vec{k}_\perp) + \psi_{+\frac{1}{2}-1}^{\uparrow\,*}(x',\vec{k}_\perp')\psi_{+\frac{1}{2}-1}^{\uparrow}(x,\vec{k}_\perp) \right.$$

$$\left. + \psi_{-\frac{1}{2}+1}^{\uparrow\,*}(x',\vec{k}_\perp')\psi_{-\frac{1}{2}+1}^{\uparrow}(x,\vec{k}_\perp) \right],$$

$$\frac{1}{\sqrt{1-\zeta}}\, \frac{(\Delta^1 - i\Delta^2)}{2M}\, E_{(2\to2)}(x,\zeta,t) \quad (17)$$

$$= \int \frac{d^2\vec{k}_\perp}{16\pi^3} \left[\psi_{+\frac{1}{2}-1}^{\uparrow\,*}(x',\vec{k}_\perp')\psi_{+\frac{1}{2}-1}^{\downarrow}(x,\vec{k}_\perp) + \psi_{-\frac{1}{2}+1}^{\uparrow\,*}(x',\vec{k}_\perp')\psi_{-\frac{1}{2}+1}^{\downarrow}(x,\vec{k}_\perp) \right],$$

where

$$x' = \frac{x-\zeta}{1-\zeta}, \qquad \vec{k}_\perp' = \vec{k}_\perp - \frac{1-x}{1-\zeta}\, \vec{\Delta}_\perp\,. \quad (18)$$

Analogous formulae hold in the domain $\zeta - 1 < x < 0$, where the struck parton in the target is an antiquark instead of a quark. In the domain $0 \leq x \leq \zeta$, the parton number changing $n + 1 \rightarrow n - 1$ ($\Delta n = -2$) off-diagonal transition matrix elements contribute. The formulae for the domain $0 \leq x \leq \zeta$ are given in Ref. [5]. The same situation also occurs in the heavy meson decays since the decay form factors are timelike [3]. In general, when the initial and final hadrons have different values of the + component of momentum, there are parton number changing contributions as well as parton number conserving ones.

5. Conclusions

The light-cone Fock representation allows one to compute the matrix elements of local currents as overlap integrals of the light-cone wavefunctions which are frame independent. In particular, we can evaluate the forward and non-forward matrix elements of the electroweak currents, moments of the deep inelastic structure functions, as well as the electromagnetic form factors. Given the local operators for the energy-momentum tensor $T^{\mu\nu}(x)$ and the angular momentum tensor $M^{\mu\nu\lambda}(x)$, one can directly compute momentum fractions, spin properties, the gravitomagnetic moment, and the form factors $A(q^2)$ and $B(q^2)$ appearing in the coupling of gravitons to composite systems [4,5]. We presented the formulae which express the electromagnetic form factors and generalized parton distributions as overlap integrals of the light-cone wavefunctions. These formulae provided useful physical intuitions in the related processes.

Acknowledgments

The author wishes to thank Stan Brodsky, Markus Diehl, Bo-Qiang Ma, and Ivan Schmidt for the collaborations on the contents presented in this article. This work is supported in part by the LG Yonam Foundation.

References

1. S. J. Brodsky, H. Pauli, and S. S. Pinsky, Phys. Rept. **301** (1998) 299.
2. G. P. Lepage and S. J. Brodsky, Phys. Rev. **D22** (1980) 2157.
3. S. J. Brodsky and D. S. Hwang, Nucl. Phys. **B543** (1999) 239.
4. S. J. Brodsky, D. S. Hwang, B. Ma, and I. Schmidt, Nucl. Phys. **B593** (2001) 311.
5. S. J. Brodsky, M. Diehl, and D. S. Hwang, Nucl. Phys. **B596** (2001) 99.
6. S. J. Brodsky and S. D. Drell, Phys. Rev. **D22** (1980) 2236.
7. X. Ji, Phys. Rev. Lett. **78** (1997) 610; Phys. Rev. **D55** (1997) 7114.
8. A. V. Radyushkin, Phys. Rev. **D56** (1997) 5524.

GENERALIZED PARTON DISTRIBUTIONS FOR WEAKLY BOUND SYSTEMS FROM LIGHT-FRONT QUANTUM MECHANICS

B. C. TIBURZI

Department of Physics
University of Washington
Box 351560
Seattle, WA 98195-1560
E-mail: bctiburz@u.washington.edu

We present generalized parton distributions for weakly bound systems on the light cone in order to build intuition about the light-front formalism. Physics at the crossover is reviewed in terms of the light-cone Fock space representation. Furthermore, we link light-cone Fock components to the equal (light-cone) time projection of the covariant Bethe-Salpeter amplitude. Continuity of the distributions arises naturally in this weak binding model.

1. Introduction

Recently much activity has been geared to investigating the next generation of exclusive processes at high momentum transfer[1]. Deeply virtual Compton scattering (DVCS) represents perhaps the cleanest of these to describe theoretically. The DVCS amplitude is a convolution of a hard scattering part and a set of new structure functions, which, at leading twist, are the generalized parton distributions (GPDs). The GPDs are off-diagonal matrix elements of bilocal field operators which interpolate between the inclusive physics of parton distributions and the exclusive limit of electromagnetic form factors. Geometric and optical interpretations of GPDs have elucidated some of the new physics they encode[2].

The purpose of this talk is two-fold. Firstly we wish to provide an intuitive basis for the light-front formalism. Light-cone correlations appear in hard processes and thus these correlators have a simple interpretation in terms of light-cone wavefunctions. Secondly we present only simple models of light-cone wavefunctions in order to demonstrate some general properties of GPDs, for example. Comparatively little work has been done to show the light-cone Fock representation of GPDs is consistent with known symmetries and limits.

The organization is as follows. We begin with the simplest example of a light-cone wavefunction: a valence quark model for the proton. This model is shown to be insufficient for describing beam-spin asymmetries which are sensitive to GPDs at the crossover between kinematic regions. Higher Fock components are needed and are derived in a weak binding, spinless model. These GPDs are continuous at the crossover due to simple relations between Fock components at vanishing plus-momentum.

2. Valence quark models

One can construct a relativistic valence quark model of a hadron by simply modeling the lowest light-cone Fock component. We analyzed GPDs from *ad hoc* valence two-body wavefunctions earlier[3]. For the proton, one would truncate the Fock space to exclude all but the three-body Fock component. To determine this constituent quark wavefunction of the proton, one would solve a quantum mechanical bound-state equation of the form

$$\left[\sum_{i=1}^{3} \frac{\mathbf{k}_i^{\perp 2} + m_i^2}{2x_i} + \hat{V} \right] \psi_3(x_i, \mathbf{k}_i^{\perp}) = M^2 \, \psi_3(x_i, \mathbf{k}_i^{\perp}), \tag{1}$$

where the m_i are constituent quark masses, and $\hat{V}$ is some effective interaction. Since solving a three-body equation is often quite difficult, the form of the wavefunction is usually postulated. The functional form used is not completely arbitrary; there are minimal physical restrictions to be satisfied. The restriction which concerns us below is at a vanishing plus-momentum fraction: $x_j = 0$. From Eq. (1), the kinetic term as well as the interaction $\hat{V}$ force the valence wavefunction to vanish, i.e. $\psi_3(\ldots, x_j = 0, \ldots) = 0$.

As is known, such valence models lead to reasonable success at low energy phenomenology, particularly hadronic spectroscopy and form factors. We shall see below that such models are insufficient for modeling the GPDs.

3. At the crossover

DVCS probes a light-cone correlation of the quark fields. Not surprisingly the GPDs can be expressed most simply as a sum of Fock component overlaps[4]. The most direct way to experimentally access the GPDs from the Bethe-Heitler background is through beam asymmetries, such as the beam-spin asymmetry[5]. Additionally the beam-spin asymmetry directly probes the Fock component overlaps since only the imaginary part of the amplitude is needed[a]

$$\Im\mathcal{M} \propto \frac{\sqrt{1-\zeta}}{1-\frac{\zeta}{2}} H(\zeta,\zeta,t) - \frac{\zeta^2}{4(1-\frac{\zeta}{2})\sqrt{1-\zeta}} E(\zeta,\zeta,t) \qquad (2)$$

$$= \sum_{n,\lambda_i} \sqrt{1-\zeta}^{2-n} \int_{\{n\}} \delta(\zeta - x_j)\psi_n^{\uparrow*}(x_i',\mathbf{k}_i'^{\perp},\lambda_i)\psi_n^{\uparrow}(x_i,\mathbf{k}_i^{\perp},\lambda_i) \qquad (3)$$

where $\int_{\{n\}} = \int \prod_{i=1}^{n} \frac{dx_i d\mathbf{k}_i^{\perp}}{16\pi^3} 16\pi^3 \delta(1 - \sum x_i)\delta(\sum \mathbf{k}_i^{\perp})$ and the primed variables are given by: $x_j' = 0$, $\mathbf{k}_j'^{\perp} = \mathbf{k}_j^{\perp} - \mathbf{\Delta}^{\perp}$ for the struck quark and $x_i' = \frac{x_i}{1-\zeta}$, $\mathbf{k}_i'^{\perp} = \mathbf{k}_i^{\perp} + \frac{x_i}{1-\zeta}\mathbf{\Delta}^{\perp}$ for the spectators. Above the final state Fock components are evaluated at a vanishing momentum fraction. Thus modeling the physics in the lowest Fock state forces the Compton amplitude to have a vanishing imaginary part. Higher Fock components are needed for a non-vanishing imaginary part, since they generally do not vanish for zero plus-momentum.[b]

Not surprisingly valence quark models that match electromagnetic form factors fail to capture the physics of GPDs. As such, valence models are poor interpolators between the exclusive limit of form factors and the inclusive physics of quark distribution functions. A popular, simple *Ansatz* accomplishes this interpolation trivially:

$$H(x,\zeta = 0,t;Q^2) \sim q(x;Q^2)F(t), \qquad (4)$$

where $F(t)$ is the dipole fit to the form factor, and $q(x;Q^2)$ is a fit to the quark distribution function from deep inelastic scattering data[c]. This form is too simple, although it respects the properties of GPDs and is hence difficult to improve upon[d]. The full Fock space expansion, however, contains the physics of both limits and the light-cone wavefunctions are hence the true interpolators (from which no factorized *Ansätze* are discernible).

---

[a]This is in contrast to the charge asymmetry which is sensitive to the real part of the Compton amplitude. The real part involves an integral of the GPDs weighted by the hard scattering amplitude.
[b]This point is clear if we utilize the forward limit: $\{\zeta,t\} \to 0$, for which the sum in Eq. (3) reduces to the quark distribution function at zero plus-momentum. In a valence model, the quark distribution function vanishes at this point whereas realistically the higher Fock states lead to a non-zero value (which is of course sizeable at a high scale).
[c]Here we include the scale dependence of the GPD which can conveniently be absorbed into the *Ansatz* via the quark distribution function. In this way, the GPD will evolve while the form factor, which stems from a conserved current, will not.
[d]This *Ansatz* has been augmented by additional physics, such as pion-pole contributions *etc.*, and has been used for quite complete phenomenological estimates of cross sections including twist-three contributions[6].

4. Weak binding model

Although exact, the Fock space representation of DVCS is not entirely useful since non-perturbative solutions to the N-body wavefunctions are obviously not currently available. Since properties of GPDs are largely unexplored in terms of the Fock space representation, we adopt a two-body perturbative model to gain some insight into the problem. To deal with ambiguities (particularly the so-called non-wavefunction vertices), we start from a covariant framework and project onto the light-cone by integrating out the light-front energy dependence. Schematically $\int dk^-$ projects onto the surface $x^+ = 0$. For scattering states, it has been shown[7] that projecting covariant perturbation theory onto the light-cone results in light-cone time-ordered perturbation theory, including the delicate issue of renormalization.

On the other hand, for two-body bound states, a three-dimensional reduction scheme has been found[8] that reproduces light-front time-ordered kernels in the Bethe-Salpeter formalism. We recently applied this reduction scheme to bound-state matrix elements of the electromagnetic current[9] from which one can extract GPDs.

The model consists of a two-body bound state (completely devoid of spin) in the ladder approximation. The bound state can be described co-variantly by the Bethe-Salpeter wavefunction $\Psi(k, P)$, where P labels the total four-momentum of the system. The two-body Fock component is merely the projection of the Bethe-Salpeter wavefunction onto the light-cone

$$\psi_2(x, \mathbf{k}_{rel}^\perp) \sim \int dk^- \Psi(k, P), \tag{5}$$

where the relative transverse momentum is $\mathbf{k}_{rel}^\perp = \mathbf{k}^\perp - x\mathbf{P}^\perp$ and $x = k^+/P^+$. The GPD is then extracted from the integrand of the form factor F, which is a matrix element of the electromagnetic current J^μ between Bethe-Salpeter vertices.

$$F \sim \int d^4k \cdots \to \int dk^+ \delta(xP^+ - k^+) dk^- d^2\mathbf{k}^\perp \cdots \tag{6}$$

The inserted delta function above keeps the plus-momentum of the struck quark fixed; the GPD is what remains of the integral. Additionally performing the k^- integral allows us to write expressions in terms of light-front wavefunctions in light-front time-ordered perturbation theory.

To zeroth order in the coupling, the GPD is an overlap of two-body wavefunctions that generalizes the Drell-Yan formula to a frame in which $\Delta^+ \neq 0$. This expression is non-vanishing only in the region $x > \zeta$. The contributions at first-order for $x > \zeta$ are depicted in Figure 1. Each contains

122

a mediator that cannot be absorbed into the initial or final state vertices. Including the diagram in which the spectator quark has a one-loop self-interaction, we see the next-to-leading contributions have the form of a three-to-three overlap (consistent with $x > \zeta$ contributions being diagonal in Fock space). Indeed the form of the three-body wavefunction is exactly as we would write down directly from time-ordered perturbation theory. In the region $x < \zeta$, the leading contribution is at first-order in the coupling. The relevant diagrams are shown in Figure 2 and are four-to-two overlaps. These diagrams cannot be summed into an effective non-wavefunction vertex[e] since the expressions do not have the same functional dependence (and hence have differing covariance properties).

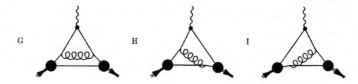

Figure 1.   Next-to-leading contributions to the GPD in the region $x > \zeta$.

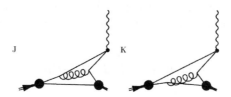

Figure 2.   Leading contributions to the GPD for $x < \zeta$.

Continuity of the GPD at $x = \zeta$ is required by factorization[11]. If the GPDs were discontinuous, the Compton amplitude would be logarithmically divergent. This weak binding model's GPD is continuous. As expected, the valence overlap vanishes at the crossover (similarly does diagram H). The Born terms (G and J) match up at $x = \zeta$ due to the dominance of the rebounding quark's energy in the time-ordering. For this reason, the final-state iteration diagrams (I and K) also match at the crossover. Furthermore, the GPD does not vanish at the crossover, which naturally

---

[e]If the four-dimensional kernel were really three-dimensional[10], i.e. independent of light-front energy (instantaneous in light-cone time), one could appeal to crossing since true higher Fock states would be absent.

stems from higher Fock states. In this model, continuity at the crossover generalizes easily to all orders in time-ordered perturbation theory.

5. Conclusion

Above we have seen the structure of GPDs in a simple weak binding model. Continuity of the distributions was maintained naturally by simple relations between Fock components evaluated at a vanishing plus-momentum fraction. In QCD, however, the relations between light-cone Fock components at small-x are far richer[12]. We have yet to see how the sum rule, which relates the zeroth moment of the GPD to the electromagnetic form factor, arises from relations between Fock components. This should first be tackled perturbatively. As to constructing phenomenological models of GPDs from light-front wavefunctions, we have demonstrated that valence models are insufficient which suggests either that constituent quark substructure needs to be added or that higher Fock components need to be modeled. Proceeding on either course is not simple: continuity and polynomiality conditions are serious constraints.

Acknowledgments

G. A. Miller provided much insight throughout the course of this project. This work was funded by the U. S. DOE, grant: DE-FG03 − 97ER41014.

References

1. X.-D. Ji, J. Phys. G **24**, 1181 (1998); A. V. Radyushkin, hep-ph/0101225; K. Goeke, et al. Prog. Part. Nucl. Phys. **47**, 401 (2001).
2. M. Burkardt, Phys. Rev. D **62**, 071503 (2000); hep-ph/0105324; J. P. Ralston and B. Pire, hep-ph/0110075.
3. B. C. Tiburzi and G. A. Miller, Phys. Rev. C **64**, 065204 (2001).
4. M. Diehl, et al. Nucl. Phys. B **596**, 33 (2001) [Erratum-ibid. B **605**, 647 (2001)]; S. J. Brodsky, et al. Nucl. Phys. B **596**, 99 (2001).
5. M. Diehl, et al. Phys. Lett. B **411**, 193 (1997).
6. A. V. Belitsky, D. Müller and A. Kirchner, Nucl. Phys. B **629**, 323 (2002).
7. N. E. Ligterink and B. L. Bakker, Phys. Rev. D **52**, 5954; 5917 (1995).
8. J. H. Sales, et al. Phys. Rev. C **61**, 044003 (2000); **63**, 064003 (2001).
9. B. C. Tiburzi and G. A. Miller, hep-ph/0205109.
10. B. C. Tiburzi and G. A. Miller, Phys. Rev. D **65**, 074009 (2002).
11. A. V. Radyushkin, Phys. Rev. D **56**, 5524 (1997).
12. F. Antonuccio, S. J. Brodsky and S. Dalley, Phys. Lett. B **412**, 104 (1997).

SEPARATION OF SOFT AND HARD PHYSICS IN DVCS

A. GÅRDESTIG, A.P. SZCZEPANIAK AND J.T. LONDERGAN

Indiana University Nuclear Theory Center,
2401 Milo B. Sampson Lane,
Bloomington, IN 47408, USA
E-mail: agardest@indiana.edu

A model for deeply virtual Compton scattering, based on analytical light-cone hadron wave functions is presented and studied at energies currently accessible at Jefferson Laboratory and DESY. It is shown that poles and perpendicular vector components play an important role at $Q^2 < 10$ GeV^2. A Q^2 suppressed diagram has to be included at these low energies, but becomes negligible above 10 GeV^2. Future prospects and developments of this model are discussed.

1. Introduction

For many years, deeply inelastic scattering (DIS) has been a major source of our knowledge of parton distributions in nucleons and nuclei. This is because it can be shown that, at sufficiently high energies and momentum transfers, the amplitude for this process factorizes into a 'hard' part which can be calculated from QCD, and a 'soft' part which can be extracted from experimental data. The soft part can be proved to be related to the probability of finding a quark with a particular flavor carrying a given fraction of the nucleon's momentum.

In recent years, much interest has been focused on studying deeply virtual Compton scattering and the electroproduction of mesons, following the proof by Collins, Frankfurt and Strikman[1]. This proof demonstrated that, under quite general conditions, the leading amplitude for hard exclusive photo-production of mesons could also be factorized into a calculable 'hard' part, and a 'soft' part. All other amplitudes were smaller than the leading amplitude by powers of $1/Q$. The soft part corresponds to the process by which a parton with a certain momentum fraction is removed from a nucleon, and replaced by a parton with a different momentum fraction. These parton distributions have been given a variety of names, but we will refer to them as 'generalized parton distributions' (GPD's). In the case of quark helicity conservation, there are four independent GPD's; $H(x, \zeta, t)$,

$E(x,\zeta,t)$, $\widetilde{H}(x,\zeta,t)$, and $\widetilde{E}(x,\zeta,t)$, where x and ζ are the light-cone momentum fractions of the struck quark and real photon, while $t = \Delta^2$ is the momentum transfer squared. Three physically different regions could be distinguished for x and ζ. The domain $0 < \zeta < x < 1$ ($\zeta - 1 < x < 0$) corresponds to the removal and replacement of a quark (antiquark) with momentum fractions x ($\zeta - x$) and $x - \zeta$ ($-x$), respectively. In the remaining region $0 < x < \zeta$, the photon scatters on a virtual quark-antiquark pair, extracted from the proton. With this notation, $\zeta \to x_B$ (Bjorken x) in the limit $Q^2 \to \infty$ (Δ fixed). In the limit of forward scattering (DIS), the H's reduce to the quark density and quark helicity distributions:

$$H(x,0,0) = q(x) \tag{1}$$

$$\widetilde{H}(x,0,0) = \Delta q(x). \tag{2}$$

The E's do not appear in DIS, they are unique to the off-forward exclusive processes and provide information not accessible through other means. The GPD's are related to the nucleon form factors by the integrals

$$\int_{\zeta-1}^{1} \frac{dx}{1-\frac{\zeta}{2}} H^q(x,\zeta,t) = F_1^q(t), \tag{3}$$

$$\int_{\zeta-1}^{1} \frac{dx}{1-\frac{\zeta}{2}} E^q(x,\zeta,t) = F_2^q(t), \tag{4}$$

and similarly for $\widetilde{H}$ and $\widetilde{E}$, all independent of ζ. The factor $1-\frac{\zeta}{2}$ is included to comply with the normalization of form factors used by Ji[2]. The GPD's might also shed some light on the nucleon spin decomposition, since Ji[2] has related them to the quark spin:

$$J_q = \frac{1}{2} \int_{\zeta-1}^{1} \frac{dx}{1-\frac{\zeta}{2}} x[H^q(x,\zeta,t=0) + E^q(x,\zeta,t=0)]. \tag{5}$$

The quark angular momentum is decomposed into $J_q = S_q + L_q$, where S_q is measured in polarized DIS experiments. If the above sum-rule is measured in DVCS, the quark spin and orbital angular momentum parts could be separated.

An alternative approach to DVCS is offered by the light-cone quark wave functions as suggested by Brodsky, Diehl, and Hwang[3] (BDH). They have used this idea to calculate DVCS on an electron in QED for large Q^2.

There are ambitious efforts under way to measure DVCS (and the hard exclusive meson photo-production) at a number of facilities, DESY and JLab in particular[4]. At JLab energies, the competing Bremsstrahlung or Bethe-Heitler (BH) process is larger than DVCS. However, by doing interference measurements (e^+/e^- beam charge asymmetry and various spin

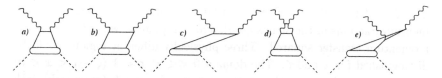

Figure 1.   The five different time-ordered 'handbag' diagrams.

asymmetries) the BH amplitude cancels out and only a BH×DVCS interference remains. At DESY the energy is large enough for DVCS to become
larger than BH, though the measurements have low statistics, not allowing for differential cross sections. Data for the asymmetries are, however,
at both facilities collected for relative low beam energies and momentum
transfers. It is thus interesting to investigate to what extent the 'leading'
amplitude is actually dominating in this kinematic region. We will here
present the first few steps toward such an understanding, by developing a
simple model using effective analytic quark-diquark wave functions, which
allows us to study the amplitudes excited at these low energies. This model
is similar to the one of BDH, but we calculate DVCS on a proton and focus
especially on the features at low Q^2, keeping higher twist terms.

2. Formalism

We start from the Fourier transform of the $\gamma^* p \to \gamma p'$ amplitude (in light-
cone coordinates)

$$T^{++} = \int d^4 y e^{iq' \cdot y} \langle p' | T J^+(y) J^+(0) | p \rangle, \qquad (6)$$

where $J^+(y) = \bar{\psi}(y) \gamma^+ \gamma_5 \frac{\tau}{\sqrt{2}} \psi(y)$ is the electromagnetic current and $p(p')$
and $q(q')$ are the four-momenta of the initial (final) hadron and photon.
This expression could be expanded to give the five different one-loop diagrams (assuming light-cone gauge $q^+ = 0$) shown in Fig. 1. These diagrams
are essentially the same as those used by BDH for DVCS on an electron.
The diagrams are represented by the integrals (because of space limitations,
only a and b are given)

$$T_a^{++} = 2(p^+)^2 \int_{\zeta < x < 1} \frac{dx d^2 \mathbf{k}_\perp}{16\pi^3} \phi^\dagger(z, \mathbf{l}_\perp)$$

$$\times \frac{i}{\left[M^2 + \left(\frac{1}{x_B} - 1 \right) Q^2 - \frac{m^2 + \mathbf{m}_{\perp a}^2}{x} - \frac{m_R^2 + \mathbf{m}_{\perp a}^2}{1-x} + i\epsilon \right]} \phi(x, \mathbf{k}_\perp), \quad (7)$$

$$T_b^{++} = 2(p^+)^2 \int_{0<x<\zeta} \frac{dx d^2\mathbf{k}_\perp}{16\pi^3} \frac{\left(m_R^2 - \frac{m^2+\mathbf{l}_\perp'^2}{z'} - \frac{M^2+\mathbf{l}_\perp'^2}{1-z'}\right)\varphi(z',\mathbf{l}_\perp')}{-\frac{\zeta(m^2+\mathbf{m}_{\perp b}^2)}{x(\zeta-x)}}$$

$$\times \frac{i}{\left[M^2 + \left(\frac{1}{x_B}-1\right)Q^2 - \frac{m^2+\mathbf{m}_{\perp a}^2}{x} - \frac{m_R^2+\mathbf{m}_{\perp a}^2}{1-x} + i\epsilon\right]}\phi(x,\mathbf{k}_\perp), \quad (8)$$

where $z = (x - \zeta)/(1 - \zeta)$, $z' = (\zeta - x)/(1 - x)$, and the relative momenta are defined as $\mathbf{l}_\perp = \mathbf{k}_\perp + (1 - z)\Delta_\perp$, $\mathbf{l}_\perp' = -(1 - z')\mathbf{k}_\perp - \Delta_\perp$, $\mathbf{m}_{\perp a} = \mathbf{k}_\perp + (1 - x)\mathbf{q}_\perp$, and $\mathbf{m}_{\perp b} = \mathbf{k}_\perp + \frac{x}{\zeta}\Delta_\perp + \frac{\zeta-x}{\zeta}\mathbf{q}_\perp$. Here $x = k^+/p^+ > 0$ and $\zeta = q'^+/p^+$ are the longitudinal momentum fractions of the struck quark and the real photon, m, m_R, and M the quark, remnant (diquark), and hadron masses, and $\Delta = q - q'$. The expressions for the other diagrams have similar structures. In the limit $Q^2 \to \infty$, the denominator of Eq. (8) is proportional to $x - \zeta + i\epsilon$, i.e., the leading twist expression of Ji[2]. Note that the integrals are over different parts of phase space and include scattering on a quark or quark-antiquark pair only, not scattering on an antiquark. The wave function is represented by the analytical form

$$\phi(x,\mathbf{k}_\perp) = N \exp\left[-\frac{1}{\beta^2}\left(\frac{m^2}{x} + \frac{m_R^2}{1-x} + \frac{\mathbf{k}_\perp^2}{x(1-x)}\right)\right], \quad (9)$$

where $\beta = 0.69$ GeV is chosen such that $F(q^2) = \int \frac{dx d^2\mathbf{k}_\perp}{16\pi^3}\phi(x,(1-x)\mathbf{q}_\perp + \mathbf{k}_\perp)\phi(x,\mathbf{k}_\perp)$ agrees with the dipole form factor for $Q^2 < 1$ GeV^2. The skew diagrams require the knowledge of the wave function $\phi(z',\mathbf{l}_\perp')$ for the diquark splitting into a hadron and quark (lower right-hand corner of diagrams b, c, and e). The form of this wave function will eventually be restricted by exclusive data, but is here arbitrarily chosen to be of the form of Eq. (9), with m_R and x replaced by M and z'.

3. Results

The T^{++} matrix elements have been calculated for present JLab and DESY kinematics and are shown in Fig. 2. The laboratory angle $\theta_{\gamma\gamma'}$ between the virtual and real photons is defined for in-plane kinematics such that it is positive for $\phi = 0$ and negative for $\phi = 180°$, where ϕ is the azimuth angle between the final electron and proton, with $\hat{\mathbf{q}}$ as the polar axis. These calculations incorporates the principal value parts of the integrals, i.e., the imaginary part of the T^{++}.

The forward peaking of the skew diagrams (b, c, and e) is a consequence of the shift of the momentum $\mathbf{k}_\perp$ in the extra denominator. The maximum of the integral is dislocated to momenta where this denominator becomes very small.

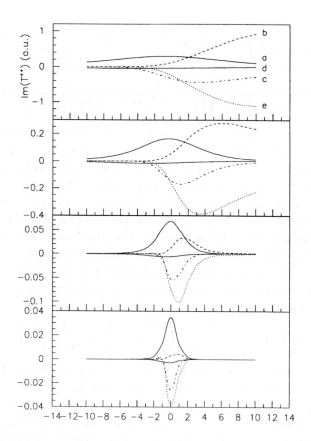

Figure 2. The $\mathrm{Im}T^{++}$ matrix elements as functions of the laboratory angle between the virtual and the real photon for in-plane scattering. The labels correspond to the ones of Fig. 1 (the lower solid line always correspond to diagram d). The calculations are for $x_B = 0.35$ and $Q^2 = 2, 4, 10, 20$ GeV^2 from top to bottom.

Because of its two hard propagator, diagram b is expected to be suppressed in the high Q^2 limit[3], and our calculation indicates that this suppression becomes significant for $Q^2 > 10$ GeV^2, i.e., well above present JLab energies.

While diagrams a and b exhibit a physical pole (for massless final photon), the other diagrams never get vanishing denominators for physical values of kinematic parameters. Thus the real part of the DVCS amplitude, which comes from the δ-function part of the propagators, has contributions from these first two diagrams only. The real part calculations will not be presented here.

This difference in pole structure also explains why diagram d is smaller than diagram a, since Q^2 terms add up in the denominator of d, while they could cancel for a. This feature is closely related to the behavior of the two leading-twist propagators of Ji[2]; $1/(x - \xi + i\epsilon)$ and $1/(x + \xi - i\epsilon)$, where x and ξ are momentum fractions related to $\frac{1}{2}(p^+ + p'^+)$ instead of p^+. In this notation there is a cancellation for $x = \xi$ (scattering on quark) or $x = -\xi$ (antiquark). These simplified propagators neglect four-vector components, e.g., $\Delta_\perp$, that do not give large scalars in the Bjorken limit.

4. Conclusions and outlook

We have introduced a model for DVCS, using simple quark wave functions and retaining all components of four-vectors throughout. The perpendicular momenta are shown to be very important for the location of the maxima of the integrals at low Q^2. Within the model we have been able to investigate the five single-loop diagrams and their relative importance for DVCS at JLab and HERA energies. The results indicate that at present JLab energies ($Q^2 < 4$ GeV^2), all of the diagrams (except possibly the crossed diagram d) need to be considered. In particular it is necessary to include diagram b, despite its two hard propagators, since the expected suppression is significant only for larger Q^2. Other higher-order diagrams might need to be considered as well. At $Q^2 > 10$ GeV^2, the process is completely dominated by the handbag diagrams with one hard propagator.

This work will be extended to cover a wider kinematic range in Q^2 and x_B and will include antiquark contributions to the amplitudes and the wave functions. The photo-production of mesons will also be calculated, with special consideration of the meson poles then made possible in the skew diagrams.

In order to compare with actual JLab and HERA experiments the interference with the Bremsstrahlung (Bethe-Heitler) process has to be calculated. We would then obtain cross sections and asymmetries to test against experimental results and be able to check the validity of commonly used approximations and to evaluate various wave functions that could be used.

This work was supported in part by NSF grant NSF-PHY0070368 and DOE grant DE-FG02-87ER40365.

References

1. J.C. Collins, L. Frankfurt and M. Strikman, *Phys. Rev.* **D56**, 2982 (1997).
2. X. Ji, *Phys. Rev. Lett.* **78**, 610 (1997); *Phys. Rev.* **D55**, 7114 (1997).
3. S. J. Brodsky, M. Diehl and D. S. Hwang, Nucl. Phys. **B596**, 99 (2001).
4. see e.g., A. Airapetian *et al.*, *Phys. Rev. Lett.* **87**, 182001 (2001); S. Stepanyan *et al.*, *Phys. Rev. Lett.* **87**, 182002 (2001).

SCALE DEPENDENCE OF HADRONIC WAVE FUNCTIONS AND PARTON DENSITIES

FENG YUAN, MATTHIAS BURKARDT, XIANGDONG JI

Department of Physics, University of Maryland, College Park, Maryland 20742

We study how the components of hadronic wave functions in light-cone quantization depend on the ultraviolet cut-off by relating them in a systematic way to the matrix elements of a class of quark-gluon operators between the QCD vacuum and the hadrons. From this, we derive an infinite set of scale-evolution equations for the individual contributions to parton distributions from the Fock expansion. When summed over all the contributions, we recover the well-known DGLAP equation.

In light-cone quantization and light-cone gauge[1], the hadronic states in QCD are expressed as an expansion of various quark and gluon Fock components [2,3]. Here, we are interested in how the light-cone Fock components depend on the momentum cut-off. We approach this problem by systematically relating the Fock expansion to the matrix elements of a certain class of quark-gluon operators between the QCD vacuum $|0\rangle$ and the hadron states. The scale-dependence of the wave function amplitudes can then be traced to the wave function renormalization constants of quark and gluon fields. Following this, we derive the scale dependence of the parton densities from individual Fock components. The scale evolution of these contributions obeys an infinite set of coupled, linear differential-integral quations. When summed over all Fock contributions, we recover the well-known DGLAP equation for parton densities.

The light-cone time x^+ and coordinate x^- are defined as $x^{\pm} = 1/\sqrt{2}(x^0 \pm x^3)$. Likewise we define Dirac matrices $\gamma^{\pm} = 1/\sqrt{2}(\gamma^0 \pm \gamma^3)$. The projection operators for Dirac fields are defined as $P_{\pm} = (1/2)\gamma^{\mp}\gamma^{\pm}$. Any Dirac field ψ can be decomposed into $\psi = \psi_+ + \psi_-$ with $\psi_{\pm} = P_{\pm}\psi$. ψ_+ is a dynamical degrees of freedom. For the gluon fields in the light-cone gauge $A^+ = 0$, $A_{\perp}$ is dynamical.

The key observation is that the light-cone Fock expansion of a hadron state is *completely* defined by the matrix elements of a special class equal light-cone time quark-gluon operators between the QCD vacuum and the hadron. These operators are specified as follows: Take the + component of the Dirac field ψ_+ and the $+\perp$-component of the gauge field $F^{+\perp}$. Assume all these fields are at light-cone time $x^+ = 0$, but otherwise with arbitrary dependence on other spacetime coordinates. Products of these fields with the right quantum numbers (spin, flavor, and color) define a set of operator basis. To simplify the discussion, we assume all fields in the operators are normal ordered, i.e., the annihilation operators appearing at the right of the creations. We believe that the matrix elements of all these operators between the hadron state and the QCD vacuum yield *complete* information about the hadron wave function.

As an example, let us consider π^+ meson with momentum P^μ along the z-direction. The leading light-cone Fock states consist of a pair of up and anti-down quarks. The light-cone helicity of the π^+ meson is zero, but the light-cone helicity of the quark-antiquark pair can either be zero or ± 1. Use $u_+(\xi^-, \xi_\perp)$ to represent the up-quark field in the coordinate space and $\bar{d}_+(0)$ the anti-down-quark field. In the massless limit, the operator $\bar{d}_+ \gamma^+ \gamma_5 u_+$ yields a helicity-0 pair, and $\bar{d}_+ \sigma^{+\perp} \gamma_5 u_+$ a helicity-1 pair. The first operator defines a coordinate amplitude,

$$\langle 0|\bar{d}_+(0)\gamma^+\gamma_5 u_+(\xi^-, \xi_\perp)|\pi^+(P)\rangle = \phi_0(\xi^-, \xi_\perp)2P^+ . \tag{1}$$

Introducing the Fourier tranformation of the amplitude, we find

$$|\pi^+(P)\rangle = \int \frac{d^2 k_\perp}{(2\pi)^3} \frac{dx}{2\sqrt{x(1-x)}} \phi_0(x, k_\perp)[b^\dagger_{u\uparrow} d^\dagger_{u\downarrow} - b^\dagger_{u\downarrow} d^\dagger_{d\uparrow}]|0\rangle + ... \tag{2}$$

The operator with helicity-1 quark-anti-quark pair defines the amplitude

$$\langle 0|\bar{d}_+(0)\sigma^{+i}\gamma_5 u_+(\xi^-, \xi_\perp)|\pi^+(P)\rangle = \partial^i \phi_1(\xi^-, \xi_\perp)2P^+ . \tag{3}$$

Similily, we find a light-cone Fock component

$$|\pi^+(p)\rangle = \int \frac{d^2 k_\perp}{(2\pi)^3} \frac{dx}{2\sqrt{x(1-x)}} \phi_1(x, k_\perp) \left[(k_1 - ik_2)b^\dagger_{u\uparrow} d^\dagger_{d\uparrow} \right.$$
$$\left. + (k_1 + ik_2)b^\dagger_{u\downarrow} d^\dagger_{d\downarrow} \right] |0\rangle + ... \tag{4}$$

It is now straightfoward to study the cut-off dependence of $\phi_{0,1}(x, k_\perp)$. Besides the explicit cut-off dependence in the wave function amplitudes, the $k_\perp$ integration in Eq. (1) is implicitly bounded by Λ. In any cut-off scheme, the quark and gluon fields in QCD as well as the strong coupling constant α_s depend on the cut-off. For large Λ, the dependence of quantum fields on Λ can be calculated in perturbation theory,

$$\psi_\Lambda(\xi) = Z_F^{1/2}(\Lambda)\tilde{\psi}(\xi), \quad A^\mu_\lambda(\xi) = Z_A^{1/2}(\Lambda)\tilde{A}^\mu(\xi), \tag{5}$$

where $\tilde{\psi}(\xi)$ and $\tilde{A}^\mu(\xi)$ are indepedent of Λ, and $Z_{F,A}(\Lambda)$ are the wave function renormalization constants.

Going back to Eqs. (1,3), it is now clear that the cut-off dependence of the wave function amplitudes ϕ_i comes entirely from the field renormalization,

$$\phi_i^\Lambda(x, k_\perp) = Z_F(\Lambda)\tilde{\phi}_i(x, k_\perp) , \tag{6}$$

where $\tilde{\phi}_i(x, k_\perp)$ is independent of Λ.

We claim that the above feature holds for all components of the pion wave function in the Fock expansion. For instance, the most general two-quark, one-gluon wave function amplitudes can be defined through the matrix elements of the operators,

$$\bar{d}_+(0)\gamma^+\gamma_5 F^{+j}(\eta_-, \eta_\perp)u_+(\xi^-, \xi_\perp) ,$$
$$\bar{d}_+(0)\sigma^{+i}\gamma_5 F^{+j}(\eta_-, \eta_\perp)u_+(\xi^-, \xi_\perp) . \tag{7}$$

Their scale dependence comes entirely from the wave function renormalization constants $Z_F(\Lambda)Z_A^{1/2}(\Lambda)$. In general, an n-particle Fock wave function amplitude with n_q quark and antiquark and n_g gluon creation operators has an explicit cut-off dependence through $Z_F^{n_q/2}(\Lambda)Z_A^{n_g/2}(\Lambda)$.

Knowing the scale dependence of individual components of the hadron wave function, we can calcualte the scale dependence of their contributions to hadronic matrix elements. As an example, we consider the Feynman parton distributions, although the discussion applies to generalized parton distributions as well [5]. Deriving the parton evolution equation from light-cone wave functions has been considered before[6]. Our approach here allows to uncover a set of new equations.

Consider, for example, the two-particle wave-function contribution to the u quark distribution in the pion. We have

$$u_2(x, \Lambda) = \int_0^\Lambda \frac{d^2 k_\perp}{(2\pi)^3} \left[|\phi_0^\Lambda(x, k_\perp)|^2 + k_\perp^2 |\phi_1^\Lambda(x, k_\perp)|^2 \right] . \tag{8}$$

Since at large $k_\perp$, $\phi_0(x, k_\perp)$ goes like $1/k_\perp^2$ and $\phi_1(x, k_\perp)$ like $1/k_\perp^4$ modulo logaritms [2,3], the $k_\perp$ integration is convergent and hence Λ can be taken to infinity. Thus the only Λ dependence in $u_2(x, \Lambda)$ comes from the wave function renormalization factor Z_F,

$$\frac{d}{d\ln\Lambda^2}u_2(x, \Lambda) = -2\gamma_F \frac{\alpha_s(\Lambda)}{4\pi}u_2(x, \Lambda) , \tag{9}$$

where γ_F is the anomalous dimension of Z_F and is gauge-dependent.

Equation (9) indicates that the two-particle Fock state contribution to the up-quark distribution graduately diminishes as $\Lambda \to \infty$. The physics is simple: As Λ gets larger, the state is probed at shorter distance, and it becomes increasingly difficult for the meson to remain in the two-particle Fock component because of the radiation.

Consider now the three-particle Fock component contribution to the u-quark distribution. We write schematically,

$$u_3(x, \Lambda) = \int \frac{d^2 k_\perp}{(2\pi)^3} \frac{d^2 k'_\perp dx'}{(2\pi)^3} |\phi^\Lambda(x, k_\perp, x', k'_\perp, \Lambda)|^2 \,, \tag{10}$$

where we have not considered individual quark-antiquark-gluon helicities and orbital angular momentum projections although this can be done straightforwardly. As discussed before, the wave function $\phi^\Lambda(x, k_\perp, x', k'_\perp)$ has an explicit dependence on Λ through the wave function renormalization constant $Z_F Z_A^{1/2}$. Additional dependence comes from integrations over the transverse momenta $k_\perp$ and $k'_\perp$.

We take derivative with respect to Λ using the chain rule. The derivative with respect to the wave function renormalization yields,

$$\frac{d}{d \ln \Lambda^2} u_3(x, \Lambda) = -(2\gamma_F + \gamma_A) \frac{\alpha_s(\Lambda)}{4\pi} u_3(x, \Lambda) + \dots \,. \tag{11}$$

where γ_A is the anomalous dimension of the gluon wave function renormalization. The physics of this part of the scale evolution is the same as the two-particle Fock component case: The splitting of the partons leads to the decrease of the probability for the pion to remain in the three-particle Fock state.

The integrations over $\vec{k}_\perp$ and $\vec{k}'_\perp$ do not yield divergences in general. In fact, there is no overall divergence (divergence arising from when all transverse momenta going to infinity at the same rate) because the power counting indicates that when two-momenta going to infinity at the same time, the integrals have negative superficial degree of divergence. However, there are subdivergences. These subdivergences arise from one-loop diagrams shown in Fig 1. The physics of these diagrams is that there are three-particle Fock amplitudes which are generated from the radiation of the two-particle Fock amplitude. Therefore, the result is proportional to $u_2(x, \Lambda)$.

Considering the loop intergral from the d-quark line shown in Fig. 1a, where the $\vec{k}'_\perp$ integration is divergent but the $\vec{k}_\perp$ integration is finite. It is easy to see that when Λ changes, the resulting 3-particle distributions also changes,

$$\frac{d}{d \ln \Lambda^2} u_3(x, \Lambda) = \dots + \gamma_F \frac{\alpha_s(\Lambda)}{4\pi} u_2(x, \Lambda) \,. \tag{12}$$

where the plus sign indicates that the three-particle Fock component receives a contribution from the radiation of the two-parton states.

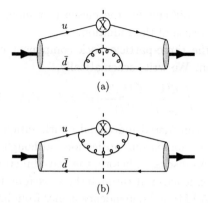

Figure 1. Contributions to the u-quark distribution from the three-particle component generated by the two-particle Fock component.

For Fig. 1b, with the gluon radition from the u-quark line, where the integration over $\vec{k}_\perp$ is now divergent but the one over $\vec{k}'_\perp$ is finite. The $\vec{k}_\perp$ integration can be done using the standard light-cone perturbation theory,

$$\frac{d}{d\ln\Lambda^2}u_3(x,\Lambda) = ... + \frac{\alpha_s(\Lambda)}{2\pi}C_F \int_x^1 \frac{dy}{y}\frac{1+y^2}{1-y}u_2\left(\frac{x}{y},\Lambda\right) , \qquad (13)$$

where the divergence at $y = 1$ must be regulated. Adding everything together, we obtain the complete evolution equation for u_3 is

$$\frac{d}{d\ln\Lambda^2}u_3(x,\Lambda) = \frac{\alpha_s(\Lambda)}{4\pi}\left[-(2\gamma_F+\gamma_A)u_3(x,\Lambda) + \gamma_F u_2(x,\Lambda) \right.$$
$$\left. +2C_F\int_x^1 \frac{dy}{y}\frac{1+y^2}{1-y}u_2\left(\frac{x}{y},\Lambda\right)\right] . \qquad (14)$$

This is an inhomogeneous equation with a driving term u_2.

Going to wave function amplitudes with four and more partons posts no special difficulty[7]. If we define the valence up-quark distribution $u_v = u - \bar{u}$, it is not difficult to see that the evolution equation for u_{nv} from Fock states with n partons is [7]

$$\frac{d}{d\ln\Lambda^2}u_{nv}(x,\Lambda) = \frac{\alpha_s(\Lambda)}{4\pi}\left[-\sum_{i=1}^{n}\gamma_i u_{nv}(x,\Lambda) + \sum_{i=1}^{n-2}\gamma_i u_{n-1v}(x,\Lambda) + \right.$$
$$\left. 2C_F\int_x^1 \frac{dy}{y}\frac{1+y^2}{1-y}u_{n-1v}(\frac{x}{y},\Lambda)\right] , \qquad (15)$$

where the sum over $n - 2$ γ_i excludes one γ_F and one γ_A. The first term describes the depletion of the n-particle Fock component due to the gluon radiation into $n+1$-particle component; the second and third terms describe the increase of the n-particle component due to the gluon emission of the $n-1$-particle component. The difference between the later two comes from whether the gluon is radiated from the active particle or the spectators. The total $u_v(x)$ distribution is a sum over all possible Fock components,

$$u_v(x) = \sum_{i=2}^{\infty} u_{iv}(x) \ . \tag{16}$$

Summing over all the equations for the individual Fock components, we recover the standard DGLAP equation,

$$\frac{d}{d\ln\Lambda^2} u_v(x,\Lambda) = \frac{\alpha_s(\Lambda)}{2\pi} \int_x^1 \frac{dy}{y} \frac{1+y^2}{(1-y)_+} u_v\left(\frac{x}{y},\Lambda\right) \tag{17}$$

which is an important check.

In summary, we find that the light-cone wave functions of hadrons in QCD can be entirely determined by the matrix elements of a class of quark-gluon operators. From this, we derive an infinite set of evolution equations for the parton distributions contributed by individual n-particle Fock components. These equations are consistent with the well-known DGLAP equation.

The authors thank S. Brodsky for a number of useful discussions and encouragement. This work was supported by the U. S. Department of Energy via grants FG-03-95ER-40965 and DE-FG02-93ER-40762.

References

1. J. B. Kogut and D. E. Soper, Phys. Rev. D1, 2901 (1971). S. Brodsky and P. Lepage, *Introduction to Perturbative QCD*, Ed. by A. Mueller, World Scientific, Singapore, 1989.
2. S. J. Brodsky, Invited Lectures presented at the Cracow School of Theoretical Physics, Zakopane, Poland, June 2001, hep-ph/0111340.
3. G. P. Lepage and S. J. Brodsky, Phys. Rev. D22, 2157 (1980).
4. S. J. Brodsky, P. Hoyer, N. Marchal, S. Peigne, F. Sannino, hep-ph/0104291.
5. X. Ji, J. Phys. G 24, 1181 (1998); A. V. Radyushkin, hep-ph/0101225; K. Goeke, M. V. Polyakov, M. Vanderhaeghen, Prog. Part. Nucl. Phys. 47, 401 (2001).
6. S. J. Brodsky, T. Huang, and G. P. Lepage, talk presented at the Summer Institute on Particle Physics, SLAC, Aug. 1981, SLAC-PUB-2857.
7. M. Burkardt, X. d. Ji, and F. Yuan, hep-ph/0205272.

LEADING-TWIST TWO GLUON DISTRIBUTION AMPLITUDE AND EXCLUSIVE PROCESSES INVOLVING η AND η' MESONS

K. PASSEK*

Fachbereich Physik, Universität Wuppertal, 42097 Wuppertal, Germany
E:mail: passek@theorie.physik.uni-wuppertal.de

I briefly review the formalism for treating the leading-twist two-gluon states appearing in processes which involve η and η' mesons. The constraints on the size of the lowest order Gegenbauer coefficients of the two-gluon distribution amplitude are obtained from the fit to the η and η' transition form factor data. The results are applied to $\chi \to \eta\eta(\eta'\eta')$ decays and deeply virtual electroproduction of η and η' mesons.

1. Introduction

The description of the hard exclusive processes involving light mesons is based on the factorization of the short- and long-distance dynamics and the application of the perturbative QCD[1]. The former is represented by the process-dependent and perturbatively calculable elementary hard-scattering amplitude, in which meson is replaced by its (valent) Fock states, while the latter is described by the process-independent meson distribution amplitude (DA), which encodes the soft physics. The lowest Fock state of flavour-nonsinglet mesons consists of quark and antiquark, while for flavour-singlet meson the two-gluon state appears additionally. In this paper I give a status report on work of Ref. [2], and discuss the proper treatment and the importance of these two-gluon states.

On the basis of recent results[3,4] on η and η' mixing, we adopt the following representation of η and η' in a octet-singlet basis:

$$|\eta \rangle = \cos\theta_8 |\widetilde{\eta}_8\rangle - \sin\theta_1 |\widetilde{\eta}_1\rangle ,$$
$$|\eta'\rangle = \sin\theta_8 |\widetilde{\eta}_8\rangle + \cos\theta_1 |\widetilde{\eta}_1\rangle , \tag{1}$$

---

[a]On leave of absence from the Rudjer Bošković Institute, Zagreb, Croatia.

where the pure octet $(\widetilde{\eta}_8)$ and singlet $(\widetilde{\eta}_1)$ states are given by

$$|\widetilde{\eta}_1\rangle = \frac{f_1}{2\sqrt{6}}\left[\phi_1(x)|(u\bar{u} + d\bar{d} + s\bar{s})/\sqrt{3}\rangle + \phi_g(x)|gg\rangle\right],$$

$$|\widetilde{\eta}_8\rangle = \frac{f_8}{2\sqrt{6}}\,\phi_8(x)|(u\bar{u} + d\bar{d} - 2s\bar{s})/\sqrt{6}\rangle, \tag{2}$$

and higher Fock states are neglected. In this representation, the mixing dependence is solely embedded in the θ_8 and θ_1 angles, while in more general approach different distribution amplitudes ϕ_8^P and ϕ_1^P could be assumed for $P = \eta, \eta'$. The numerical values[4] $f_8 = 1.26 f_\pi$, $f_1 = 1.17 f_\pi$, $\theta_8 = -21.2°$, and $\theta_1 = -9.2°$ are used in this work. Alternatively, one could use the recently suggested quark-flavour basis[4], but the analysis of DA evolution is more straightforward in the above given octet-singlet basis.

2. Two gluon distribution amplitude and the transition form factor for the flavour-singlet meson

Employing for this section more transparent notation $\phi_q \equiv \phi_1$, the DA evolution equation for $\widetilde{\eta}_1$ takes the matrix form

$$\mu_F^2 \frac{\partial}{\partial \mu_F^2} \begin{pmatrix} \phi_q(x, \mu_F^2) \\ \phi_g(x, \mu_F^2) \end{pmatrix} = V(x, u, \alpha_S(\mu_F^2)) \otimes \begin{pmatrix} \phi_q(u, \mu_F^2) \\ \phi_g(u, \mu_F^2) \end{pmatrix}, \tag{3}$$

where $\otimes$ denotes the usual convolution symbol. The kernel V is 2x2 matrix with a well defined expansion in α_S. The evolution of the flavour-singlet pseudoscalar meson distribution amplitude (DA) has been investigated in a number of papers[5,6,7]. Most of the results[5,6] are in agreement up to differences in conventions. On the other hand, the consistent set of conventions has to be used in calculation of both the hard-scattering and the distribution amplitude, and these are not easy to extract from the literature.

Following the recent analysis of the pion transition form factor[8], we have performed a detailed next-to-leading order (NLO) analysis of the $\widetilde{\eta}_1$ transition form factor taking into account both hard-scattering and perturbatively calculable DA part, and this enabled us to fix and test the convention we are using, and to make a connection with other conventions.

The hard-scattering amplitude one obtains by evaluating the $\gamma^* + \gamma \to q\bar{q}$ and $\gamma^* + \gamma \to gg$ amplitudes which we denote by $T_{q\bar{q}}(u, Q^2)$ and $T_{gg}(u, Q^2)$, respectively. Owing to the fact that final state quarks and gluons are taken to be massless and onshell, $T_{q\bar{q}}$ and T_{gg} contain collinear singularities, which have to be factorized out in order to obtain the finite quantities $T_{H,q\bar{q}}$ and $T_{H,gg}$:

$$T(u, Q^2) = \left(T_{H,q\bar{q}}(x, Q^2, \mu_F^2) \quad T_{H,gg}(x, Q^2, \mu_F^2)\right) \otimes Z^{-1}(x, u, \mu_F^2). \tag{4}$$

On the other hand, the unrenormalized quark and gluon distribution amplitudes $\phi_q(u)$ and $\phi_g(u)$ are defined in terms of $\langle 0|\bar{\Psi}(-z)\gamma^+\gamma_5\Omega\Psi(z)|\tilde{\eta}_1\rangle$ and $\langle 0|G^{+\alpha}(-z)\,\Omega\,\tilde{G}_\alpha{}^+(z)|\tilde{\eta}_1\rangle$, respectively. The renormalization introduces the mixing of these quark and gluon composite operators and

$$\phi(u) = \begin{pmatrix} \phi_q(u) \\ \phi_g(u) \end{pmatrix} = Z(u,x,\mu_F^2) \otimes \begin{pmatrix} \phi_q(x,\mu_F^2) \\ \phi_g(x,\mu_F^2) \end{pmatrix}. \tag{5}$$

We note that Z represents a 2x2 matrix. Perturbation theory cannot be used for a direct evaluation of $\Phi(u)$, but replacing $|\tilde{\eta}_1\rangle$ by $|q\bar{q}\rangle$ or $|gg\rangle$ enables us to obtain the perturbatively calculable DA part and to determine matrix Z. Finally, the $\tilde{\eta}_1$ transition form factor is given by

$$F_{\gamma^*\gamma\tilde{\eta}_1}(Q^2) = \frac{f_1}{2\sqrt{6}}\,T(u,Q^2)\otimes\phi(u) = \frac{f_1}{2\sqrt{6}}\,T_H(x,Q^2,\mu_F^2)\otimes\phi(u,\mu_F^2). \tag{6}$$

Hence, the singularities, which appear in the calculation of the hard-scattering (4) and the DA part (5) should cancel, and we have used this requirement to check the consistency of our calculation.

By differentiating (5) with respect to μ_F^2 one obtains the DA evolution equation (3) with evolution potential V expressed in terms of Z $V = -Z^{-1} \otimes (\mu_F^2 \partial/\partial \mu_F^2 \, Z)$. The solutions of the leading-order (LO) evolution equation are given by

$$\phi_q(x,\mu_F^2) = 6x(1-x)\left[1 + \sum_{n=2}^{\infty}{}' B_n^q(\mu_F^2)\, C_n^{3/2}(2x-1)\right]$$

$$\phi_g(x,\mu_F^2) = x^2(1-x)^2 \sum_{n=2}^{\infty}{}' B_n^g(\mu_F^2)\, C_{n-1}^{5/2}(2x-1), \tag{7}$$

where

$$B_n^q(\mu_F^2) = B_n^+(\mu_0^2)\left(\frac{\alpha_S(\mu_0^2)}{\alpha_S(\mu_F^2)}\right)^{\frac{\gamma_+^n}{\beta_0}} + \rho_n^-\, B_n^-(\mu_0^2)\left(\frac{\alpha_S(\mu_0^2)}{\alpha_S(\mu_F^2)}\right)^{\frac{\gamma_-^n}{\beta_0}}$$

$$B_n^q(\mu_F^2) = \rho_n^+\, B_n^+(\mu_0^2)\left(\frac{\alpha_S(\mu_0^2)}{\alpha_S(\mu_F^2)}\right)^{\frac{\gamma_+^n}{\beta_0}} + B_n^-(\mu_0^2)\left(\frac{\alpha_S(\mu_0^2)}{\alpha_S(\mu_F^2)}\right)^{\frac{\gamma_-^n}{\beta_0}}. \tag{8}$$

The coefficients $B_n^\pm(\mu_0^2)$, i.e., $B_n^{q,(g)}(\mu_0^2)$, represent nonperturbative input at scale μ_0^2, while $\gamma_n^\pm = 1/2\left[(\gamma_n^{qq} + \gamma_n^{gg}) \pm \sqrt{(\gamma_n^{qq} - \gamma_n^{gg})^2 + 4\gamma_n^{qg}\gamma_n^{gq}}\right]$ and

$$\rho_n^+ = 6\,\frac{\gamma_n^{gq}}{\gamma_n^+ - \gamma_n^{gg}}, \qquad \rho_n^- = \frac{1}{6}\,\frac{\gamma_n^{qg}}{\gamma_n^- - \gamma_n^{qq}} \tag{9}$$

are defined in terms of anomalous dimensions $\gamma_n^{qq} = \gamma_n^{(0)}8$, $\gamma_n^{gg}5$,

$$\gamma_n^{qg} = \sqrt{n_f C_F} \, \frac{n(n+3)}{3(n+1)(n+2)} \qquad \gamma_n^{gq} = \sqrt{n_f C_F} \, \frac{12}{(n+1)(n+2)} . \quad (10)$$

The DA ϕ_q is normalized to 1, but, since $\int_0^1 dx \; \phi_g(x, \mu_F^2) = 0$, there is no such natural way to normalize ϕ_g. It is important to emphasize that any change of the normalization of the gluon DA is accompanied by the corresponding change in the hard-scattering part. Namely, for $\phi_g \to \sigma \, \phi_g$, the projection of gg state on the $\widetilde{\eta}_1$ state, which can be derived from the definition of the gluon distribution amplitude, gets modified by factor $1/\sigma$, i.e.,

$$\frac{i}{2} \, \epsilon^{\mu\nu\alpha\beta} \, \frac{n_\alpha P_\beta}{n \cdot P} \, \frac{1}{x(1-x)} \to \frac{1}{\sigma} \frac{i}{2} \, \epsilon^{\mu\nu\alpha\beta} \, \frac{n_\alpha P_\beta}{n \cdot P} \, \frac{1}{x(1-x)} , \quad (11)$$

and, thus, the hard-scattering amplitude changes while the physical quantity ($F_{\gamma^*\gamma\widetilde{\eta}_1}$, in this case) remains independent of the choice of convention. By inspecting Eqs. (7-9), it is easy to see that the change of the normalization of ϕ_g can be translated into the change of the off-diagonal anomalous dimensions

$$\gamma_n^{qg} \to \frac{1}{\sigma} \gamma_n^{qg} \qquad \gamma_n^{gq} \to \sigma \, \gamma_n^{gq} \quad (12)$$

and of the coefficient $B_n^-(\mu_0^2) \to \sigma B_n^-(\mu_0^2)$. The former can be easily understood in the "operator" language, i.e., by considering the impact of the change of the normalization of the gluon operator on the anomalous dimensions. Hence, by employing (12) along with (10) and (11), we can consistently change our conventions in order to compare our results with other calculations from the literature. For historical reasons (comparison with the forward case), in what follows $\sigma = \sqrt{n_f/C_F}$ is used.

3. Applications

Using the mixing scheme defined in Eq. (1), we have obtained the NLO leading-twist prediction for the η and η' transition form factors. For the treatment of $\phi_8(x, \mu_F^2)$, we use the well-known LO result for the flavour-nonsinglet meson distribution amplitude[1,8]. We truncate the Gegenbauer series at $n = 2$, and fit our results to the experimental data[9]. The fits are carried through with $\mu_R = Q/\sqrt{2}$ and $\mu_F = Q$, with α_S evaluated from the two-loop expression with $n_f = 4$ and $\Lambda_{\overline{MS}}^{(4)} = 305$ MeV. For $Q^2 \geq 2$ GeV^2 and $\mu_0 = 1$ GeV, the results of the fits read

$$B_2^8(\mu_0^2) = -0.04 \pm 0.04 \quad B_2^1(\mu_0^2) = -0.08 \pm 0.04 \quad B_2^g(\mu_0^2) = 9 \pm 12 . \quad (13)$$

The existing experimental data and their quality allow us to obtain not more than a constraint on the value of B_2^g. As expected, we have observed a strong correlation between B_2^1 and B_2^g. The quality of the fit is shown in Fig. 1.

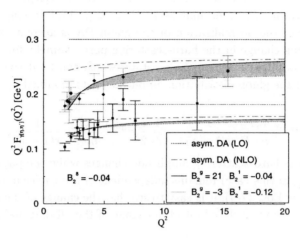

Figure 1. η (below) and η' (above) transition form factors. The shaded area corresponds to the in Eq. (13) given range for $B_2^1(\mu_0^2)$ and $B_2^g(\mu_0^2)$.

As a first application of the above given results, we have analyzed the $\chi_{c0} \to \eta\eta, \eta'\eta'$ decays. Following previous work[10], we take χ_{c0} as a non-relativistic $c\bar{c}$ bound state, and obtain the decay amplitudes for $c\bar{c} \to (q\bar{q})(q\bar{q})$ and $c\bar{c} \to (gg)(gg)$. The ratio $\Gamma_{\eta'\eta'}/\Gamma_{\eta\eta}$ is suitable for investigating the sensitivity to the gluon contributions. Despite possibly large value of B_2^g, we have observed only modest dependence on the variations of B_1 and B_g in the allowed range (13) (up to 20 % difference).

Finally, I report on our result for the two gluon contribution to the deeply virtual electroproduction of η, η' mesons. In this case, the subprocess amplitudes $q+\gamma_L^* \to (q\bar{q})+q$ and $q+\gamma_L^* \to (gg)+q$ have to be evaluated. Our result for the former amplitude $H_{0+,0+}^{P(q)}$ is in agreement with the results from the literature[11,12]. In terms of subprocess Mandelstam variables $(\hat{s},\hat{u},\hat{t}=t)$, the latter amplitude reads

$$H_{0+,0+}^{P(g)} = \frac{4\pi\alpha_s}{Q} \frac{C_F}{N_C} \frac{1}{2} \frac{1}{\sqrt{n_f}} \frac{Q\sqrt{-\hat{u}\hat{s}}}{Q^2+\hat{s}} \int_0^1 d\tau \frac{\phi_g(\tau)}{\tau(1-\tau)} \frac{t}{\hat{u}\hat{s}} \left(\frac{1}{\tau} - \frac{1}{1-\tau}\right). \quad (14)$$

In deeply virtual electroproduction of mesons (DVEM), the limit $t \to 0$ has to be considered. For $t = 0$ and $\hat{s} + \hat{u} = -Q^2$, our result (14) gives $H_{0+,0+}^{P(g)} = 0$. We conclude that the two-gluon contribution is suppressed.

4. Conclusions

We have performed a detailed analysis of the proper inclusion of the two gluon states in the (octet-singlet scheme based) description of the hard exclusive processes involving η and η' mesons. Normalization and conventions have been fixed and discrepancies found in the literature have been resolved. From the fit of the η, η' transition form factors to experimental data, we have obtained the range for the B_2^1, B_2^g and B_2^8 coefficients of the ϕ_1, ϕ_g and ϕ_8 DAs, respectively. Expected strong correlation between B_2^1 and B_2^g has been observed. The results have been applied to $\chi_{c0} \to \eta\eta(\eta'\eta')$ decay, where only a moderate dependence on B_2^g has been found. For DVEM, the $\gamma^*q \to (gg)q$ subprocess has been found to be suppressed for small momentum transfer t. We conclude that for considered processes the theoretical and experimental uncertainties do not allow further restriction of the parameter range since only a modest dependence on the value of B_2^g has been observed. This contrasts the findings for $\eta'g^*g^*$ vertex[13].

Acknowledgments

This work is supported by Deutsche Forschungs Gemeinschaft and partially supported by the Ministry of Science and Tech., Croatia (No. 0098002).

References

1. S.J. Brodsky and G.P. Lepage in *Perturbative QCD*, ed. A.H. Mueller (World Scientific, Singapore, 1989), and references therein.
2. P. Kroll, K. Passek-Kumerički, *"The leading-twist two gluon distribution amplitude"*, WU B 02-22, work in progress.
3. H. Leutwyler, Nucl. Phys. Proc. Suppl. **64**, 223 (1998).
4. T. Feldmann, P. Kroll and B. Stech, Phys. Rev. D **58**, 114006 (1998); Phys. Lett. B **449**, 339 (1999).
5. M. V. Terentev, Sov. J. Nucl. Phys. **33**, 911 (1981);
6. M. A. Shifman and M. I. Vysotsky, Nucl. Phys. B **186**, 475 (1981); V. N. Baier and A. G. Grozin, Nucl. Phys. B **192**, 476 (1981); A. V. Belitsky, D. Müller, L. Niedermeier and A. Schäfer, Nucl. Phys. B **546**, 279 (1999).
7. T. Ohrndorf, Nucl. Phys. B **186**, 153 (1981).
8. B. Melić, B. Nižić and K. Passek, Phys. Rev. D **65**, 053020 (2002).
9. J. Gronberg *et al.* [CLEO Collaboration], Phys. Rev. D **57**, 33 (1998); M. Acciarri *et al.* [L3 Collaboration], Phys. Lett. B **418**, 399 (1998).
10. J. Bolz, P. Kroll and G. A. Schuler, Phys. Lett. B **392**, 198 (1997); Eur. Phys. J. C **2**, 705 (1998) and references therein.
11. H. W. Huang and P. Kroll, Eur. Phys. J. C **17**, 423 (2000).
12. L. Mankiewicz, G. Piller and T. Weigl, Eur. Phys. J. C **5**, 119 (1998); M. Vanderhaeghen, P. A. Guichon and M. Guidal, Phys. Rev. D **60**, 094017 (1999).
13. A. Ali and A. Y. Parkhomenko, Phys. Rev. D **65**, 074020 (2002).

NUCLEON COMPTON SCATTERING WITH TWO
SPACE–LIKE PHOTONS

A. AFANASEV

Jefferson Lab, Newport News, VA 23606, USA
E-mail: afanas@jlab.org

I. AKUSHEVICH*

Duke University, Durham, NC 27708, USA
E-mail: aku@jlab.org

N.P. MERENKOV

NSC Kharkov Institute of Physics and Technology, Kharkov 63108, Ukraine
E-mail: merenkov@kipt.kharkov.ua

We calculated two–photon exchange effects for elastic electron–proton scattering at high momentum transfers. The corresponding nucleon Compton amplitude is defined by two space–like virtual photons that appear to have significant virtualities. We make predictions for a) a single–spin beam asymmetry, and b) a single–spin target asymmetry or recoil proton polarization caused by an unpolarized electron beam.

1. Introduction

The two-photon exchange mechanism in elastic electron-nucleon scattering can be observed experimentally by a) measuring the C-odd difference between electron–proton and positron–proton scattering cross sections; b) analyzing deviations from the Rosenbluth formula and c) studying T-odd parity–conserving single–spin observables. This paper concentrates on the latter.

Early measurements of the parity–conserving single–spin observables include induced polarization of the recoil proton in elastic ep-scattering[1] and the target asymmetry[2]. These experiments were able to set only upper

---

* On leave from the National Center of Particle and High Energy Physics, 220040, Minsk, Belarus

bounds that appeared to be at one per cent level. Corresponding theoretical calculations were given in Refs.[3] and Ref.[4], with deep–inelastic intermediate states considered in the latter.

The transverse beam asymmetry of spin-$\frac{1}{2}$ particle scattering on a nuclear target was first calculated by N.F. Mott[5] in 1932, providing, for example, an operating principle for low–energy (about 1 MeV) electron beam polarimeters [6]. The measurement of this asymmetry at higher energies of several hundred MeV was reported recently by SAMPLE Collaboration [7]. The observed magnitude of this effects is about 10^{-5} and it appears to be the only nonzero parity-conserving single–spin effect measured so far for elastic ep–scattering. Here we present the first (to the best of our knowledge) theoretical evaluation of this asymmetry that takes nucleon structure into consideration. We also present results of our calculations of parity–conserving single–spin effects due to initial or final proton polarization in elastic ep–scattering. Since we are dealing with large transferred momenta, we describe excitation of intermediate hadronic states (Fig.1) in terms of deep–inelastic structure functions of the non–forward Compton amplitude with two space–like photons.

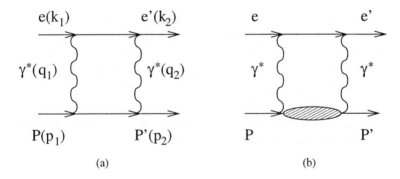

Figure 1.   Two-photon exchange mechanism responsible for single–spin asymmetries in elastic ep–scattering. a) Elastic intermediate state. b) Inelastic intermediate states.

2. Formalism

In QED, beam and target parity–conserving single–spin asymmetries and polarizations are caused by the two–photon exchange mechanism (Fig.1). In the leading order of electromagnetic coupling constant, the imaginary (absorptive) part of the two–photon exchange amplitude interferes with a (real) amplitude of the lowest–order one–photon exchange to produce a single–spin effect due to transverse (namely, normal to the scattering plane) polarization of either the electron or the proton. As was noticed by De Rujula and collaborators over 30 years ago[4], the quantity which governs transverse polarization effects is *the absorptive part of the non–forward Compton amplitude for off-shell photons scattering from nucleons.*

Using parity- and time-reversal invariance, one can demonstrate that a) the beam asymmetry is zero in the ultrarelativistic beam energies, b) beam and target asymmetries are independent observables and c) the target asymmetry and recoil proton polarization are equal.

There are two basic contributions to the two–photon exchange mechanism shown in Fig.1 which differ by the intermediate hadronic state. In the first case (Fig.1a) the intermediate state is purely elastic, containing only a proton and electron. In the second case (Fig.1b) the target proton is excited producing continuum of particles in the intermediate state.

A general formula for the transverse single–spin asymmetries includes integration over the loop 4–momentum k. It can be written as[4]

$$A_{l,p}^{el,in} = \frac{8\alpha}{\pi^2} \frac{Q^2}{D(Q^2)} \int dW^2 \frac{S + M^2 - W^2}{S + M^2} \frac{dQ_1^2}{Q_1^2} \frac{dQ_2^2}{Q_2^2} \frac{1}{\sqrt{K}} B_{l,p}^{el,in}, \qquad (1)$$

where $S = 2k_1 p_1$, $Q_{1,2}^2 = -q_{1,2}^2$, M is the proton mass and notation for particle momenta is shown in Fig.1. For the elastic intermediate state, the integration is two dimensional because of two Dirac deltas that put the intermediate lepton and proton on the mass shell. It does not apply to the proton in the inelastic case, so the additional integration over W^2 has to be done (from M^2 to $S + M^2$), resulting in a triple integral. The quantity $D(Q^2)$ comes from Born (*i.e.*, one–photon exchange) contribution. The formulae for the relevant quantities read

$$D(Q^2) = 8[Q^4(F_1 + F_2)^2 + 2S_m(F_1^2 + \tau F_2^2)],$$
$$S_m = S^2 - SQ^2 - M^2Q^2, \tau = Q^2/4M^2, \qquad (2)$$
$$K = \sqrt{1 - z_1^2 - z^2 - z_2^2 + 2zz_1z_2}.$$

Here z, z_1 and z_2 are cosines of the scattering angles in c.m.s. They are related to Q^2, Q_1^2 and Q_2^2 as follows:

$$Q^2 = \frac{S^2}{2(S + M^2)}(1 - z), \quad Q_{1,2}^2 = \frac{S(S - W^2 + M^2)}{2(S + M^2)}(1 - z_{1,2}). \qquad (3)$$

The above formula (3) also sets the limits of the integration region for Q_1^2 and Q_2^2, which is shown in Fig. 2 for the representative electron beam energy $E_b = 5$ GeV. It can be seen from Fig. 2 that the virtualities of the exchanged photons, albeit limited, can become significantly larger than the overall transferred momentum Q^2. Experimentally, by selecting the electron scattering angles and beam energies, one can control the limits of photon virtualities contributing to the single–spin asymmetries.

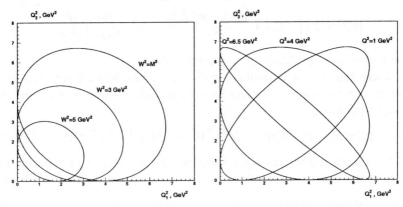

Figure 2. Integration region over Q_1^2 and Q_2^2 in Eq.(2) for elastic ($W^2 = M^2$) and inelastic contributions. The latter (left) is given for Q^2=4 GeV^2 and two values of W^2, which is an integration variable in this case. The elastic case is shown on the right as a function of external Q^2. The electron beam energy is E_b= 5 GeV.

The quantities B in Eq.(1) result from contraction of leptonic and hadronic tensors. Explicitly, they read

$$B_l^{el,in} = \frac{i}{4}Tr[(\hat{k}_2 + m)\gamma_\alpha(\hat{k} + m)\gamma_\beta\gamma_5\hat{\eta}_l(\hat{k}_1 + m)\gamma_\mu] \times$$

$$\frac{1}{4}Tr[(\hat{p}_2 + M)W_{\alpha\beta}^{el,in}(\hat{p}_1 + M)\Gamma_\mu^*] \qquad (4)$$

$$B_p^{el,in} = \frac{1}{4}Tr[\hat{k}_2\gamma_\alpha\hat{k}\gamma_\beta\hat{k}_1\gamma_\mu]\frac{i}{4}Tr[(\hat{p}_2 + M)\gamma_5\hat{\eta}_p W_{\alpha\beta}^{el,in}(\hat{p}_1 + M)\Gamma_\mu^*(q^2)],$$

where $k(p)$ is the 4–momentum of the intermediate electron (proton), m is the electron mass, $\eta_{l,p}$ is the electron (proton) polarization vector and the nucleon electromagnetic current is parameterized in the form

$$\Gamma_\mu(q^2) = (F_1(q^2) + F_2(q^2))\gamma_\mu - F_2(q^2)\frac{(p_1 + p_2)_\mu}{2M}. \qquad (5)$$

The only quantities than remain to be defined are elastic and inelastic non–forward tensors $W_{\alpha\beta}$[a]. For the elastic case, it is known exactly:

$$W_{\alpha\beta}^{el} = 2\pi\delta(W^2 - M^2)\Gamma_\alpha(q_2^2)(\hat{p} + \hat{q}_1 + M)\Gamma_\beta(q_1^2). \qquad (6)$$

---

[a]Strictly speaking, we deal only with imaginary parts of these tensors

To compute the non–forward inelastic hadronic tensor, we need model assumptions. We use an expression for it inspired by Blumlein and Robaschik calculation [9],

$$W_{\alpha\beta}^{in} = 2\pi \left(-g_{\alpha\beta} + \frac{p_\alpha^b q_\beta^b + p_\beta^b q_\alpha^b}{q^b p^b} W_1 + p_\alpha^b p_\beta^b \frac{W_2}{M^2} \right) \frac{\hat{q}^b}{q^b p^b} \tag{7}$$

where $q^b = (q_1 - q_2)/2$ and $p^b = (p_1 + p_2)/2$.

These tensors are normalized such that in the forward direction they reproduce conventional relations between inclusive structure functions and elastic form factors,

$$W_1 = 2M\tau (F_1 + F_2)^2 \delta(W^2 - M^2), \quad W_2 = 2M(F_1^2 + \tau F_2^2)\delta(W^2 - M^2). \tag{8}$$

For the elastic intermediate state, we use unitarity to obtain model–independent analytic expressions for the target and beam asymmetries B^{el},

$$B_p^{el} = \frac{\sqrt{S_m}}{4M\sqrt{Q^2}} \left[Q_1^2 Q_2^2 \left(S_q F_1 (2\mathcal{F}_{22} + \mathcal{F}_+) - \frac{1}{2}(4F_1 M^2 - F_2 Q^2)(2\mathcal{F}_{11} + \mathcal{F}_+) \right) \right.$$
$$\left. -SS_q Q^2 (2F_2 \mathcal{F}_{11} - F_1 \mathcal{F}_+ - F_1 \mathcal{F}_-) + \frac{(Q^2 - Q_+^2)S}{8S_m} \sum_{ij} C_{ij}^p F_i \mathcal{F}_j) \right] \tag{9}$$

$$B_l^{el} = \frac{m\sqrt{S_m}}{4\sqrt{Q^2}} \left[(-2Q_1^2 Q_2^2 (F_1 + F_2)(2\mathcal{F}_{11} + \mathcal{F}_+) + \frac{Q^2 - Q_+^2}{4M^2 S_m} \sum_{ij} C_{ij}^l F_i \mathcal{F}_j) \right]$$

where $Q_+^2 = Q_1^2 + Q_2^2$, $S_q = S + M^2 - \frac{Q^2}{2}$ and expressions for the coefficients C_{ij} are given in the Appendix.

The index $i = 1, 2$ corresponds to form factors $F_{1,2} = F_{1,2}(Q^2)$, and the index j takes values $11, 22, +, -$ where $\mathcal{F}_j$ are quadratic combinations of elastic form factors with the arguments $Q_{1,2}^2$:

$$\mathcal{F}_{11} = F_1(Q_1^2)F_1(Q_2^2), \quad \mathcal{F}_{22} = F_2(Q_1^2)F_2(Q_2^2),$$
$$\mathcal{F}_+ = F_1(Q_1^2)F_2(Q_2^2) + F_1(Q_2^2)F_2(Q_1^2), \tag{10}$$
$$\mathcal{F}_- = \frac{Q_1^2 - Q_2^2}{Q^2}(F_1(Q_1^2)F_2(Q_2^2) - F_1(Q_2^2)F_2(Q_1^2)).$$

The asymmetries $B_{l,p}^{el}$ are given in the symmetric form with respect to the transformation $Q_1^2 \leftrightarrow Q_2^2$. It can be seen also that

$$B_{l,p}^{el}(Q^2 = Q_1^2) = B_{l,p}^{el}(Q^2 = Q_2^2) = 0$$

so there are no infrared singularities in Eq.(1).

An inelastic contribution to the asymmetries reads

$$B_l^{in} = \frac{mQ^2 W_1}{16\nu_b N_s} [R_1 + 2R_2 + 8R_3] + \frac{mQ^2 W_2}{32M^2 \nu_b N_s} R_4$$

$$B_p^{in} = \frac{(Q_+^2 - Q^2)(SF_2 - 2M^2F_1) + 2F_2 S_q w_m}{128\nu_b^2 N_s M^3}(4M^2 W_1 T_1 + \quad (11)$$

$$\nu_b W_2 T_2)Q^2$$

where

$$
\begin{aligned}
R_1 &= (2\nu_b(2S - 2\nu_b - Q^2)F_1 + (F_1 + F_2)Q_+^2 Q^2)(w_m Q^2 - 2\nu_b S + \\
&\quad 2M^2(Q^2 - Q_+^2)), \\
R_2 &= (w_m Q^2 - (Q^2 - Q_+^2)S)\nu_b Q^2(F_1 + F_2), \\
R_3 &= (S^2 - M^2 Q^2 - Q^2 S)\nu_b^2 F_1, \\
R_4 &= (\nu_b(4F_1 M^2 - F_1 Q^2 - 2F_2 Q^2) + 2(F_1 + F_2)Q^2 S_q)(w_m Q^2 - \\
&\quad 2\nu_b S + 2M^2(Q^2 - Q_+^2)), \\
T_1 &= 2Q_+^2 S w_m + 2Q^2(S - w_m)^2 - 2Q^2 Q_+^2(S + M^2) - \qquad\qquad (12) \\
&\quad Q^2 Q_+^2(S - w_m) + Q_+^4 S, \\
T_2 &= 4M^2(Q^2 Q_+^2 - Q^2 S_w - Q_+^2 S) - 4Q_+^2 S^2 + Q^4 S_w + \\
&\quad 3Q^2 Q_+^2 S + 8(S - Q^2)SS_w, \\
w_m &= W^2 - M^2, \quad \nu_b = w_m + \frac{Q_+^2 - Q^2}{2}, \\
N_s &= \frac{1}{2}\sqrt{Q^2(S^2 - M^2 Q^2 - SQ^2)}, \quad S_w = S + M^2 - W^2.
\end{aligned}
$$

In general, the structure functions $W_{1,2}$ are functions of four invariant variables, Q^2, $Q_{1,2}^2$ and W^2. These structure functions were neither measured nor calculated theoretically. A possibility to construct a model for them was discussed in Ref.[8], where upper bounds were obtained from the positivity conditions. Following Ref.[8], we can write

$$W_{1,2}(W^2, Q_1^2, Q_2^2, Q^2) = [W_{1,2}^{DIS}(W^2, Q_1^2)W_{1,2}^{DIS}(W^2, Q_2^2)]^{1/2} F(Q^2), \quad (13)$$

where we assumed additional form factor–like dependence $F(Q^2)$ on the overall 4–momentum transfer. If the deep–inelastic conditions ($W > 2$ GeV, $Q^2 > 1$ GeV^2) take place, the non–forward Compton form factor $F(Q^2)$ may be related to the integral of Generalized Parton Distributions (GPD)[10] at large transverse momenta t (with the Mandelstam variable t equal to $-Q^2$ in our case). Note that since the *absorptive* part of the non–forward Compton amplitude contributes, then $x = \xi$ part of GPDs is selected similar to the single–spin asymmetry arising from interference of the Bethe–Heitler process with virtual Compton scattering[11]. The computed quantities are then described by the zeroth moments of nucleon GPDs and it is therefore natural to assume for further estimates that the introduced form factor

$F(Q^2)$ depends on Q^2 like the nucleon form factor described, with a good accuracy, by the dipole formula $F(Q^2) = (1 + \frac{Q^2}{0.71 GeV^2})^{-2}$.

This model choice for the non–forward Compton amplitude has three main properties. It is symmetric with respect to the transformation $Q_1^2 \leftrightarrow Q_2^2$, has a correct forward $(Q^2 = 0)$ limit and assumes form factor–like suppression with respect to the overall transferred momentum Q^2. The latter was not considered in the early papers[4,8], leading to dramatic over-estimates of the two–photon–exchange effects for elastic ep–scattering.

3. Numerical results and conclusions

Our results for the single–spin asymmetries are presented in Fig.3. The asymmetries are kinematically suppressed in the forward and backward directions by a factor $\sin(\Theta_{c.m.})$. The target asymmetry increases with increasing beam energies, while the beam asymmetry decreases due to the additional supression factor m/E_b. The target asymmetry (= recoil polarization) is evaluated at the per cent level. Below the pion threshold, only the nucleon intermediate state is allowed and the calculation becomes model–independent, based only on unitarity and known values of proton form factors. At higher energies, the contribution from excited intermediate hadronic states exceeds the elastic contribution. As can be seen from Fig.4, the integration region where both exchanged photons are highly virtual plays an important role. Measurements of the single–spin asymmetries due to two–photon exchange provide information about the absorptive part of the virtual Compton amplitude with two space–like photons at large values of the Mandelstam variable t.

To our knowledge, this is the first published calculation of the single–spin beam asymmetry in elastic ep–scattering for the kinematics where nucleon structure effects become important. Our results appear to be in reasonable agreement with recent SAMPLE data[7].

Acknowledgements

The work of IA and AA was supported by the U.S. Department of Energy under contract DE–AC05–84ER40150. We thank A. Radyushkin, S. Wells and S. Brodsky for useful discussions.

Appendix

Expressions for the coefficients $C_{i,j}$ in Eq.(9) are as follows

$$C_{1,11}^p = 4M^2 Q_+^2 (Q^2 + 2S)$$

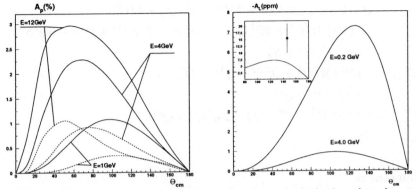

Figure 3. Proton (left) and lepton (right) asymmetries. Both elastic and total contribution are shown for proton asymmetry and only elastic one for lepton asymmetry. The plot in the insert gives comparison with SAMPLE data[7] at $E_B = 0.2$ GeV.

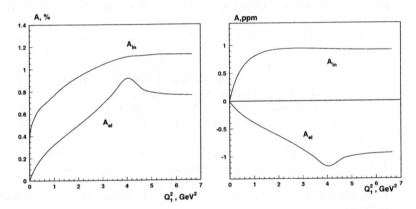

Figure 4. Elastic and inelastic contributions to the target (left) and beam (right) asymmetries versus the upper integration limit over the virtuality of one of the two exchanged photons. Importance of high virtualities is evident.

$$C^p_{1,22} = 2(-(Q^2 - Q^2_+)(2S^2 - M^2 Q^2) + Q^4 S)$$
$$C^p_{1,+} = (4Q^2 S_q (S + M^2) + Q^2_+ (4M^2 Q^2 + 4M^2 S - 2SS_q))$$
$$C^p_{1,-} = -Q^2 (2M^2 Q^2 - 2SS_q)$$
$$C^p_{2,11} = -4Q^2 (2S_q (S + M^2) + SQ^2_+)$$
$$C^p_{2,22} = -Q^2 (-(Q^2 - Q^2_+)(S - M^2) - 2S^2 + 2Q^2 S)S/M^2$$
$$C^p_{2,+} = -3Q^2 SQ^2_+$$
$$C^p_{2,-} = SQ^4$$
$$C^l_{1,11} = 2M^2 (2Q^2_+ M^2 Q^2 + Q^2_+ Q^2 S + 2Q^2_+ S^2 + 8M^4 Q^2 + 12M^2 Q^2 S)$$

$$C^l_{1,22} = Q^2((Q^2 - Q^2_+)M^2(S + 2M^2) - S^2(Q^2_+ + 4M^2))$$

$$C^l_{1,+} = M^2(4Q^2_+ M^2 Q^2 + 3Q^2_+ Q^2 S + 2Q^2_+ S^2 + 2M^2 Q^4 + Q^4 S - 6Q^2 S^2)$$

$$C^l_{1,-} = -2M^2 Q^4 (M^2 + S)$$

$$C^l_{2,11} = -4M^2(M^2 Q^4 + Q^2 S^2 + Q^4 S - Q^2_+ S^2)$$

$$C^l_{2,22} = -S^2 Q^2 (Q^2_+ - Q^2)$$

$$C^l_{2,+} = S^2(2Q^2_+ M^2 - Q^2_+ Q^2/2 - 2M^2 Q^2 + Q^4)$$

$$C^l_{2,-} = S^2 Q^4/2$$

References

1. G.V. Di Girgio *et al.*, Nuovo Cim. **39**, 474 (1965); J.C. Bizot *et al.*, Phys. Rev. **140**, B1387 (1965); D.E. Lundquist *et al.*, Phys. Rev. **168**, 1527 (1968); J.R. Chen *et al.*, Phys. Rev. Lett. **21**, 1279 (1968); H.C. Kirkman *et al.*, Phys. Lett. **32**, B519 (1970); J.A. Appel *et al.*, Phys. Rev. D **1**, 1285, (1970).
2. T. Powell *et al.*, Phys. Rev. Lett. **24**, 753 (1970)
3. F. Guerin, C.A. Picketty, Nuovo Cim. **32**, 971 (1964); J. Arafune, Y. Shimizu, Phys. Rev. D **1**, 3094 (1970); U. Günther, R. Rodenberg, Nuovo Cim. A **2**, 25 (1971); T.V. Kukhto *et al.*, Preprint JINR-E2-92-556, Feb 1993, 28pp.
4. A. De Rujula, J.M. Kaplan and E. De Rafael, Nucl. Phys. B **35**, 365 (1971).
5. N.F. Mott, Proc. R. Soc. London, Ser. A. **135**, 429 (1932).
6. T.J. Gay, F.B. Dunning, Rev. Sci. Instrum. **63**, 1635 (1992); J.S. Price, B.M. Poelker, C.K. Sinclair *et al.*, In: Proc. Protvino 1998, High Energy Spin Physics Symposium, p.554.
7. S. P. Wells *et al.* [SAMPLE collaboration], Phys. Rev. C **63**, 064001 (2001).
8. A. De Rujula, J. M. Kaplan and E. De Rafael, Nucl. Phys. B **53**, 545 (1973).
9. J. Blumlein and D. Robaschik, Nucl. Phys. B **581**, 449 (2000).
10. X. Ji, Phys. Rev. Lett. **78**, 610 (1997); A. Radyushkin, Phys. Lett. B **380**, 417 (1996).
11. M. Diehl *et al.*, Phys. Lett. B **411**, 193 (1997); I. Akushevich *et al.*, Phys. Rev. D **64**, 094010 (2001).

EXCLUSIVE PROCESSES AT HERMES

D. GASKELL (FOR THE HERMES COLLABORATION)

Nuclear Physics Laboratory, University of Colorado at Boulder,
Campus Box 390,
Boulder, CO 80309, USA
E-mail: gaskell@dilsey.colorado.edu

The open geometry of the HERMES spectrometer provides a unique opportunity to study several exclusive and nearly–exclusive processes simultaneously. At HERMES, we have measured the longitudinal cross section for ρ^0 production, target–spin asymmetries for π^+ production, and beam–spin and beam–charge asymmetries from the interference of the deeply virtual Compton scattering (DVCS) and Bethe–Heitler (BH) processes. These measurements provide first constraints on parameterizations of the generalized parton distributions (GPDs). The upcoming HERMES run will provide new data from a transversely polarized target, providing further constraints on GPD models. A recoil detector planned for the HERMES spectrometer in combination with dedicated high target–density running will improve identification of the exclusive final state and allow measurements with high statistical precision.

The primary scientific mission of the HERMES experiment at DESY has been the extraction of the polarized quark distributions via inclusive and semi–inclusive deep inelastic scattering from polarized targets. However, the open geometry and good particle identification capabilities of the HERMES spectrometer [1] also allow for the study of several exclusive processes. These include longitudinal ρ^0 production, the azimuthal dependence of the target–spin asymmetry from π^+ production, and the azimuthal dependence of beam–spin and beam–charge asymmetries from the DVCS and Bethe–Heitler interference. These different final states are sensitive to different combinations of generalized parton distributions, e.g. H and E for vector meson production, $\tilde{H}$ and $\tilde{E}$ for pseudoscalar meson production, while DVCS depends on all four. The initial HERMES measurements have already provided constraints on existing GPD models.

1. Exclusive ρ^0 electroproduction

The detection of ρ^0 mesons at HERMES is accomplished via the reconstruction of $\pi^+\pi^-$ pairs. The missing mass, M_x, of the recoiling system is recon-

152

structed, and exclusive events selected via a cut on $\Delta E = (M_x^2 - m_p^2)/2m_p$ (this is described in more detail elsewhere in these proceedings [2]).

At large Q^2 (> 5 GeV^2) and high W (> 10 GeV), the ρ^0 electroproduction cross section is dominated by multiple gluon exchange. However, recent calculations based on GPD models [3] indicate that quark exchange should play a significant role at small W.

Since factorization has only been proven for the longitudinal channel in meson electroproduction, it is necessary to extract the longitudinal ρ^0 cross section to make a fair comparison to GPD–based predictions. This is accomplished by extracting the spin–density matrix elements from the $\pi^+\pi^-$ angular distributions from the ρ^0 decay. Assuming s–channel helicity conservation (SCHC), the ratio, R, of longitudinal to transverse cross sections can be expressed in terms of the r_{00}^{04} spin–density matrix element,

$$R = \frac{\sigma_L}{\sigma_T} = \frac{r_{00}^{04}}{\epsilon(1 - r_{00}^{04})}, \qquad (1)$$

where ϵ is the longitudinal polarization of the virtual photon. The longitudinal cross section, σ_L, can then be extracted using this measurement of R, and the measured total cross section σ_{total}.

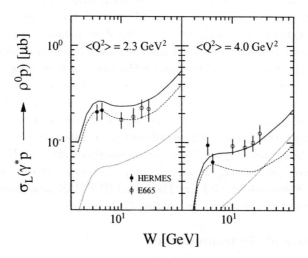

Figure 1. Longitudinal cross section for ρ^0 electroproduction from the proton. The closed circles are HERMES data while the open circles are from E665. The curves are the results of GPD–based calculations from Vanderhaeghen et al. [3]. The dotted curve represents the contribution from gluon exchange, the dashed–dotted the quark exchange contribution, and the solid line the sum of the two. Clearly, the quark exchange piece is required to achieve agreement with the data.

The HERMES results [4] for σ_L are shown in Fig. 1 with the calculations of Vanderhaeghen et al. [3]. In the context of this GPD model, it is clear that for $W < 10$ GeV, quark exchange plays a significant role.

2. Exclusive π^+ Production

Experimentally, the study of exclusive π^+ production is somewhat challenging at HERMES. The resolution of the HERMES spectrometer is unfortunately not sufficient to differentiate the exclusive π^+ sample from the semi–inclusive sample event–by–event. Hence, the semi–inclusive background is estimated (and subtracted) by using the π^- missing mass spectrum, for which exclusive production to a final proton only is impossible. This is illustrated in Fig. 2. This method of background subtraction is complicated by the fact that there may be differences for π^+ and π^- production in the resonant relative to the non–resonant channels.

The study of exclusive π^+ production at HERMES is motivated by the prediction of Frankfurt et al. [5], that, for a transversely polarized target, one expects a large asymmetry in the azimuthal ϕ distribution (where ϕ is the angle between the electron scattering plane and the π^+ production plane) of the target–spin asymmetry. This asymmetry comes about from the interference between the pseudoscalar $\tilde{E}$ and pseudovector $\tilde{H}$ GPDs.

Although the above prediction is for a transverse target asymmetry, one may also expect some asymmetry when the target is polarized longitudinally along the beam direction. When taken relative to the virtual photon direction, such a longitudinally polarized target will have a small, but non-zero perpendicular component. In this case the polarized π^+ cross section can be written [6],

$$\sigma_S \sim \left[S_\perp \sigma_L + S_\| \sigma_{LT}\right] \sin\phi, \tag{2}$$

where $S_\perp$ ($S_\|$) is the component of the target polarization perpendicular (parallel) to the virtual photon direction, σ_L is the longitudinal cross section, σ_{LT} is the interference between the longitudinal and transverse cross sections.

The $\sin\phi$ asymmetry averaged over the HERMES acceptance is shown in Fig. 2. This asymmetry is defined by,

$$A_{UL} = \frac{1}{P_{Target}} \frac{N^\uparrow - N^\downarrow}{N^\uparrow + N^\downarrow}, \tag{3}$$

where P_{Target} is the mean target polarization and $N^\uparrow$ ($N^\downarrow$) is the number of normalized π^+ events when the target is polarized anti–parallel (parallel) to the beam direction. The $\sin\phi$ amplitude of this asymmetry, $A_{UL}^{\sin\phi}$, is then

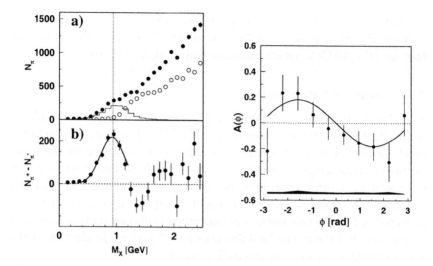

Figure 2. Left: Missing mass spectrum for exclusive π^+ production. The top panel shows the π^+ (closed circles) and π^- (open circles) yields including semi–inclusive contributions. The solid curve is a Monte Carlo calculation of the exclusive π^+ yield only. The bottom panel shows the missing mass distribution after using the (normalized) π^- spectrum to subtract the π^+ semi–inclusive background. Right: The $\sin\phi$ distribution of the target–spin asymmetry for exclusive π^+ production.

$-0.18 \pm 0.05(\text{stat}) \pm 0.02(\text{sys})$ [7]. Note that in Eqn. 2, σ_{LT} is suppressed by $1/Q$, but $S_\perp/S$ is about 0.16, so the two terms may be of similar magnitude. Hence one can make no concrete statement about the size of the asymmetry coming from the GPD interference, but this result does indicate that such measurements with a transversely polarized target are feasible.

3. Deeply Virtual Compton Scattering

HERMES measurements of the interference between the deeply virtual Compton scattering (DVCS) and Bethe–Heitler processes are discussed in detail elsewhere in these proceedings [8].

4. Future Measurements

4.1. *Transversely Polarized Target*

Starting in 2002, the HERMES experiment will begin running with a transversely polarized hydrogen target. Measurements of exclusive processes from such a target could prove to be extremely interesting. As mentioned in Sec. 2, the azimuthal asymmetry of exclusive π^+ production should be more directly sensitive to the interference between the $\tilde{H}$ and $\tilde{E}$ GPDs. Analogously, the azimuthal asymmetries of exclusive vector meson production from a transverse target would be sensitive to the interference between the E and H GPDs. The latter is particularly interesting in that, in the context of the model of Goeke et al. [6] the parameterization of the GPD E involves J_u and J_d, the total angular momentum of the up and down quarks in the nucleon. Since the interference involves E linearly, measurements of the azimuthal asymmetries of ρ^0, ρ^+, and ω mesons would provide strict constraints on E, and hopefully give some insight into the quark contribution to the nucleon angular momentum.

4.2. Recoil Detector

The existing HERMES measurements of exclusive processes would be much improved with more statistics and improved identification of the exclusive final state. A recoil detector, planned for installation in 2004, in conjunction with dedicated high luminosity running will address both these issues.

The HERMES recoil detector [9] will basically consist of a barrel of detectors around the target region. This will allow the detection of the slowly recoiling proton (at large angle) without interfering with the acceptance of high momentum mesons and photons in the HERMES spectrometer. The recoil detector will consist of two layers of silicon detectors and two layers of scintillating fiber detectors, both used for tracking as well as particle identification via energy deposition. Additional scintillators will be used for π^0 and photon detection and a longitudinal magnetic field will allow momentum analysis of the recoiling proton.

Since the recoil detector is designed to detect the (charged) proton, it can only be used to study exclusive production of high energy neutral particles, i.e., π^0, η, ρ^0 mesons, as well as photons. There are also plans to study the exclusive production of Λ particles.

The exclusive final state will be selected via "co–planarity" cuts, i.e., ensuring that the recoiling proton is in the same plane as the produced meson or photon. For the exclusive production of photons (in studying the DVCS + Bethe–Heitler interference), it is estimated that the background due to Δ production (the primary contaminating process) can be reduced to 1%.

Fig. 3 shows the projected uncertainties for measurements of the beam–spin and beam–charge asymmetries associated with the DVCS + Bethe–Heitler interference for one year of dedicated, high–luminosity running with the recoil detector. This degree of statistical precision will make it possible

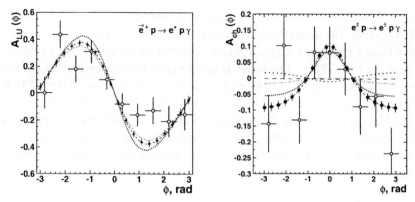

Figure 3. Projections of statistical accuracies (closed circles) for measurements of the beam–spin (left) and beam–charge (right) azimuthal asymmetries coming from the interference of the DVCS and Bethe–Heitler processes. These projections assume one year of dedicated, high–luminosity running with the HERMES recoil detector. For comparison, the HERMES results from existing data are also shown (open circles). The curves represent various GPD model predictions.

to finely bin the data to look at the asymmetries as functions of x_{Bj}, Q^2, and t.

Acknowledgments

I would like to thank my fellow members of the HERMES collaboration for their hard work in taking and analyzing the data presented in this work.

References

1. K. Ackerstaff *et al.* [HERMES Collaboration], Nucl. Instrum. Meth. A **417**, 230 (1998).
2. A. Borissov [HERMES Collaboration], "Exclusive Electroproduction of Vector Mesons at HERMES," these proceedings.
3. M. Vanderhaeghen, P. A. Guichon and M. Guidal, Phys. Rev. Lett. **80**, 5064 (1998).
4. A. Airapetian *et al.* [HERMES Collaboration], Eur. Phys. J. C **17**, 389 (2000).
5. L. L. Frankfurt, . V. Polyakov, M. Strikman and M. Vanderhaeghen, Phys. Rev. Lett. **84**, 2589 (2000).
6. K. Goeke, . V. Polyakov and M. Vanderhaeghen, Prog. Part. Nucl. Phys. **47**, 401 (2001).
7. A. Airapetian *et al.* [HERMES Collaboration], Phys. Lett. B **535**, 85 (2002)
8. J. Ely [HERMES Collaboration], "Azimuthal Asymmetries Associated with Deeply Virtual Compton Scattering," these proceedings.
9. The HERMES Collaboration, "The HERMES Recoil Detector", DESY PRC Proposal 02-01 (HERMES Internal Note 02-003).

AZIMUTHAL ASYMMETRIES ASSOCIATED WITH DEEPLY VIRTUAL COMPTON SCATTERING

J. ELY (FOR THE HERMES COLLABORATION)

Nuclear Physics Laboratory, University of Colorado at Boulder,
Campus Box 390, Boulder, CO 80309, USA
E-mail: ely@ucsu.colorado.edu

The asymmetries in the azimuthal distributions of hard exclusive electroproduced photons, with respect to the beam spin and beam charge, have been observed for the first time. Here the azimuthal angle ϕ is defined as the angle between the lepton scattering plane and the photon production plane. The data were collected by the HERMES experiment at DESY for the 1996-2000 running period, using the 27.6 GeV longitudinally polarized lepton beam at HERA and an unpolarized hydrogen gas target. The observed asymmetries are attributed to the interference of the Bethe-Heitler and deeply virtual Compton scattering processes.

Deep inelastic scattering (DIS) experiments have yielded valuable information on the parton momentum distributions and their helicity dependence. Recent theoretical work extends the QCD framework of DIS to the non-forward region of the virtual Compton process [1,2]. The resulting non-perturbative functions are expressed in terms of generalized parton distribution functions (GPD's). These functions describe dynamical correlations between partons of different momenta. The GPD's have attracted increased interest since they were shown to allow access to the total angular momentum of the partons [3].

The GPDs can be determined experimentally using exclusive reactions in which the target nucleon remains intact. Deeply virtual Compton scattering (DVCS) (see Fig. 1a) has the simplest final state: a single photon in addition to the recoiling target nucleon and scattered lepton. However DVCS has the same final state as the Bethe-Heitler(BH) process (see Fig. 1b), and in most presently accessible kinematic regimes the BH process dominates over the DVCS process. At HERMES, asymmetries with respect to the charge and helicity of the beam are used to isolate the interference term of the BH and DVCS processes, providing access to the DVCS contribution.

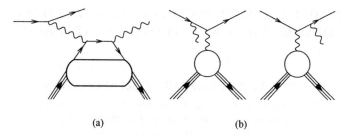

Figure 1. (a) Feynman diagram for deeply-virtual Compton scattering, and (b) photon radiation from the incoming and scattered lepton in the Bethe-Heitler process.

As demonstrated in Ref [4], the difference in the unpolarized positron and electron beam cross sections (charge asymmetry) from an unpolarized target is proportional to the real part of the interference term of the BH and DVCS processes. The imaginary part of the interference term can be measured by taking the cross section difference between polarized lepton beam helicities from an unpolarized target (single spin asymmetry).

The HERMES experiment, uses the 27.6 GeV longitudinally polarized lepton beam at DESY [5]. The results presented here include data from positrons (1996/97/2000) and electrons (1998) scattered off a hydrogen gas target [6]. Events from an unpolarized and spin-averaged polarized hydrogen target were collected using the HERMES detector[7]. Events were selected if they contained exactly one positron track with momentum larger than 3.5 GeV and exactly one photon with an energy deposition greater than 0.8 GeV in the calorimeter. The scattered positron was identified using four particle identification detectors, a threshold Čerenkov counter, a transition radiation detector, a preshower scintillator counter and a lead-glass calorimeter. Photons were identified by the detection of an energy deposition in the calorimeter and preshower detector without any associated track. This sample includes exclusive and non-exclusive events since the recoiled proton was not detected. The following requirements were imposed on the positron kinematics: $Q^2 > 1$ GeV^2, $W^2 > 4$ GeV^2, and $\nu < 24$ GeV, where Q^2 is the negative square of the virtual photon 4-momentum, ν is the virtual photon energy and W is the invariant mass of the photon-nucleon system. The opening angle between the virtual photon and the real photon ($\theta_{\gamma\gamma^*}$) was constrained to be greater than 15 mrad (due to the granularity of the calorimeter) and less than 70 mrad (for a more uniform coverage in ϕ, the azimuthal angle of the final state photon with respect to the lepton scattering plane).

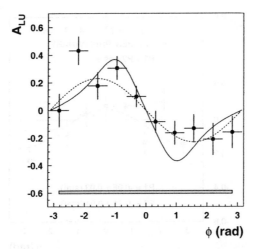

Figure 2. Single spin asymmetry as a function of the azimuthal angle ϕ from the HERMES experiment. The data correspond to the missing mass region between -1.5 and 1.7 GeV with statistical (error bars) and systematic (error band) uncertainties.

Shown in Fig. 2 is the azimuthal dependence of the single spin asymmetry:

$$A_{LU}(\phi) = \frac{1}{\langle |P_l| \rangle} \cdot \frac{N^+(\phi) - N^-(\phi)}{N^+(\phi) + N^-(\phi)}. \tag{1}$$

Here N^+ and N^- represent the luminosity-normalized yields of events in the two beam helicity states, $\langle |P_l| \rangle$ is the magnitude of the average beam polarization, and the subscripts L and U denote the use of a longitudinally polarized beam and an unpolarized target. The data in Fig. 2 are confined to the exclusive region of missing mass from -1.5 to 1.7 GeV, and show a distinct $\sin\phi$ behavior. The missing mass M_x is defined as $\sqrt{(q + P_p - k)^2}$ with q, P_p and k being the four-momenta of the virtual photon, the target proton and the produced real photon, respectively. A negative M_x^2 may arise due to the finite momentum resolution of the spectrometer in which case $M_x \equiv -\sqrt{-M_x^2}$. The data are superimposed with a simple $\sin\phi$ curve with an amplitude of 0.23 (dashed curve) as well as the GPD calculation of Ref. [8] for the average kinematic values of this measurement (solid curve).

Fig. 3 displays the HERMES preliminary measurement of the charge asymmetry A_C, which is defined as follows:

$$A_C(\phi) = \frac{N_{e^+}(\phi) - N_{e^-}(\phi)}{N_{e^+}(\phi) + N_{e^-}(\phi)}. \tag{2}$$

Here N_{e^+} and N_{e^-} represent the luminosity-normalized yields of events in the two beam charge states. The data in the exclusive region of missing

160

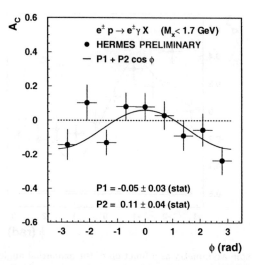

Figure 3. Charge asymmetry as a function of the azimuthal angle ϕ from the HERMES experiment. The data correspond to the missing mass region between -1.5 and 1.7 GeV with the error bars representing the statistical uncertainties only.

mass from -1.5 to 1.7 GeV displayed in Fig. 3 show a clear $\cos\phi$ behavior as illustrated by the best-fit curve of amplitude 0.11.

To display the results for exclusive as well as non-exclusive events together, the beam-spin analyzing power $A_{LU}^{\sin\phi}$ has been determined:

$$A_{LU}^{\sin\phi} = \frac{2}{N} \sum_{i=1}^{N} \frac{\sin\phi_i}{(P_l)_i}, \qquad (3)$$

where $N = N^+ + N^-$. Here, the difference of moments (single spin asymmetry) is implicitly contained in the sign of the beam polarization P_l. The analyzing power $A_{LU}^{\sin\phi}$ is shown in Fig. 4 as a function of missing mass. A negative asymmetry is observed in the exclusive region while $A_{LU}^{\sin\phi}$ approaches zero for larger M_x values. The missing mass bins left and right of $M_x = M_{proton}$ also show a non-zero value of $A_{LU}^{\sin\phi}$ in Fig. 4 due to the missing-mass resolution of the HERMES spectrometer, which amounts to ~ 0.8 GeV for these events. By combining the $A_{LU}^{\sin\phi}$ values for the exclusive region as in Fig. 2 ($-1.5 < M_x < 1.7$ GeV), a value of -0.23 ± 0.04 (stat) ± 0.03 (syst) is obtained. The average kinematic values in this region are: $\langle x \rangle = 0.11$, $\langle Q^2 \rangle = 2.6$ GeV, and $\langle -t \rangle = 0.27$ GeV^2.

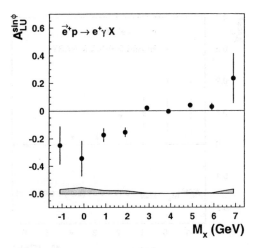

Figure 4. The beam spin analyzing power $A_{LU}^{\sin\phi}$ as a function of missing mass for the electroproduction of photons from HERMES.

Similarly, the moment $A_C^{\cos\phi}$ of the beam charge asymmetry has been determined using the formula

$$A_C^{\cos\phi} = \frac{1}{2}\left[\frac{2}{N_{e^+}}\sum_{i=1}^{N_{e^+}}\cos\phi_i - \frac{2}{N_{e^-}}\sum_{i=1}^{N_{e^-}}\cos\phi_i\right] \qquad (4)$$

and is shown in Fig. 5 as a function of missing mass. Here, an explicit difference in moments is taken to produce the beam charge asymmetry. A positive asymmetry is observed in the exclusive region while $A_C^{\cos\phi}$ approaches zero for larger M_x values. Combining the events in the exclusive region of $-1.5 < M_x < 1.7$ GeV as in Fig. 3, a value of 0.11 ± 0.04 (stat) has been obtained for $A_C^{\cos\phi}$. The systematical uncertainty is presently being studied.

In conclusion, the beam-spin [9] and beam-charge azimuthal asymmetries for the hard electroproduction of photons have been measured for the first time at HERMES. Significant asymmetries have been observed in the exclusive region, and approach zero in the non-exclusive region. The expected beam-helicity and beam-charge dependence of the asymmetries from the interference between the DVCS and BH processes is present in the data. As these charge and helicity depedences can not be caused by the BH process alone, the present data constitute evidence for the observation of the DVCS process.

162

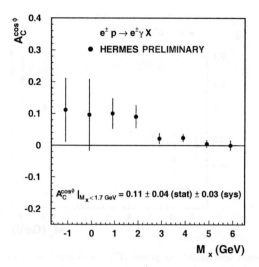

Figure 5. The beam charge asymmetry $A_C^{\cos\phi}$ as a function of missing mass for the electroproduction of photons from HERMES.

Acknowledgments

I gratefully acknowledge the U.S. Department of Energy for financial support to attend this conference. I would thank my HERMES colleagues, especially D. Gaskell, for useful discussions.

References

1. J.C. Collins and A. Freund, Phys. Rev. **D59** (199) 074009.
2. A.V. Radyushkin and C. Weiss, Phys. Rev. **D63** (2001) 114012.
3. X. Ji, Phys. Rev. **D55** (1997) 7114.
4. A.V. Belitsky *et al.*, Nucl. Phys. **B593** (2001) 289.
5. D.P. Barber *et al.*, Phys. Lett. **B343** (1995) 436.
6. J. Stewart. Proc. of the Workshop "Polarized gas targets and polarized beams" eds. R.J. Holt and M.A. Miller, Urbana-Champaign, AIP Conf. Proc. **421** (1997) 69.
7. HERMES Collaboration, K.Ackerstaff et al., Nucl. Instr. Meth. **A417** (1998) 230.
8. N. Kivel, M. V. Polyakov, and M. Vanderhaeghen, Phys. Rev. **D63** (2001) 114014.
9. HERMES Collaboration, A. Airapetian, *et al*, Phys. Rev. Lett. **87** (2001) 182001.

THE GENERALIZED PARTON DISTRIBUTIONS PROGRAM AT JEFFERSON LAB

F. SABATIÉ

C.E.A. Saclay
DAPNIA/SPhN
91191 Gif sur Yvette, France
E-mail: fsabatie@cea.fr

The Generalized Parton Distributions (GPD) have drawn a lot of interest from the theoretical community since 1997, but also from the experimental community and especially at Jefferson Lab. First, the results for Deeply Virtual Compton Scattering (DVCS) at 4.2 GeV beam energy have recently been extracted from CLAS data. The single spin asymmetry shows a remarkably clean sine wave despite the rather low Q^2 achievable at this energy. Two new dedicated DVCS experiments using 6 GeV beam will run in 2003 and 2004 in Hall A and Hall B respectively. Both experiments will yield very accurate results over a wide range of kinematics, and allow for the first time a precise test of the factorization of the DVCS process. Assuming the Bjorken regime is indeed reached, these experiments will allow the extraction of linear combinations of GPD's and put strong constraints on the available phenomenological models. Upon successful completion of both experiments, a wider experimental program at 6 GeV can be envisioned, using for instance Deuterium targets trying to nail down the neutron and deuteron GPD's. In addition, resonances can be probed using ΔVCS where one produces a resonance in the final state. In a way similar to DVCS, such a process depends on Transition GPD's which describe in a unique way the structure of resonances. Finally, the 12 GeV upgrade of Jefferson Lab extends the available kinematical range, and will allow us to perform a complete, high precision GPD program using various reactions among which, Deeply Virtual Meson Electroproduction (DVMP) and DVCS.

1. Introduction

The understanding of the structure of the nucleon is a fundamental topic. Despite having been studied during the past forty years, there are still many questions left unanswered. An example of such is the extensive debate over the spin structure of the nucleon ground state. Two kinds of electromagnetic observables linked to the nucleon structure have been considered so far. Electromagnetic form factors, first measured on the proton by Hoftstader in the 1950's, then more recently on the neutron. Weak form factors have been measured in parity violating experiments. Another approach initiated in the late 60's studies parton distribution functions via Deep Inelastic Scattering (DIS), and Drell-Yan processes.

Recently a new theoretical framework has been proposed, namely the Generalized Parton Distributions (GPD) [1,2]. These provide an intimate connection between the ordinary parton distributions and the elastic form factors and therefore contain a wealth of information on the quark and gluon structure of the nucleon. Moreover, not only do they depend on the usual DIS variables, the skewdness ξ (linked to x_B) and Q^2, but also on the average parton momentum fraction in the loop x and the momentum transfer between the initial and recoil protons $t = (p'-p)^2$, giving the GPD's much more degrees of freedom than the regular parton distributions.

In addition, it has been shown recently [3] that the t dependence of the GPD's provide information on the transverse position of partons inside the nucleon. One could imagine the possibility to take "pictures" [4] of the nucleon with a very high definition (fixed by the virtuality of the incoming virtual photon), which would revolutionize our understanding of the nucleon structure and confinement in general. Such a femto-picture applied to the deuteron case for instance would yield a much clearer understanding of where quarks are actually located inside of the deuteron.

2. DVCS with CLAS at 4.2 GeV

The first accurate result on DVCS has come from Jefferson Lab with the CLAS detector in Hall B [5]. We have measured a globally exclusive beam-spin asymmetry in the reaction $\vec{e}p \to ep\gamma$, using a 4.25 GeV longitudinally polarized electron beam on a liquid hydrogen target. As usual with low-energy experiments, the $ep \to ep\gamma$ process is dominated by Bethe-Heitler (BH) where photons are emitted from the incoming or scattered electron lines. While the interesting process is actually DVCS where the photon is emitted by the proton in response to the electromagnetic excitation by the virtual photon, the interference between the BH and DVCS amplitude boosts the effect of DVCS and produces a large cross-section difference for electrons of opposite helicities. In this difference, the large BH contribution drops out and only the helicity dependent interference term remains, parametrized as $\alpha . \sin\phi + \beta . \sin 2\phi$, where ϕ is the angle between the leptonic and hadronic planes. Note that the α coefficient is linear in the leading twist GPD's. Experimentally, the Hall B experiment measured the relative asymmetry $A = (d^4\sigma^+ - d^4\sigma^-)/(d^4\sigma_{tot})$, which has a more complex ϕ dependence than the difference in cross-sections, but is much simpler to measure in a large acceptance spectrometer such as CLAS.

The reaction $ep \to ep\gamma$ was identified by examining $ep \to epX$ events and requiring the mass of the missing particle to be zero. Unfortunately, CLAS is not able to separate π^0 electroproduction from photon electroproduction using an event-by-event missing mass technique. The number of photon events was determined using a fitting technique that analyzed the shape of the missing mass distribution. The exclusivity of the reaction is therefore demonstrated globally rather than event-by-event. Figure 1 shows the resulting asymmetry A as a function of ϕ. The data points are fitted with the function $A(\phi) = \alpha . \sin \phi + \beta . \sin 2\phi$ where $\alpha = 0.202 \pm 0.028^{stat} \pm 0.013^{syst}$ and $\beta = -0.024 \pm 0.021^{stat} \pm 0.009^{syst}$. Up to the accuracy of the data, β seems to be very small at Q^2 as low as 1.25 GeV^2, indicating that the tranverse part of the process dominates as expected. More accurate data are clearly needed to test the factorization in a systematic way.

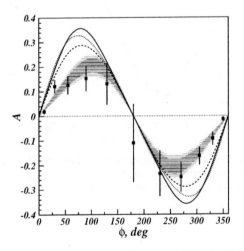

Figure 1. ϕ dependence of the beam spin asymmetry A. The dark shaded region is the range of the fitted function defined by statistical errors only. The kinematics averages to $< Q^2 >= 1.25$ GeV^2, $< x_B >= 0.19$ and $< -t >= 0.19$ GeV^2.

3. DVCS in Hall A at 6 GeV

The Hall A DVCS experiment [6] proposes to check the factorization of the DVCS process by performing an accurate measurement of the (properly) weighted cross-section difference for three values of Q^2 from 1.5 up to 2.5 GeV^2 at fixed $x_B = 0.35$. The high resolution and high luminosity achievable in Hall A allows one to make a very clean interpretation of the data, since the exclusivity will be checked on an event-by-event basis. The experimental apparatus is composed of a High Resolution Spectrometer for the detection of the scattered electron, a high resolution PbF$_2$ calorimeter for the detection of the emitted photon and an array of plastic scintillators for the detection of the recoil proton. An example of the quality of the data achievable by this experiment is shown on Figure 2 for the $Q^2 = 2.5$ GeV^2 setting.

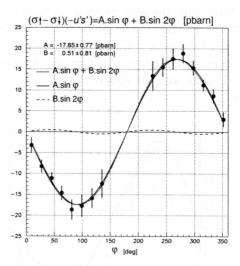

Figure 2. Expected cross-section difference mutiplied by a kinematical factor corresponding to the BH propagators as a function of ϕ for the $Q^2 = 2.5$ GeV^2 setting. Note that the convention for ϕ is different compared to Fig. 1.

4. DVCS in Hall B at 6 GeV

Once the factorization for the DVCS process has been confirmed by the Hall A experiment, DVCS in Hall B will allow one to look at various kinematical dependences of the beam spin asymmetry [7]. Indeed, CLAS with its large acceptance will allow to scan this observable in function of x_B, t and Q^2, for a total of 378 bins with good staticstics. In order to address the issue of the full exclusivity of DVCS events, the CLAS DVCS collaboration is designing a forward PbWO$_4$ calorimeter to detect low angle photons typical of DVCS at small t. In order to achieve a higher luminosity, a Moller shield composed of a superconducting solenoid surrounding the target is also under design. Figure 3 shows the quality of the data which is expected from this experiment.

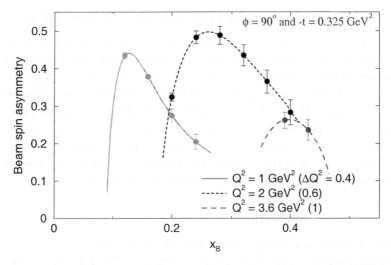

Figure 3. x_B dependence of the single spin asymmetry at $\phi = 90°$ expected in the 6 GeV CLAS DVCS experiment. Three sets of points correspond to three $< Q^2 >$.

5. New experiments at 6 GeV

Although DVCS and Deeply Virtual Meson Production are not easy experiments, one could think of the future of these measurements using 6 GeV beam. DVCS on the Deuterium is of remarkable interest: it allows not only to measure the coherent electroproduction of photons off the deuteron, but also, using deuteron as a quasi-free neutron target, to measure the DVCS reaction on the neutron. This type of experiments is essential if one envisions a full flavor decomposition of the GPD's. However, several difficulties arise. As far as the Deuteron DVCS (D_2VCS) is concerned, the recoil deuteron has to escape the target, which is more difficult than in the proton case. This will contribute to raising the minimum $-t$ achievable by the experiment. The cross-section has been evaluated recently [8]: at low x_B, it is comparable to the proton cross-section. The beam spin asymmetry is also sizeable much like in the proton case. For the neutron case, the experiment is much more difficult. Once again, the cross-section is only 2 to 3 times smaller, but the asymmetry is much smaller than for the proton, of the order of a few %.

The GPD formalism has been extended to transition reactions where the final particle is a nucleon resonance [9]. Just as in the neutron case, the cross-section is smaller, although not too small, and the asymmetry is a few %. Even though ΔVCS is a very difficult experiment, the prospects for a new kind of baryon resonance spectroscopy makes the study worth.

6. The GPD program with 12 GeV beam at Jefferson Lab

The extension of the planned 6 GeV GPD program up to 12 GeV is rather straightforward. Higher energy beam allows one to open up the kinematical coverage: $0.1 \leq x_B \leq 0.6$ and $1 \leq Q^2 \leq 8\,\text{GeV}^2$. One can therefore imagine a complete GPD program using all kinds of Hard Exclusive reactions ($ep \to ep\gamma$, $ep \to ep\rho$, $eD \to eD\gamma$, $ep \to en\pi^+$, ...) on different targets, polarized or not, and looking at various observables such as cross-sections, single spin asymmetries or even double spin asymmetries. Both the luminosity and/or the high resolution of the Jefferson Lab equipments will make it possible to perform accurate measurements in a reasonnable amount of time.

7. Conclusion and outlook

Jefferson Lab is in a unique position to make pioneering steps towards a better understanding of the structure of the nucleon. The 4.2 GeV CLAS data has confirmed that the DVCS single spin asymmetry is indeed large and close to a pure sine wave, encouraging us to perform two new experiments at 6 GeV in Hall A and Hall B. These experiments will not only try to justify the factorization of DVCS at moderate Q^2, but will give the first strong contrains to GPD phenomenological models. Additional information can be obtained at 6 GeV using a Deuterium target and looking either at coherent deuteron DVCS, or DVCS on quasi-free neutrons. Finally, the 12 GeV of Jefferson Lab greatly enhances the kinematical coverage of Deeply Exclusive processes, and a complete GPD program can be performed, which undoubtedly, will shed a new light on the understanding of the nucleon.

References

1. X. Ji, Phys. Rev. Lett 78 (1997) 610 (1997)
   X. Ji, Phys. Rev. D55 (1997) 7114.
2. A.V. Radyushkin, Phys. Lett. B380 (1996) 417
   A.V. Radyushkin, Phys. Lett. B385 (1996) 333.
3. M. Burkardt, Phys. Rev. D62 (2000) 071503.
4. P. Ralston and B. Pire, hep-ph/0110075.
5. S. Stepanyan et al., Phys. Rev. Lett. 87 (2001) 182002.
6. P. Bertin, C-E. Hyde-Wright, R. Ransome and F. Sabatié, cospokespersons of Experiment 00-110 in Hall A. http://www.jlab.org/~sabatie/dvcs.
7. V.D. Burkert, L. Elouadrhiri, M. Garçon and S. Stepanyan, cospokespersons of Experiment 00-113 in Hall B.
8. F. Cano and B. Pire, hep-ph/0206215.
9. L.L. Frankfurt, M. Polyakov, M. Strikman and M. Vanderhaeghen, Phys. Rev. Lett. 84 (2000) 2589.

EXCLUSIVE ELECTROPRODUCTION OF VECTOR MESONS AT HERMES: NUCLEAR EFFECTS

A.B. BORISSOV, ON BEHALF OF THE HERMES COLLABORATION

Randall Laboratory of Physics, University of Michigan, Ann Arbor, MI 48109-1120, USA. E-mail: Alexander.Borissov@desy.de

Exclusive coherent and incoherent electroproduction of the ρ^0 meson from ^{1}H and ^{14}N targets has been studied at the HERMES experiment as a function of coherence length and negative squared transfer Q^2. The ratio of ^{14}N to ^{1}H cross sections per nucleon, known as the nuclear transparency, was found to increase (decrease) with increasing coherence length for coherent (incoherent) ρ^0 electroproduction. For fixed coherence length, a rise of the nuclear transparency with Q^2 is observed for coherent and incoherent ρ^0 production, which is in agreement with theoretical calculations of color transparency.

One of the fundamental predictions of QCD is the existence of a phenomenon called color transparency (CT), whose characteristic feature is that, at sufficiently high negative squared four-momentum of virtual photon (Q^2) in deep inelastic scattering, the initial state (ISI) and final state (FSI) interactions of hadrons traversing a nuclear medium supressed [1,2,3,4,5,6]. The idea is that the dominant amplitudes for exclusive reactions at high Q^2 involve hadrons of reduced transverse size; that these small color-singlet objects, or small size configurations (SSC), have reduced interactions with hadrons in the surrounding nuclear medium; and that these SSC remain small long enough to traverse the nucleus without interaction. Consequently, the signature of CT is an increase in the nuclear transparency as Q^2 increases. The nuclear transparency T is a measure of the average probability that a hadron escapes from the nucleus A with reduced interaction. It can be measured by taking the ratio of the cross section for a selected process on a nucleus with atomic number A to the atomic number of the nucleus times the corresponding cross section on a free nucleon, $T = \sigma_A/(A\sigma_N)$.

In order to study CT for exclusive electroproduction of ρ^0 mesons, one has to select a sample of ρ^0 mesons that contains SSC produced via a hard process, i.e. with large Q^2. In these processes, the hadronic structure of a high energy virtual photon gives rise to a $q\bar{q}$ pair that evolves from its initial

SSC to a normal-size ρ^0 meson, where the initial SSC has a transverse size $r_\perp \sim 1/Q$. The $q\bar{q}$ fluctuation of the virtual photon can propagate over a distance l_c known as the coherence length. It is given by $l_c = \frac{2\nu}{Q^2+M_{q\bar{q}}^2}$, where ν is the virtual photon energy and $M_{q\bar{q}}$ is the invariant mass of the $q\bar{q}$ pair. This SSC will then propagate through the nuclear medium with little interaction. After the hard scattering, the $q\bar{q}$ pair will evolve to a normal-size ρ^0 meson over a distance l_f called the formation length. It is the governing scale for CT effects and is given by $l_f = \frac{2\nu}{m_{V'}^2-m_V^2}$, where m_V is the mass of the ρ^0 meson in the ground state and $m_{V'}$ the mass of its first orbital excitation. For small values of l_f, the $q\bar{q}$ pair evolves quickly into a normal-size ρ^0 meson, and the ρ^0 meson experiences increased interactions in the nuclear medium. Therefore, the optimal condition for a clean demonstration of CT is that l_f is larger than the nuclear size ("frozen" approximation [3,4,7]).

The phenomenon of nuclear transparency is an intricate mixture of coherence and formation length effects with different signatures for coherent and incoherent ρ^0 production. For incoherent ρ^0 production, the chance for the $q\bar{q}$ pair to interact with the nuclear medium increases with l_c. Only if l_c becomes considerably smaller than the size of the nucleus, the nuclear transparency approaches unity $(T \to 1)$. This coherence length effect [8] for small values of l_c can mimic the Q^2-dependence of the nuclear transparency predicted by CT. For coherent ρ^0 production, in contrast, the Q^2-dependence of the nuclear form factor suppresses the nuclear transparency at small l_c, as $l_c \to 0$ corresponds to a large longitudinal momentum transfer $(q_c \sim 1/l_c)$, where the form factor is small. Hence T decreases with Q^2 in coherent production and can significantly modify the Q^2-dependence. In order to disentangle coherence length from CT effects, it is important to study the variation of T with Q^2, while keeping l_c fixed. In this way, a change of T with Q^2 can be associated with the onset of CT. A rigorous quantum mechanical description of the SSC evolution, based on the light-cone Green function formalism [7], naturally incorporates the interplay between the coherence length effect (ISI) and CT effect (FSI). In this formalism it is shown that the signature of CT is a positive slope of the Q^2-dependence of the nuclear transparency for both coherent and incoherent ρ^0 production. We have adopted this approach and extracted the slope of the Q^2-dependence of T at very narrow (0.1 fm) bins of l_c.

The data were obtained during the 1996-1997 running periods of the HERMES experiment [9,8] in the 27.5 GeV HERA positron storage ring at DESY. The HERMES average kinematic values well satisfy the requirements for studying CT effects. At $\langle Q^2 \rangle \simeq 1.6$ GeV^2 the transverse size of the $q\bar{q}$ pair in the ρ^0 meson is about 0.4 fm, much smaller than a typical hadron size. This ensures that the requirement for SSC is satisfied.

For nuclear targets, the diffractive interaction of the virtual photon with the target can occur incoherently from an individual nucleon or coherently from the nucleus as a whole. The exclusive ρ^0 production signal was extracted in the kinematic region $-2 < \Delta E < 0.6$ GeV and $0.6 < M_{\pi\pi} < 1$ GeV, where ΔE is the exclusivity variable [8,10] and $M_{\pi\pi}$ the invariant mass of the detected pion pair. The coherent exclusive ρ^0 mesons have been selected with $|t'| < 0.045$ GeV^2 for nitrogen and $|t'| < 0.4$ GeV^2 for hydrogen, while for incoherent production the t' cut was $0.09 < |t'| < 0.4$ GeV^2 for both data samples. It has been shown [8] that the incoherent t' slope parameter b_N for various nuclei is consistent with the hydrogen value $b_N = 7.1 \pm 0.3$ GeV^{-2}, and that the observed Q^2-dependence of b_N agrees well with world data [11]. The coherent slope parameter on nitrogen, $b_{14N} = 57.2 \pm 3.3$ GeV^{-2}, is in agreement with the values predicted by the relationship $b_A \approx R_A^2/3$ [12], where R_A is the nuclear radius.

The background under the exclusive ρ^0 sample has been described earlier [8]. It is mainly caused by hadrons from DIS fragmentation processes. Part of this background is removed by excluding the region $|t'| > 0.4$ GeV^2, where the background dominates the ρ^0 yield. The remaining background $(6 \pm 3\%)$ is estimated and subtracted as a tail of the normalized background spectrum for $\Delta E < 0.4$ GeV [8,9]. The double-diffractive contribution to the incoherent ρ^0 production cross section is found to be $(4 \pm 2\%)$ [10] for which the data were corrected. For coherent ρ^0 production the contamination due to double-diffractive dissociation is found to be negligible.

For incoherent ρ^0 production the nuclear transparency has been evaluated as $T = \sigma_A/(A\sigma_H) = N_A \mathcal{L}_H/(A N_H \mathcal{L}_A)$ [8], where H refers to ^{1}H, A to ^{14}N, $N_{A,H}$ is the number of incoherent events, and $\mathcal{L}_{A,H}$ is the effective luminosity of the nitrogen or hydrogen samples, corrected for detector and reconstruction inefficiencies.

For coherent ρ^0 production, some additional correction factors have been applied to extract the nuclear transparency because different t' requirements have been applied to the nitrogen and the hydrogen data. These include the ratio of the different acceptance correction factors caused by the different t' regions selected, which has been obtained from Monte Carlo simulations of exclusive ρ^0 production in a 4π geometry and in the HERMES acceptance [10]; the radiative corrections factors, which were calculated separately for nitrogen and hydrogen [13] for each coherence length bin; and the 'Pauli' correction factor, which has been calculated according to Ref. [14].

No Q^2-dependence has been observed for any of the correction factors, when applied to the ratios.

The nuclear transparency for coherent ρ^0 production, presented in Fig. 1 (left panel, squares), increases with coherence length as expected [7]. In contrast, the nuclear transparency for incoherent ρ^0 production decreases with l_c as shown in Fig. 1 (left panel, circles). These data supersede the previously published data [8]. The present analysis is based on an improved tracking method, a different l_c binning, and a t' cut which is fixed over the entire coherence length region. Fair agreement is found between the measured nuclear transparencies, integrated over the available Q^2 region, and calculations including both the coherence length and CT effects [7,15], evaluated at fixed l_c and Q^2 values, given by the dashed lines in Fig. 1 (left panel).

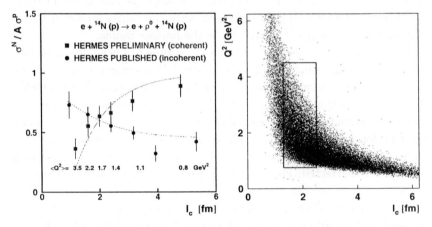

Figure 1. Left panel: Nuclear transparency as a function of coherence length for coherent (squares) and incoherent (curcles) ρ^0 production on nitrogen, compared to predictions with CT effects included [15] (corresponding lines). Error bars include statistical and point-to-point systematic uncertainties added in quadrature.
Right panel: Raw distribution of coherence length versus Q^2 for exclusive ρ^0 production with $\langle l_c \rangle \simeq 2.0$ fm and $\langle Q^2 \rangle \simeq 1.64$ GeV^2. The region surrounded by the square represents the subset that is used for the two-dimensional analysis of the nuclear transparency.

The systematic uncertainties are separated into kinematics-dependent and kinematics-independent contributions. The ratio of the integrated luminosities represents the largest kinematics-independent uncertainty. Additional contributions come from corrections due to 'Pauli blocking', from double-diffractive dissociation, and from the efficiency of the ΔE exclusivity cut. The estimated systematic uncertainty from all normalization factors is 11%. The main contributions of kinematics-dependent systematic uncertainties come from DIS background subtraction, from acceptance corrections, the t' cut reconstruction efficiency, and the application of radiative corrections. Thus, the contribution of the kinematics-dependent systematic uncertainty varies between 8% and 14%. This results in a total systematic uncertainty of 14% to 18%. Note, that it is smallest in the region around $l_c \simeq 2$ fm, where the two-dimensional analysis of the nuclear transparency as a function of coherence length and Q^2 has been performed.

This two-dimensional analysis, representing a new approach in the search for CT, is constrained by the phase space boundaries displayed in Fig. 1 (right panel). Since the combination of statistical significance and Q^2 coverage is largest near $l_c \simeq 2.0$ fm, the region of $1.3 < l_c < 2.5$ fm has been chosen for this two-dimensional analysis. To deconvolute the CT and coherence length effects, narrow coherence length bins were used with bin widths of 0.1 fm. In order to extract the Q^2-dependence, each l_c bin was split into three or four Q^2 bins. The Q^2 bin width, determined by the available Q^2 statistics in each l_c bin, was optimized to have approximately equal statistics in each Q^2 bin.

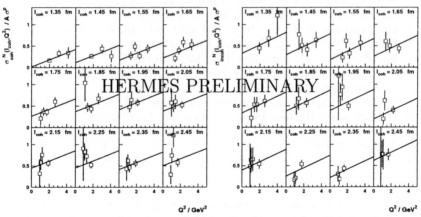

Figure 2. Nuclear transparency as a function of Q^2 in specific coherence length bins (as indicated in each panel) for coherent in the left panel and incoherent in the right panel ρ^0 production on nitrogen. The straight line is the result of the common fit of the Q^2-dependence. The error bars include only statistical uncertainties.

The nuclear transparency was extracted in each (l_c, Q^2) sub-bin, and is shown in Fig. 2 (left and right panels) for twelve l_c bins for coherent and incoherent ρ^0 production, respectively. The low statistics in each (l_c, Q^2) sub-bin makes it difficult to fit the slope of the Q^2-dependence for each coherence length bin, separately. Instead, the data have been fitted with a common Q^2-dependent slope of the transparency ratio, presented as lines in Fig. 2, resulting in reduced chi-square values near unity. The slope has been extracted by fitting $T_{c(inc)} = \sigma_{c(inc)}^{14N}(l_c, Q^2)/A\sigma_H = P_0 + P_1 \cdot Q^2$ to the twelve l_c bins, treating P_0 and P_1 as free parameters. The common slope parameter P_1 of the Q^2-dependence represents the signature of the CT effect averaged over the coherence length range. This procedure was performed separately for the coherent and incoherent data. The results of these fits are presented in Tab. 1 and compared to dedicated theoretical calculations [15]. If the results are combined, the common value for the slope of the Q^2-dependence of exclusive ρ^0 production, 0.085 ± 0.025 GeV^{-2}, is in agreement with the combined theoretical prediction of about 0.058 GeV^{-2} [15].

References

1. G. Bertsch *et al.*, Phys. Rev. Lett. **47**, 297 (1981).
2. A.B. Zamolodchikov *et al.*, Pis'ma Zh. Eksp. Teor. Fiz. **33**, 612 (1981); Sov. Phys. JETP Lett. **33**, 595 (1981).
3. S.J. Brodsky and A.H. Mueller, Phys. Lett. **B206**, 685 (1988).
4. N.N. Nikolaev, Comm. Nucl. Part. Phys., **21**, 1, 41 (1992).
5. S.J. Brodsky *et al.*, Phys. Rev. **D50**, 3134 (1994).
6. L.L. Frankfurt *et al.*, Ann. Rev. Nucl. Part. Sci., **45**, 501 (1994).
7. B.Z. Kopeliovich *et al.*, Phys. Rev. **C65**, 035201 (2002).
8. K. Ackerstaff *et al.*, Phys. Rev. Lett. **82**, 3025 (1999).
9. K. Ackerstaff *et al.*, Nucl. Instr. Meth. **A417**, 230 (1998). **36**, 495 (1987).
10. A. Airapetian *et al.*, Eur. Phys. J. **C17**, 389 (2000).
11. M. Tytgat, PhD Thesis, Gent University (2001); DESY-THESIS-2001-018.
12. B. Povh and J. Huefner, Phys. Rev. Lett. **58**, 1612 (1987).
13. I. Akushevich, Eur. Phys. J. C **8**, 457 (1999).
14. T. Renk, G. Piller, W. Weise, Nucl. Phys. **A689**, 869 (2001).
15. B.Z. Kopeliovich and J. Nemchik, private communication.

Table 1. Fitted common slope parameter of the Q^2-dependence of the nitrogen to hydrogen ratio with statistical and systematic uncertainties given separately as a HERMES preliminary result. The results are compared to the theoretical predictions [15].

Data sample	Measured Q^2 slope (GeV^{-2})	Prediction (GeV^{-2})
coherent	$0.081 \pm 0.027 \pm 0.011$	0.060
incoherent	$0.097 \pm 0.048 \pm 0.008$	0.053

MEASUREMENT OF LIGHT-CONE WAVE FUNCTIONS BY DIFFRACTIVE DISSOCIATION

DANIEL ASHERY *

School of Physics and Astronomy, Raymond and Beverly Sackler Faculty of Exact Sciences, Tel Aviv University, Israel

Diffractive dissociation of particles can be used to study their light-cone wave functions. Results from Fermilab experiment E791 for diffractive dissociation of 500 GeV/c π^- mesons into di-jets show that the $|q\bar{q}\rangle$ light-cone asymptotic wave function describes the data well for $Q^2 \sim 10$ $(\text{GeV/c})^2$ or more.

1. Light-Cone Wave Functions

A very powerful description of the hadronic structure is obtained through the light-cone wave functions. These are expanded in terms of a complete basis of Fock states having increasing complexity [1]. The negative pion has the Fock expansion:

$$
\begin{aligned}
|\psi_{\pi^-}\rangle &= \sum_n < n|\pi^- > |n> \\
&= \psi_{d\bar{u}/\pi}^{(\Lambda)}(u_i, \vec{k}_{\perp i})|\bar{u}d> \\
&\quad + \psi_{d\bar{u}g/\pi}^{(\Lambda)}(u_i, \vec{k}_{\perp i})|\bar{u}dg> + \cdots
\end{aligned}
\tag{1}
$$

They have longitudinal light-cone momentum fractions:

$$
u_i = \frac{k_i^+}{p^+} = \frac{k_i^0 + k_i^z}{p^0 + p^z}, \qquad \sum_{i=1}^n u_i = 1
\tag{2}
$$

and relative transverse momenta

$$
\vec{k}_{\perp i} \ , \qquad \sum_{i=1}^n \vec{k}_{\perp i} = \vec{0}_\perp.
\tag{3}
$$

*supported in part by the israel science foundation and the us-israel binational science foundation. representing fermilab e791 collaboration.

The first term in the expansion is referred to as the valence Fock state, as it relates to the hadronic description in the constituent quark model. This state must have a very small size and hence a large mass. The higher terms are related to the sea components of the hadronic structure, are larger and lighter. It has been shown that once the valence Fock state is determined, it is possible to build the rest of the light-cone wave function [2,3].

The essential part of the wave function is the hadronic distribution amplitude $\phi(u, Q^2)$. It describes the probability amplitude to find a quark and antiquark of the respective lowest-order Fock state carrying fractional momenta u and $1 - u$. For pions it is related to the light-cone wave function of the respective Fock state ψ through [4]:

$$\phi_{q\bar{q}/\pi}(u, Q^2) \sim \int_0^{Q^2} \psi_{q\bar{q}/\pi}(u, \tilde{k}_\perp) d^2\tilde{k}_\perp \tag{4}$$

$$Q^2 = \frac{k_\perp^2}{u(1 - u)} \tag{5}$$

2. The Pion Light-Cone Wave Function

For many years two functions were considered to describe the momentum distribution amplitude of the quark and antiquark in the $|q\bar{q}\rangle$ configuration. The asymptotic function was calculated using perturbative QCD (pQCD) methods [4,5,6], and is the solution to the pQCD evolution equation for very large Q^2 ($Q^2 \to \infty$):

$$\phi_{Asy}(u) = \sqrt{3}u(1 - u). \tag{6}$$

Using QCD sum rules, Chernyak and Zhitnitsky (CZ) proposed [7] a function that is expected to be correct for low Q^2:

$$\phi_{CZ}(u) = 5\sqrt{3}u(1 - u)(1 - 2u)^2. \tag{7}$$

As can be seen from Eqns. 6 and 7 and from Fig. 1, there is a large difference between the two functions. Measurements of the electromagnetic form factors of the pion are related to the integral over the wave function and the scattering matrix element and their sensitivity to the shape of the wave function is low [8]. Open questions are what can be considered to be high enough Q^2 to qualify for perturbative QCD calculations, what is low enough to qualify for a treatment based on QCD sum rules, and how to handle the evolution from low to high Q^2.

The concept of the measurement presented here is the following: a high energy pion dissociates diffractively while interacting with a heavy nuclear target. The first (valence) Fock component dominates at large Q^2; the other terms are suppressed by powers of $1/Q^2$ for each additional parton, according to counting rules [8,9]. This is a coherent process in which the quark and antiquark break apart and hadronize into two jets. If in this fragmentation process the quark momentum is transferred to the jet, measurement of the jet momentum gives the quark (and antiquark) momentum. Thus: $u_{measured} = \frac{p_{jet1}}{p_{jet1}+p_{jet2}}$. It has been shown by Frankfurt et $al.$ [10] that the cross section for this process is prportional to ϕ^2.

An important assumption is that quark momenta are not modified by nuclear interactions; i.e., that color transparency [6] is satisfied. The valence Fock state is far off-shell with mass M_n of several GeV/c^2. This corresponds to a transverse size of about 0.1 fm., much smaller than the regular pion size. Under these conditions, the valence quark configuration will penetrate the target nucleus freely and exhibit the phenomenon of color transparency. The lifetime of the configuration is given by $2P_{lab}/(M_n^2 - M_\pi^2)$ so at high beam momenta the configuration is kept small throughout its journey through the nucleus. Bertsch et $al.$ [6] proposed that the small $|q\bar{q}\rangle$ component will be filtered by the nucleus. Frankfurt et $al.$ [10] show that for $k_t > 1.5 GeV/c$ the interaction with the nucleus is completely coherent and $\sigma(|q\bar{q}\rangle N)$ is small. This leads to an A^2 dependence of the forward amplitude squared. When integrated over transverse momentum the signature for color transparency is a cross section dependence of $A^{4/3}$.

The basic assumption that the momentum carried by the dissociating $q\bar{q}$ is transferred to the di-jets was examined by Monte Carlo (MC) simulations of the asymptotic and CZ wave functions (squared). The MC samples were allowed to hadronize through the LUND PYTHIA-JETSET model [11] and then passed through simulation of the experimental apparatus. In Fig. 1, the initial distributions at the quark level are compared with the final distributions of the detected di-jets. As can be seen, while the fragmentation process and experimental effects smear the distributions, their qualitative features are retained.

3. Experimental Results

Results of experimental studies of the pion light-cone wave function were recently published by the Fermilab E791 collaboration [12]. In the experiment, diffractive dissociation of 500 GeV/c negative pions interacting with carbon and platinum targets was measured. Diffractive di-jets were required to carry the full beam momentum. They were identified through the $e^{-bq_t^2}$ dependence of their yield (q_t^2 is the square of the transverse momentum transferred to the nucleus and $b = \frac{<R^2>}{3}$ where R is the nuclear radius).

178

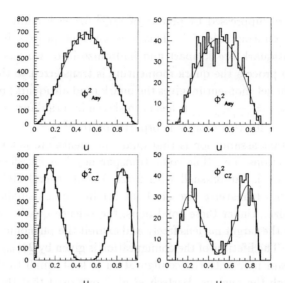

Figure 1. Monte Carlo simulations of squares of the two wave functions at the quark level (left) and of the reconstructed distributions of di-jets as detected (right).

Fig. 2 shows the q_t^2 distributions of di-jet events from platinum and carbon. The different slopes in the low q_t^2 coherent region reflect the different nuclear radii. Events in this region come from diffractive dissociation of the pion.

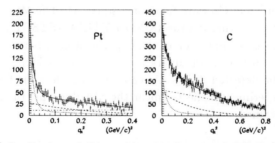

Figure 2. q_t^2 distributions of di-jets with $1.5 \leq k_t \leq 2.0$ GeV/c for the platinum and carbon targets. The lines are fits of the MC simulations to the data: coherent dissociation (dotted line), incoherent dissociation (dashed line), background (dashed-dotted line), and total fit (solid line).

For measurement of the wave function the most forward events ($q_t^2 < 0.015$ GeV/c^2) from the platinum target were used. For these events, the value of u was computed from the measured longitudinal momenta of the jets. The analysis was carried out in two windows of k_t: 1.25 GeV/c $\leq$ k_t $\leq$ 1.5 GeV/c and 1.5 GeV/c $\leq$ k_t $\leq$ 2.5 GeV/c.

The resulting u distributions are shown in Fig. 3.

In order to get a measure of the correspondence between the experimental results and the calculated light-cone wave functions, the results were fit with a linear combination of squares of the two wave functions (after smearing - right side of Fig. 1). This assumes an incoherent combination of the two wave functions and that the evolution of the CZ function is slow (as stated in [7]). The results for the higher k_t window show that the asymp-

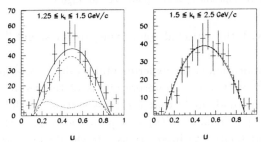

Figure 3. The u distribution of diffractive di-jets from the platinum target for $1.25 \leq k_t \leq 1.5$ GeV/c (left) and for $1.5 \leq k_t \leq 2.5$ GeV/c (right). The solid line is a fit to a combination of the asymptotic and CZ wave functions. The dashed line shows the contribution from the asymptotic function and the dotted line that of the CZ function.

totic wave function describes the data very well. Hence, for $k_t > 1.5$ GeV/c, which translates to $Q^2 \sim 10$ (GeV/c)^2, the pQCD approach that led to construction of the asymptotic wave function is reasonable. The distribution in the lower window is consistent with a significant contribution from the CZ wave function or may indicate contributions due to other non-perturbative effects. As the measurements are done within k_t windows, the results actually represent the square of the light-cone wave function averaged over k_t in the window: $\psi^2_{q\bar{q}}(x, \overline{k_\perp})$.

The k_t dependence of diffractive di-jets is another observable that can show how well the perturbative calculations describe the data. As shown in [10] it is expected to be: $\frac{d\sigma}{dk_t} \sim k_t^{-6}$. The results, shown in Fig. 4, are consistent with this dependence only in the region above $k_t \sim 1.8$ GeV/c, in agreement with the conclusions from the u-distributions. For lower k_t (Q^2) values, non-perturbative effects are expected to be significant.

Finally, verification that the $|q\bar{q}\rangle$ configuration is small and does not suffer from final state nuclear interaction was done by observing the color transparency effect for the diffractive dijets. The A-dependence of the diffractive di-jet yield was measured and found to have $\sigma \propto A^\alpha$ with $\alpha \sim$

180

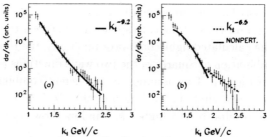

Figure 4. Comparison of the k_t distribution of acceptance-corrected data with fits to cross section dependence (a) according to a power law, (b) based on a nonperturbative Gaussian wave function for low k_t and a power law for high k_t.

1.5, consistent with the expected color transparency signal [10].

Following publication of the experimental results [12] several papers were published [13,14,15] with various interpretations of the results. In particular, the statement made in [10] that the cross section for diffractive Di-jet production is proportional to $\phi^2(u)$ was questioned due to effects of gluon distributions and final-state interactions. It should be noted that some effects due to gluon distributions are included in the studies discussed in section 1. These subjects were addressed in other talks in this session [16].

4. The Photon Light-Cone Wave Function

The photon light-cone wave function can be described in a similar way except that it has two major components: the electromagnetic and the hadronic states:

$$\psi_\gamma = a|\gamma_p\rangle + b|l^+l^-\rangle + c|l^+l^-\gamma\rangle + (other\ e.m.)$$
$$+ d|q\bar{q}\rangle + e|q\bar{q}g\rangle + (other\ hadronic) + \dots . \tag{8}$$

where $|\gamma_p\rangle$ describes the point bare-photon and $|l^+l^-\rangle$ stands for $|e^+e^-\rangle$, $|\mu^+\mu^-\rangle$ etc. Each of these states is a sum over the relevant helicity components. The wave function is very rich: it can be studied for real photons, for virtual photons of various virtualities, for transverse and longitudinal photons and the hadronic component may be decomposed according to the quark's flavor. An experimental program based on diffractive dissociation of real and virtual photons into di-leptons or di-jets is presently on-going at HERA.

I would like to thank Drs. S. Brodsky and L. Frankfurt for many helpful discussions. I also want to acknowledge the efforts of the E791 collaboration, of which I am a member, and particularly my graduate student, R. Weiss-Babai, for the data presented in this work.

References

1. S.J. Brodsky, D.S. Hwang, B.Q. Ma and I. Schmidt, Nucl.Phys. **B593**, 311 (2001).
2. A. H. Mueller, Nucl. Phys. **B415**, 373 (1994), F. Antonuccio, S. J. Brodsky and S. Dalley, Phys. Lett. **B412**, 104 (1997).
3. S. Daley, proc. 10th Light Cone meeting, Heidelberg, June 2000, hep-ph/0007081 and this workshop.
4. S.J. Brodsky and G.P. Lepage, Phys. Rev. **D22**, 2157 (1980); S.J. Brodsky and G.P. Lepage, Phys. Scripta **23**, 945 (1981); S.J. Brodsky, Springer Tracts in Modern Physics **100**, 81 (1982).
5. A.V. Efremov and A.V. Radyushkin, Theor. Math. Phys. **42**, 97 (1980).
6. G. Bertsch, S.J. Brodsky, A.S. Goldhaber, and J. Gunion, Phys. Rev. Lett. **47**, 297 (1981).
7. V.L. Chernyak and A.R. Zhitnitsky, Phys. Rep. **112**, 173 (1984).
8. G. Sterman and P. Stoler, Ann. Rev. Nuc. Part. Sci. **43**, 193 (1997).
9. S.J. Brodsky and G.R. Farrar, Phys. Rev. Lett. **31**, 1153 (1973).
10. L.L. Frankfurt, G.A. Miller, and M. Strikman, Phys. Lett. **B304**, 1 (1993).
11. H.-U. Bengtsson and T. Sjöstrand, Comp. Phys. Comm. **82**, 74 (1994); T. Sjöstrand, PYTHIA 5.7 and JETSET 7.4 Physics and Manual, CERN-TH.7112/93, (1995).
12. E791 Collaboration, E.M. Aitala *et al.*, Phys. Rev. Lett. **86**, 4768 and **86**, 4773 (2001).
13. N.N. Nikolaev, W. Schafer and G. Schwiete, Phys. Rev. **D63**, 014020 (2000).
14. V.M. Braun *et al.* Phys. Lett. **B509**, 43 (2001).
15. V. Chernyak, Phys. Lett. **B516**, 116 (2001).
16. See contributions by V. Chernyak, D. Ivanov and L. Frankfurt, this session.

THE PION DIFFRACTIVE DISSOCIATION
INTO TWO JETS

V.L. CHERNYAK

Budker Institute for Nuclear Physics,
630090 Novosibirsk, Russia
E-mail: chernyak@inp.nsk.su

The method used and results obtained are described for calculation of the cross section for the pion diffractive dissociation into two jets. The main new qualitative result is that the distribution of longitudinal momenta for jets is not simply proportional to the profile of the pion wave function $\phi_\pi(x)$, but depends on it in a complicated way. In particular, it is shown that under conditions of the E791 experiment, the momentum distribution of jets is similar in its shape for the asymptotic and CZ wave functions. It is concluded therefore that, unfortunately, the process considered is really weakly sensitive to the profile of the pion wave function, and the accuracy of data is insufficient to distinguish clearly between different models of $\phi_\pi(x)$. Comparison with the results of other papers on this subject is given.

1. The E791 experiment at Fermilab [1] has recently measured the cross section of the hard diffractive dissociation of the pion into two jets. In particular, the distribution of the total pion longitudinal momentum into fractions y_1 and y_2, $(y_1 + y_2) = 1$, between jets has been measured. The main purpose was to obtain in this way the information about the leading twist pion wave function $\phi_\pi(x_1, x_2)$, which describes the distribution of quarks inside the pion in the longitudinal momentum fractions x_1 and $x_2 = 1 - x_1$.

The hope was based on theoretical calculations of this cross section, see Refs. [2],[3],[4],[5]. It has been obtained in these papers that the cross section is simply proportional to the pion wave function squared: $d\sigma/dy_1 \sim |\phi_\pi(y_1)|^2$. In such a case, it would be sufficient to measure only the gross features of $d\sigma/dy$ to reveal the main characteristic properties of $\phi_\pi(x)$, and to discriminate between various available models of $\phi_\pi(x)$.

Our main new qualitative result [6,7] is that this is not the case. The real situation is much more complicated, with $d\sigma/dy$ depending on $\phi_\pi(x)$ in a highly nontrivial way. We describe below in a short form our approach and the results obtained for this cross section.

2. The kinematics of the process is shown in Fig.1. We take the nucleon as a target, and the initial and final nucleons are substituted by two soft transversely polarized gluons with momenta $q_1 = (u + \xi)\overline{P}$ and $q_2 = (u - \xi)\overline{P}$, $\overline{P} = (P+P')/2$, $\Delta = (q_1-q_2)$ [a]. The lower blob in Fig.1 represents the generalized gluon distribution of the nucleon, $G_\xi(u) = G_\xi(-u)$ [8],[9],[10].

The final quarks are on shell, carry the fractions y_1 and y_2 of the initial pion momentum, and their transverse momenta are $(\mathbf{k}_\perp + (\mathbf{q}_\perp/2))$ and $(-\mathbf{k}_\perp+(\mathbf{q}_\perp/2))$, $q_\perp \ll k_\perp$, where $q_\perp$ is the small final transverse momentum of the target, while $k_\perp$ is large. The invariant mass of these two quark jets is $M^2 = k_\perp^2/y_1 y_2$, and $\xi = M^2/\nu$, $\nu = (s - u)$.

The upper blob M in Fig.1 represents the hard kernel of the amplitude. According to the well developed approach to description of hard exclusive processes in QCD [11],[12],[13],[14] (see Ref. [15] for a review), all hard gluon and quark lines in all diagrams have to be written down explicitly and substituted by their perturbative propagators. In other words, the hard momentum flow have to be made completely explicit and these hard lines of diagrams constitute the hard kernel M. They should not be hidden (if it is possible at all) as (derivatives of) "the tails" of the unintegrated pion wave function $\Psi_\pi(x, l_\perp)$, or of the "unintegrated gluon distribution". This is, first of all, what differs our approach from previous calculations of this process in Refs. [4],[5].

For calculation of M in the leading twist approximation and in the lowest order in α_s, the massless pion can be substituted by two massless on-shell quarks with the collinear momenta $x_1 p_\pi$ and $x_2 p_\pi$ and with zero transverse momenta, as account of primordial virtualities and transverse momenta results only in higher twist corrections to M. The leading twist pion wave function $\phi_\pi(x, \mu_o)$ describes the distribution of these quarks in momentum fractions x_1 and x_2 [b].

So, the hard kernel M is proportional to the scattering amplitude of

---

[a]The small skewedness, $\xi \ll 1$, is always implied. It is typically: $\xi \sim 10^{-2}$, in the Fermilab experiment.

[b]As usual, on account of leading logs from loops, the soft pion wave function $\phi_\pi(x,\mu_o)$ evolves to $\phi_\pi(x, \mu)$, where μ is the characteristic scale of the process. In other words, in the leading twist component of the pion wave function the two pion quarks prepare themselves for a hard collision by exchanging gluons and increasing their virtualities and transverse momenta from the initial scale $\mu_o \sim \Lambda_{QCD}$ up to the characteristic scale of the process $\mu \lesssim k_\perp$, so that this quark pair enters the nucleon (nuclei) already having the small transverse size $r_\perp \sim 1/k_\perp$ (while the smallness of the longitudinal size is ensured by the Lorenz contraction).

184

two initial collinear and on-shell quarks of the pion on the on-shell gluon:

$$M \sim \{d(x_1 p_\pi) + \bar{u}(x_2 p_\pi) + g(q_1) \to d(p_1) + \bar{u}(p_2) + g(q_2)\} \quad .$$

In the lowest order in $\alpha_s(k_\perp)$, M consists of 31 connected Born diagrams, each one is $\sim O(\alpha_s^2(k_\perp))$.

The general structure of the amplitude is, therefore (symbolically):

$$T \sim \langle P'|A^\perp \cdot A^\perp|P \rangle \otimes (\bar{\psi}_1 M \psi_2) \otimes \langle 0|\bar{u} \cdot d|\pi^- \rangle,$$

where the first matrix element introduces the generalized gluon distribution of the nucleon $G_\xi(u)$, $\bar{\psi}_1$ and ψ_2 are the free spinors of final quarks, "M" is the hard kernel, i.e., the product of all vertices and hard propagators, the last matrix element introduces the pion wave function $\phi_\pi(x)$, and $\otimes$ means the appropriate convolution.

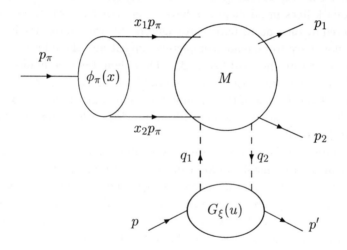

Figure 1.: Kinematics and notations

3. Proceeding in the above described way and summing up contributions of all Born diagrams, one obtains for the cross section [c] :

$$d\sigma_N = \frac{1}{8(2\pi)^5} \frac{1}{s^2} |T|^2 \frac{dy_1}{y_1 y_2} d^2 k_\perp d^2 q_\perp; \quad \mathrm{Im}\, T_g = \frac{2\pi s \omega_o}{k_\perp^2} G_\xi(\xi)\, \Omega, \quad (1)$$

---

[c] Only the imaginary part of the amplitude, $\mathrm{Im}\, T_g$ (from the gluon distribution in the target), is shown explicitly in Eqs.(1)-(6), as it is expected to be the main one at high energies. For obtaining $\mathrm{Im}\, T$, the terms $i\epsilon$ were introduced into all denominators through: $s \to s + i\epsilon$, i.e. $\xi \to \xi - i\epsilon$ [6]. The contribution of the quark distribution to $\mathrm{Im}\, T$ is given explicitly in Ref.[7]. Let us note also that all integrals over "x_1" entering Eqs.(2)-(6) are convergent at the end points $x_1 \to 0$ or $x_1 \to 1$.

$$\Omega = \int_0^1 dx_1 \, \phi_\pi(x_1) \, (\Sigma_1 + \Sigma_2 + \Sigma_3 + \Sigma_4) \,, \tag{2}$$

$$\Sigma_1 = \left[\frac{4}{x_1 x_2 \, |x_1 - y_1|} \frac{G_\xi(\bar{u})}{G_\xi(\xi)} \, \Theta(|x_1 - y_1| > \delta) \right] + (y_1 \leftrightarrow y_2), \tag{3}$$

$$\Sigma_2 = \frac{1}{x_1^2 \, x_2^2 \, y_1 \, y_2} \left\{ -(x_1 x_2 + y_1 y_2) + \right.$$

$$\left. + \left[|x_1 - y_1|(x_1 - y_2)^2 \frac{G_\xi(\bar{u})}{G_\xi(\xi)} \, \Theta(|x_1 - y_1| > \delta) + (y_1 \leftrightarrow y_2) \right] \right\}, \tag{4}$$

$$\Sigma_3 = \frac{x_1 x_2 + y_1 y_2}{9 x_1^2 x_2^2 y_1 y_2} \left\{ -1 + \left[|x_1 - y_1| \frac{G_\xi(\bar{u})}{G_\xi(\xi)} \Theta(|x_1 - y_1| > \delta) \right.\right.$$

$$\left.\left. + (y_1 \leftrightarrow y_2) \right] \right\} \tag{5}$$

$$\Sigma_4 = \frac{16}{9} \frac{1}{x_1 x_2 y_1 y_2} \xi \frac{dG_\xi(u)/du|_{u=\xi}}{G_\xi(\xi)}; \qquad \delta = \frac{k_\perp^2}{s}, \tag{6}$$

$$\omega_o = \frac{\delta_{ij}(4\pi\alpha_s)^2 f_\pi}{96} (\bar{\psi}_1 \Delta_\mu \gamma_\mu \gamma_5 \psi_2) \frac{(y_1 y_2)^2}{k_\perp^4}; \quad \bar{u} = \xi \left(\frac{x_1 y_2 + x_2 y_1}{x_1 - y_1} \right). \tag{7}$$

It is seen from the above equations that $d\sigma/dy$ is not $\sim |\phi_\pi(y)|^2$, but depends on the profile of $\phi_\pi(x)$ in a very complicated way.

4. In this section we present some numerical estimates of the cross section, based on the above expressions (1)-(6). Our main purpose is to trace the distribution of jets in the longitudinal momentum fractions y_1, y_2 depending on the profile of the pion wave function $\phi_\pi(x)$.

a) For the skewed gluon distribution $G_\xi(u, t, \mu)$ of the nucleon at $t \simeq -q_\perp^2 \simeq 0$ we use the simple form

$$G_\xi(u, t = 0, \mu \simeq k_\perp \simeq 2\,\text{GeV})|_{u \geq \xi} \simeq u^{-0.3}(1-u)^5 \tag{8}$$

(as we need it at $|u| \geq \xi$ only, and because $G_\xi(u) \to G_o(u)$ at $|u| \gg \xi$). This form agrees numerically reasonably well with the ordinary, $G_o(u, \mu \simeq 2\,\text{GeV})$, and skewed, $G_\xi(u, t = 0, \mu \simeq 2\,\text{GeV})$, gluon distributions of the nucleon calculated in Ref.[16] and Ref. [17], respectively (in the typical region of the E791 experiment: $|u| \geq \xi \sim 10^{-2}$).

c) As for the pion leading twist wave function, $\phi_\pi(x, \mu)$, we compare two model forms: the asymptotic form, $\phi_\pi^{asy}(x, \mu) = 6x_1 x_2$, and the CZ-model

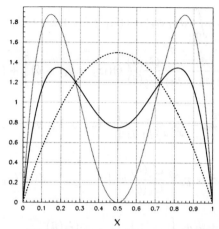

Figure 2.: Profiles of the pion wave functions: a) $\phi_\pi^{CZ}(x, \mu \simeq 0.5\,GeV) = 30\,x_1 x_2 (x_1 - x_2)^2$ - dotted line; b) $\phi_\pi^{CZ}(x, \mu \simeq 2\,GeV) = 15\,x_1 x_2 [0.2 + (x_1 - x_2)^2]$ - solid line; c) $\phi_\pi^{asy}(x) = 6\,x_1 x_2$ - dashed line.

[18]. The latter has the form: $\phi_\pi^{CZ}(x, \mu_o \simeq 0.5\,\text{GeV}) = 30\,x_1 x_2 (x_1 - x_2)^2$, at the low normalization point. Being evolved to the characteristic scale of this process, $\mu \simeq k_\perp \simeq 2\,\text{GeV}$, it looks as: $\phi_\pi^{CZ}(x, \mu \simeq 2\,\text{GeV}) = 15\,x_1 x_2 \left[(x_1 - x_2)^2 + 0.2 \right]$, see Fig. 2. [b]

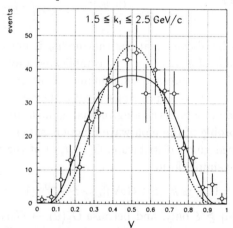

Figure 3.: The y-distribution of jets calculated for $k_\perp = 2\,GeV$, $E_\pi = 500\,GeV$ and with the pion wave functions: $\phi_\pi^{CZ}(x, \mu \simeq 2\,GeV)$ - solid line, $\phi_\pi^{asy}(x)$ - dashed line. The overall normalization is arbitrary, but the relative normalization of two curves is as calculated. The data points are from the E791 experiment [1].

The results of these numerical calculations are then compared with the E791-data, see Fig. 3.

It is seen that while the two model pion wave functions are quite different, the resulting distributions of jets in longitudinal momenta are similar and, it seems, the present experimental accuracy is insufficient to distinguish clearly between them. Moreover, even the ratio of the differential cross sections is not much different from unity: $d\sigma^{asy}/d\sigma^{CZ} \simeq 1.2$ at $y_1 = 0.5$, and the same ratio is $\simeq 0.7$ at $y_1 = 0.25$ [d].

5. The purpose of this section is to compare (in a short form) the above results and those obtained by other authors, see Refs. [4],[5] and Refs. [19],[20].

a) In comparison with Eqs.(1)-(6), the approximations used in Ref. [5] correspond to neglecting the terms Σ_2, Σ_3 and Σ_4, and supposing that the region of small $|x_1 - y_1|$ dominates the only remaining contribution: $\int dx_1 \phi_\pi(x_1)\Sigma_1 \simeq \phi_\pi(y_1) \int dx_1 \Sigma_1$. In this approximation, indeed, $\mathrm{Im}\, T_g \sim \phi_\pi(y)$ and $d\sigma/dy \sim |\phi_\pi(y)|^2$.

Unfortunately, with the real conditions of the E791 experiment, this is a poor approximation, and on account of neglected terms in Eqs.(1)-(6) the form of $d\sigma/dy$ changes *qualitatively* in comparison with $|\phi_\pi(y)|^2$, see Figs.2 and 3 [e].

b) The authors of papers in Ref. [4] argued that from the whole set of 31 Feynman diagrams only a small especially chosen subset dominates $\mathrm{Im}\, T_g$, while contributions of all other diagrams will be suppressed by Sudakov form factors. From this subset of diagrams they obtained also: $\mathrm{Im}\, T_g(y) \sim \phi_\pi(y)$.

First, I am unable to understand this last result, as the direct calculation of this subset of diagrams according to the rules described above in Section 2 shows that, here also, the resulting $\mathrm{Im}\, T_g(y)$ depends on $\phi_\pi(x)$ through a complicated integrated form, like those in Eqs.(4)-(6) above.

As for the Sudakov suppression of all other diagrams, I also disagree. The line of reasoning in Ref. [4] was the following. Consider, for instance, the diagrams like those in Fig.2 in Ref. [6]. Exchanging the hard gluon, the two pion quarks undergo abrupt change of their direction of motion and will

---

[d]On account of the quark contributions to $\mathrm{Im}\, T$ the resulting curves for $d\sigma/dy$ remain very similar to those in Fig.3, see Ref.[7]. For the numerical calculation of $\mathrm{Re}\, T$ see Ref.[20]. The estimate of $\mathrm{Re}\, T$ can be obtained from the well known formula: $\mathrm{Re}\, T(\nu) \simeq (\pi/2)\, d\ln(\nu^{-1}\mathrm{Im}\, T(\nu))/d\ln\nu$. Because $\nu^{-1}\mathrm{Im}\, T(\nu) \sim \nu^{0.3}$ here, $\mathrm{Re}\, T \simeq 0.45\,\mathrm{Im}\, T$ and $|T|^2 \simeq 1.2\,(\mathrm{Im}\, T)^2$.

[e]The forms of $d\sigma/dy$ for $\phi_\pi^{asy}(x)$ and $\phi_\pi^{CZ}(x, \mu \simeq 2\,\mathrm{GeV})$ and with the pion energy ten times larger, $E_\pi = 5\,\mathrm{TeV}$, $k_\perp = 2\,\mathrm{GeV}$, are shown in Fig.6 in Ref.[6]. It is seen from therein that even this does not help much (as the form of the distribution in y of $d\sigma/dy$ weakly depends on the pion energy), and so the approximation used in Ref.[5] remains poor even at such energies.

tend to radiate gluons in their previous direction of motion. Because the events with such radiation are excluded from the data set, what remains will be suppressed by the Sudakov form factors.

Applying this line of reasoning to nearly any hard exclusive amplitude, one will conclude that all such amplitudes will be suppressed, *as a whole*, by Sudakov effects. For instance, let us consider the large angle Compton amplitude. In the c.m.s. and at high energy, the three proton quarks change abruptly their direction of motion, deviating by large angle θ. So, following the line of reasoning in Ref. [4], the whole large angle Compton amplitude will be suppressed by the Sudakov form factors.

Really, this is not the case. The reason is that the radiations of three collinear quarks of the proton cancel each other due to the color neutrality of the proton [f].

   c) Unlike Ref. [4] and Ref. [5], the results from Refs.[19],[20] are very similar to those from Refs. [6],[7]. In particular, the analytic expressions for $(T_g + T_q)$ in terms of corresponding integrals over $\phi_\pi(x)$ and the gluon (quark) distributions are the same, the only difference is that the signs of $i\epsilon$ in denominators of a few terms are opposite. For T_g for instance (and similarly for T_q), this difference can be represented as:

$$T_g = \delta T_g(y,\, \nu + i\epsilon,\, M^2)$$
$$+ \int_0^1 dx_1 \phi_\pi(x_1)[f_1(x,y)g^+(\nu, M^2) + f_2(x,y)g^\pm(\nu, M^2)],$$

$$g^\pm(\nu, M^2) = \int_{-1}^1 du \frac{G_\xi(u)}{(u + \frac{M^2}{\nu \pm i\epsilon})(u - \frac{M^2}{\nu \pm i\epsilon})} \quad , \tag{9}$$

where δT_g is the same in Ref. [6] and Refs. [19],[20], and $f_1(x,y)$ and $f_2(x,y)$ are definite simple functions. The upper sign in Eq.(9) is from Ref. [6] (where, as pointed out above, the terms $i\epsilon$ were introduced into all denominators through: $\nu \to \nu + i\epsilon$), while the lower sign is from Refs. [19],[20].

The result for $Im\, T_g$ was obtained in Refs. [19],[20] by a direct calculation of one-loop Feynman diagrams, and so it can be checked only by a similar independent calculation. But there at least one simple argument

---

[f]This cancellation becomes ineffective only at the end points of the quark phase space, where, for instance, one quark carries nearly all the proton momentum, while other ones are wee. The contributions from these regions to the total amplitude are, indeed, suppressed by the Sudakov effects. But in any case, the contributions from these end point regions give only power corrections to the total amplitude, even ignoring the additional Sudakov suppression.

following from general analyticity properties, in favour of the approach used in Ref. [6].

Let us first consider our amplitude $T_g(y, \nu, M^2)$ in the Euclidean region $M^2 < 0$. There is no M^2 - discontinuity in this region, and the only discontinuity is due to the s-cut. So, $i\epsilon$ terms enter unambiguously into *all* denominators as: $\nu = (s - u) \to \nu + i\epsilon$, and the terms f_1 and f_2 in Eq.(9) will be multiplied by *the same analytic function* $g(\nu, M^2)$:

$$[f_1(x, y) + f_2(x, y)] g(\nu + i\epsilon, M^2) . \tag{10}$$

Let us continue now this expression into the Minkowski region $M^2 > 0$. How it may be that after the analytic continuation the functions $f_1(x, y)$ and $f_2(x, y)$ will be multiplied by two different functions, as in Refs. [19],[20] (see Eq.(9)):

$$[f_1(x, y) g(\nu + i\epsilon, M^2) + f_2(x, y) g^\star(\nu + i\epsilon, M^2)] \quad ? \tag{11}$$

In conclusion, I would like to emphasize that in spite of some "internal" differences between the results from Refs. [6],[7] and Refs.[19],[20], the final curves for $d\sigma/dy$ are practically the same, and Fig.3 gives a good representation of both answers. So, in any case, the qualitative conclusions agree: under the conditions of the E791 experiment, the process considered appeared to be weakly sensitive to the profile of the pion wave function $\phi_\pi(x)$ and, unfortunately, these data cannot discriminate between, say, $\phi_\pi^{asy}(x)$ and $\phi_\pi^{CZ}(x)$.

Acknowledgements

I am grateful to the Theory Group of JLab and Organizing Committee and, in particular, to A.V. Radyushkin for kind hospitality and support.

References

1. E.M. Aitala et. al. (E791 Collaboration), *Phys. Rev. Lett.* **86**, 4768 (2001); hep-ex/**0010043**
   D. Ashery, hep-ex/**9910024**
2. S.F. King, A. Donnachie and J. Randa, *Nucl. Phys.* **B167**, 98 (1980)
   J. Randa, *Phys. Rev.* **D22**, 1583 (1980)
3. G. Bertsch, S.J. Brodsky, A.S. Goldhaber and J.F. Gunion,
   Phys. Rev. Lett. **47**, 297 (1981)
4. L. Frankfurt, G.A. Miller and M. Strikman, *Phys. Lett.* **B304**, 1 (1993);
   Found. of Phys. **30**, 533 (2000); hep-ph/**9907214**; hep-ph/**0010297**
5. N.N. Nikolaev, W. Schafer and G. Schwiete, *Phys. Rev.* **D63**, 014020 (2001);
   hep-ph/**0009038**
6. V.L. Chernyak, *Phys. Lett.* **B516**, 116 (2001); hep-ph/**0103295**

190

7. V.L. Chernyak and A.G. Grozin, *Phys. Lett.* **B517**, 119 (2001); hep-ph/**0106162**
8. D. Muller, D. Robaschik, B. Geyer, F.-M. Dittes and J. Horejsi, *Forts. Phys.* **42**, 101 (1994)
9. A.V. Radyushkin, *Phys. Lett.* **B385**, 333 (1996); *Phys. Rev.* **D56**, 5524 (1997)
10. X. Ji, *Phys. Rev. Lett.* **78**, 610 (1997); *J. Phys.* **G24**, 1181 (1998)
11. V.L. Chernyak and A.R. Zhitnitsky, *JETP Lett.* **25**, 510 (1977); *Sov. J. Nucl. Phys.* **31**, 544 (1980)
12. V.L. Chernyak, V.G. Serbo and A.R. Zhitnitsky, *JETP Lett.* **26**, 594 (1977); *Sov. J. Nucl. Phys.* **31**, 552 (1980)
13. A.V. Efremov and A.V. Radyushkin, *Phys. Lett.* **B94**, 245 (1980); *Teor. Math. Phys.* **42**, 97 (1980)
14. G.P. Lepage and S.J. Brodsky, *Phys. Lett.* **B87**, 359 (1979); *Phys. Rev.* **D22**, 2157 (1980)
15. V.L. Chernyak and A.R. Zhitnitsky, *Phys. Rep.* **112**, 173 (1984)
16. M. Gluck, E. Reya and A. Vogt, *Eur. Phys. J.* **C5**, 461 (1998)
17. K.J. Golec-Biernat, A.D. Martin and M.G. Ryskin, *Phys. Lett.* **B456**, 232 (1999); hep-ph/**9903327**
18. V.L. Chernyak and A.R. Zhitnitsky, *Nucl. Phys.* **B201**, 492 (1982)
19. V.M. Braun, D.Yu. Ivanov, A. Schafer and L. Szymanowski, *Phys. Lett.* **B509**, 43 (2001); hep-ph /**0103275**
20. V.M. Braun, D.Yu. Ivanov, A. Schafer and L. Szymanowski, hep-ph/**0204191**

QCD FACTORIZATION FOR THE PION DIFFRACTIVE DISSOCIATION INTO TWO JETS

D. YU. IVANOV[1,2]

[1] *Institut für Theoretische Physik, Universität Regensburg,*
D-93040 Regensburg, Germany,
[2] *Institute of Mathematics, 630090 Novosibirsk, Russia,*
E-mail: Dmitri.Ivanov@physik.uni-regensburg.de

We report a detailed study of the process of pion diffraction dissociation into two jets with large transverse momenta. We find that the standard collinear factorization does not hold in this reaction. The structure of non-factorizable contributions is discussed and the results are compared with the experimental data. Our conclusion is that the existing theoretical uncertainties do not allow, for the time being, for a quantitative extraction of the pion distribution amplitude. (Talk presented at the Workshop on Exclusive Processes at High Momentum Transfer, Jefferson News, VA, May 15-18, 2002)

1. To our knowledge, the pion (and photon) diffraction dissociation into a pair of jets with large transverse momentum on a nucleon target was first discussed in [1]. Then the possibility to use this process to probe the nuclear filtering of pion components with a small transverse size was suggested in [2]. The A-dependence and the $q_\perp^2$-dependence of the coherent dijet cross section was first calculated in [3]. In the same work it was argued that the jet distribution with respect to the longitudinal momentum fraction has to follow the quark momentum distribution in the pion and hence provides a direct measurement of the pion distribution amplitude. Recent experimental data by the E791 collaboration [4,5] indeed confirm the strong A-dependence which is a signature for color transparency, and are consistent with the predicted $\sim 1/q_\perp^8$ dependence on the jet transverse momentum. Moreover, the jet longitudinal momentum fraction distribution turns out to be consistent with the $\sim z^2(1-z)^2$ shape corresponding to the asymptotic pion distribution amplitude which is also supported by an independent measurement of the pion transition form factor $\pi\gamma\gamma^*$ [6].

After these first successes, one naturally asks whether the QCD description of coherent dijet production can be made fully quantitative. Two recent studies [7] and [8] address this question, with contradictory conclusions.

Therefore we attempt to clarify the situation and develop a perturbative QCD framework for the description of coherent dijet production that would be in line with other known applications of the QCD factorization techniques. The results reported here have been obtained in collaboration with V.M. Braun, A. Schäfer and L. Szymanowski [9,10].

2. The kinematics of the process is shown in Fig. 1. The momenta of

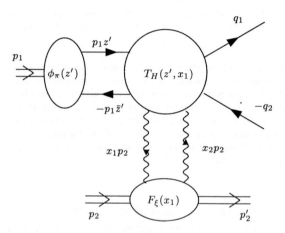

Figure 1. Kinematics of the coherent hard dijet production $\pi \to 2$ jets. The hard scattering amplitude T_H contains at least one hard gluon exchange.

the incoming pion, incoming nucleon and the outgoing nucleon are p_1, p_2 and p_2', respectively. The pion and the nucleon masses are both neglected, $p_1^2 = 0$, $p_2^2 = (p_2')^2 = 0$. We denote the momenta of the outgoing quark and antiquark (jets) as q_1 and q_2, respectively. They are on the mass shell, $q_1^2 = q_2^2 = 0$, and can be decomposed

$$q_1 = zp_1 + \frac{q_{1\perp}^2}{zs}p_2 + q_{1\perp} \, , \, q_2 = \bar{z}p_1 + \frac{q_{2\perp}^2}{\bar{z}s}p_2 + q_{2\perp} \qquad (1)$$

such that z is the longitudinal momentum fraction of the quark jet in the lab frame. We use the shorthand notation: $\bar{u} \equiv (1 - u)$ for any longitudinal momentum fraction u.

We are interested in the forward limit, when the transferred momentum $t = (p_2 - p_2')^2$ is equal to zero and the transverse momenta of jets compensate each other $q_{1\perp} \equiv q_\perp$, $q_{2\perp} \equiv -q_\perp$. In this kinematics the invariant mass of the produced $q\bar{q}$ pair is equal to $M^2 = q_\perp^2/z\bar{z}$. The invariant c.m. energy $s = (p_1 + p_2)^2 = 2p_1p_2$ is taken to be much larger than the transverse jet momentum $q_\perp^2$.

3. The possibility to constrain the pion distribution amplitude $\phi_\pi(z', \mu_F^2)$ in the dijet diffractive dissociation experiment assumes that the amplitude of this process can be calculated in the collinear approximation as suggested by Fig. 1:

$$M_{\pi \to 2\,\text{jets}} = \sum_{p=q,\bar{q},g} \int_0^1 dz' \int_0^1 dx_1 \, \phi_\pi(z', \mu_F^2) \, T_H^p(z', x_1, \mu_F^2) \, F_\zeta^p(x_1, \mu_F^2). \quad (2)$$

Here $F_\zeta^p(x_1, \mu_F^2)$ is the generalized parton distribution (GPD) $p = q, \bar{q}, g$ [13,14,15] in the target nucleon; x_1 and $x_2 = x_1 - \zeta$ are the momentum fractions of the emitted and the absorbed partons, respectively. The asymmetry parameter ζ is fixed by the process kinematics: $\zeta = M^2/s = q_\perp^2/z\bar{z}$. $T_H(z', x_1, \mu_F^2)$ is the hard scattering amplitude involving at least one hard gluon exchange and μ_F is the (collinear) factorization scale. By definition, the pion distribution amplitude only involves small momenta, $k_\perp < \mu_F$, and the hard scattering amplitude is calculated neglecting the parton transverse momenta.

We calculated [10] both the leading-order gluon and quark contributions to the amplitude and find that in both cases the corresponding hard kernels T_H^q, T_H^g diverge as $1/z'^2$ and $1/\bar{z}'^2$ in the $z' \to 0$ and $z' \to 1$ limit, respectively. This implies that the integration of the pion momentum fraction diverges at the end-points and the collinear factorization is, therefore, broken.

At high energies the contribution of gluon GPD dominates. We found that up to kinematical factors

$$T_H^g = C_F \left(\frac{\bar{z}}{z'} + \frac{z}{\bar{z}'} \right) \left(\frac{\zeta}{[x_1 - i\epsilon]^2} + \frac{\zeta}{[x_2 + i\epsilon]^2} - \frac{\zeta}{[x_1 - i\epsilon][x_2 + i\epsilon]} \right)$$

$$+ \left(\frac{z\bar{z}}{z'\bar{z}'} + 1 \right) \left[C_F \left(\frac{z\bar{z}}{z'\bar{z}'} + 1 \right) + \frac{1}{2N_c} \left(\frac{z}{z'} + \frac{\bar{z}}{\bar{z}'} \right) \right]$$

$$\times \left(\frac{1}{[(z - z')x_1 - z\bar{z}'\zeta + i\epsilon]} + \frac{1}{[(z' - z)x_2 - z\bar{z}'\zeta + i\epsilon]} \right) \quad (3)$$

$$- \left[C_F \frac{z\bar{z}}{z'\bar{z}'} \left(\frac{\bar{z}}{z'} + \frac{z}{\bar{z}'} \right) + \frac{1}{2N_c z'\bar{z}'} \left(\frac{z\bar{z}}{z'\bar{z}'} + 1 \right) \right] \frac{\zeta}{[x_1 + i\epsilon][x_2 - i\epsilon]}.$$

The differential cross section summed over the polarizations and the color of quark jets is given by

$$\frac{d\sigma_{\pi \to 2\,\text{jets}}}{dq_\perp^2 \, dt \, dz} = \frac{\alpha_s^4 f_\pi^2 \pi (1 - \zeta)}{8 N_c^3 q_\perp^8} |M|^2, \quad (4)$$

where M is calculated as in (2) with T_H^g given in (3), $f_\pi = 133$ MeV is the pion decay constant.

4. According to Eq. (3) the leading end-point behavior of the gluon amplitude at $z' \to 0$ and $z' \to 1$ is given by the following expression

$$M\Big|_{\text{end-points}} = -i\pi \left(N_c + \frac{1}{N_c} \right) z\bar{z} \int\limits_0^1 dz' \, \frac{\phi_\pi(z',\mu^2)}{z'^2} F_\zeta^g(\zeta,\mu^2) \,. \qquad (5)$$

Since $\phi_\pi(z') \sim z'$ at $z' \to 0$, the integral over z' diverges logarithmically. Remarkably, the integral containing the pion distribution amplitude does not involve any z-dependence. Therefore, the longitudinal momentum distribution of the jets in the nonfactorizable contribution is calculable and, as it turns out, has the shape of the asymptotic pion distribution amplitude $\phi_\pi^{\text{as}}(z) = 6z\bar{z}$.

In technical terms, the appearance of the end point divergency is due to pinching of the x_1 contour in the point $x_1 = \zeta(x_2 = 0)$ in the case when variable z' is closed to the end points, cf. Eqs. (2,3). One can trace [10] that this pinching occurs between soft gluon (the gluon with momentum $x_2 \to 0$) interactions in the initial and in the final state, and is related with the existence of the unitarity cuts of the amplitude in different, s and M^2, channels.

The other important integration region in (2) is the one when $\zeta \ll |z' - z| \ll 1$, i.e. when the longitudinal momentum fraction carried by the quark is close (for high energies) to that of the quark jet in the final state

$$M\Big|_{\zeta \ll |z'-z| \ll 1} = -4i\pi N_c \, \phi_\pi(z) \int\limits_z^1 \frac{dz'}{z'-z} F_\zeta^g\left(\zeta\frac{z\bar{z}}{z'-z}, q_\perp^2\right)$$

$$\simeq -4i\pi N_c \, \phi_\pi(z) \int\limits_\zeta^1 \frac{dy}{y} \, F_\zeta^g(y, q_\perp^2) \,. \qquad (6)$$

This logarithmic integral is nothing but the usual energy logarithm that accompanies each extra gluon in the gluon ladder. Its appearance is due to the fact the hard gluon which supplies jets by the high transverse momentum can be emitted in a broad rapidity interval and is not constrained to the pion fragmentation region. The integral on the r.h.s. of (6) can be identified with the unintegrated generalized gluon distribution. And, therefore, in the region $z' \sim z$ hard gluon exchange can be viewed as a large transverse momentum part of the gluon distribution in the proton, cf. [7]. This contribution is proportional to the pion distribution amplitude $\phi_\pi(z, q_\perp^2)$ and contains the enhancement factor $\ln 1/\zeta \sim \ln s/q_\perp^2$.

5. We performed numerical calculations for the kinematics of E791 experiment: the transverse momentum range $1.5 \le q_\perp \le 2.5$ GeV and

$s = 1000 \, \text{GeV}^2$. We found that the diffractive (6) and the end-point (5) contributions are numerically comparable to each other. In the later case in order to regularize the end-point divergence we used the simplest prescription, an explicit cutoff on the quark momentum fraction in the pion $z' \geq \mu_{\text{IR}}^2/q_\perp^2$, where μ_{IR} is of order of intrinsic quark transverse momentum in the pion, see [10] for more details. For GPD's the parametrization [16] was used. Fig. 2 shows the comparison of the calculated dijet momentum fraction distribution with the data [4]. The two solid curves correspond to the asymptotic, $\phi_\pi^{\text{as}}(z)$, and 'two-humped", $\phi_\pi^{\text{CZ}}(z, \mu = 0.5 \, \text{GeV}) = 30z(1-z)(1-2z)^2$, forms of the pion distribution amplitude. The dashed curve corresponds to the Chernyak-Zhitnitsky model evolved to the scale $\mu = 2 \, \text{GeV}$, $\phi_\pi^{\text{CZ}}(z, \mu = 2 \, \text{GeV}) = 15z(1-z)[0.20 + (1-2z)^2$. The overall normalization is arbitrary, but is the same for all three choices of the distribution amplitude. It is seen that experimental uncertainties do not allow for the separation between the distribution amplitudes $\phi_\pi^{\text{as}}(z)$ and $\phi_\pi^{\text{CZ}}(z, \mu = 2 \, \text{GeV})$ while the extreme choice $\phi_\pi^{\text{CZ}}(z, \mu = 0.5 \, \text{GeV})$ is not favored. This general conclusion is in agreement with the analysis in [12].

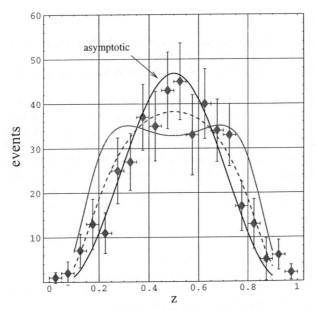

Figure 2. The longitudinal momentum fraction distribution of the dijets from the platinum target [4] in comparison with our predictions, see text.

6. Our calculation is close in spirit to [11,12] although the conclusion about the factorization is different. We examined the transition to the light-cone limit carefully and found that the approximation used in [11,12] breaks down for soft gluons (and quarks). In the double logarithmic approximation our result in (6) is similar to [7] obtained using different methods. We, therefore, agree with the interpretation suggested in [7] that in the true diffraction limit, for very large energies, the dijet production can be considered as a probe of the hard component of the pomeron. We note, however, that this interpretation breaks down beyond the double logarithmic approximation and is not sufficient for the energy region of the E791 experiment. Finally, we have to mention an approach to coherent diffraction suggested in [8] that attributes hard dijets to a hard component of the pion wave function. This technique is interesting, but apparently complicated for the discussion of factorization. We found that the general argumentation in [8] appears to be in contradiction with our explicit calculations.

Acknowledgments

This work was supported by Alexander von Humboldt Foundation.

References

1. S. F. King, A. Donnachie and J. Randa, *Nucl. Phys.* **B167** 98 (1980); J. Randa, *Phys. Rev.* **D22** 1583 (1980).
2. G. Bertsch, S. J. Brodsky, A. S. Goldhaber and J. F. Gunion, *Phys. Rev. Lett.* **47** 297 (1981).
3. L. Frankfurt, G. A. Miller and M. Strikman, *Phys. Lett.* B **304** 1 (1993).
4. E. M. Aitala *et al.* [E791 Collaboration], *Phys. Rev. Lett.* **86** 4768 (2001).
5. E. M. Aitala *et al.* [E791 Collaboration], *Phys. Rev. Lett.* **86** 4773 (2001).
6. J. Gronberg *et al.* [CLEO Collaboration], *Phys. Rev.* **D57** 33 (1998).
7. N. N. Nikolaev, W. Schäfer and G. Schwiete, *Phys. Rev.* **D63** 014020 (2001).
8. L. Frankfurt, G. A. Miller and M. Strikman, *Found. Phys.* **30** 533 (2000) [hep-ph/9907214]; *Phys. Rev.* **D65** 094015 (2002) [hep-ph/0010297].
9. V. M. Braun, D. Yu. Ivanov, A. Schäfer and L. Szymanowski, *Phys. Lett.* **B509** 43 (2001).
10. V. M. Braun, D. Yu. Ivanov, A. Schafer and L. Szymanowski, arXiv:hep-ph/0204191.
11. V. Chernyak, *Phys. Lett.* **B516** (2001) 116.
12. V. L. Chernyak and A. G. Grozin, *Phys. Lett.* **B517** 119 (2001).
13. D. Müller, D. Robaschik, B. Geyer, F. M. Dittes and J. Horejsi, *Fortsch. Phys.* **42** 101 (1994); [arXiv:hep-ph/9812448].
14. A. V. Radyushkin, *Phys. Lett.* **B385** 333 (1996); *Phys. Rev.* **D56** 5524 (1997).
15. X. Ji, *Phys. Rev. Lett.* **78** 610 (1997); *J. Phys. G.* **24** 1181 (1998).
16. A. Freund and M. F. McDermott, *Phys. Rev.* **D65**, 074008 (2002); http://durpdg.dur.ac.uk/hepdata/dvcs.html

NUCLEAR TRANSPARENCY AND LIGHT CONE PION WAVE FUNCTION IN THE COHERENT PION-NUCLEON AND PION-NUCLEUS PRODUCTION OF TWO JETS AT HIGH RELATIVE MOMENTA

LEONID FRANKFURT

Tel Aviv University,
Israel

G. A. MILLER

University of Washington,
U.S.A.

M. STRIKMAN

Pennsylvania State University,
USA

In the target rest frame at fixed $x = 2\kappa_t^2/s$, with $\kappa_t^2 \to \infty$, the space-time evolution of coherent high κ_t di-jet production is dominated by the formation of a high κ_t $q\bar{q}$ component (point like configuration) before the incident pion reaches the target. This point like configuration then interacts through two gluon exchange with the target, leading to disappearance of initial and final state interaction effects. These predictions are in accord with the FNAL data. We compute the cross section in the approximation of keeping the leading order(LO) in powers of α_s, and in the leading log approximation in $\alpha_s \ln(\kappa_t^2/\Lambda_{QCD}^2)$. Amplitudes for other processes are shown to be smaller by at least a power of α_s. Thus the high κ_t component of pion wave function, can be measured at sufficiently large values of κ_t^2. The resulting dominant amplitude is proportional to $z(1-z)\alpha_s(\kappa_t^2)/\kappa_t^{-4}(\ln \kappa_t^2/\Lambda^2)^{C_F/\beta}$ (z is the fraction light-cone $(+)$ momentum carried by the quark in the final state, β is the coefficient in the running coupling constant) times the skewed gluon distribution of the target.

1. Nuclear Dependence of the amplitude

The picture we have obtained is that the amplitude is dominated by a process in which the pion becomes a $q\bar{q}$ pair of essentially zero transverse extent well before hitting the nuclear target. The $q\bar{q}$ can interact with one nucleon and can pass undisturbed through any other nucleon. For zero momentum transfer q_t to the nucleus, the amplitude $M(\text{A})$ takes the form

$$M(\text{A}) = \text{A}\, M(N)\frac{G_A(x_1, x_2, m_f^2)}{AG_N(x_1, x_2, m_f^2)}\left(1 + \frac{\epsilon}{<b^2>\kappa_t^2}\text{A}^{1/3}\right) \equiv \text{A}M(N)\Gamma\,,\quad(1)$$

in which the skewedness of the gluonic distribution is made explicit, and where the real number $\epsilon > 0$. Observe the factor A which is the dominant effect here. This factor is contained in the ratio of nuclear to nucleon gluon distributions[1]. This dependence on the atomic number is a reliable prediction of QCD in the limit that the dijet mass, m_f^2, and $s \to \infty$, with fixed x_N. On the contrary, for $x_N \to 0$ with fixed m_f^2, the nuclear shadowing of the gluon distribution becomes very important [1,4]. At even larger values of l_c pQCD interaction becomes strong and absorbtion will be restored[2].

The ϵ correction term in Eq. (1) is a higher twist contribution which arises from a single rescattering which can occur as the PLC moves through the nuclear length ($R_A \propto \text{A}^{1/3}$). That $\epsilon > 0$ was a somewhat surprising feature of our 1993 calculation because the usual second order rescattering, as treated in the Glauber theory, always reduces cross sections. This highly unusual sign follows from the feature in QCD that the relative contribution of the rescattering term (screening term) decreases with increasing size of the spatially small dipole.

The differential cross section takes the form

$$\frac{d\sigma(A)}{dq_t^2} = A^2\Gamma^2\frac{d\sigma(N)}{dt}e^{tR_A^2/3}\,,\quad(2)$$

for small values of t. Note that

$$-t = q_t^2 - t_{min},\quad(3)$$

where $-t_{min}$ is the minimum value of the square of the longitudinal momentum transfer:

$$-t_{min} = \left(\frac{m_f^2 - m_\pi^2}{2p_\pi}\right)^2.\quad(4)$$

Our discussion below is applicable in the kinematics where $-t_{min}R_A^2/3 \leq 1$ so that the entire dependence of the cross section on t_{min} is contained in the factor $e^{t_{min}R_A^2/3}$.

One measures the integral

$$\sigma(A) = \int dt\frac{d\sigma(A)}{dt} = \frac{3}{R_A^2}A^2\Gamma^2\sigma(N)\,.\quad(5)$$

A typical procedure is to parametrize $\sigma(A)$ as

$$\sigma(A) = \sigma_1 A^\alpha \tag{6}$$

in which σ_1 is a constant independent of A. For the R_A corresponding to the two targets Pt (A = 195) and C(A = 12) of E791, one finds $\alpha \approx 1.45$. The experiment[3] does not directly measure the coherent nuclear scattering cross section. This must be extracted from a measurement which includes the effects of nuclear excitation. The extraction is discussed below.

Another potentially important A-dependent effect is the nuclear shadowing of the parton densities [4]. Very little direct experimental information is available for the A-dependence of the gluon structure function. The analyses of the data combined with the momentum sum rules and the calculation of gluon shadowing at small x suggest that the value $x \sim 0.01$ (which corresponds to the kinematics of [3]) is in a transition region between the regime of an enhancement of the gluon distribution at $x \sim .1$ and the strong shadowing at $x \leq 0.005$ [5,6,7]. Using the A-dependence of F_{2A} as a guide, and in particular the NMC ratio F_{2Sn}/F_{2C}, the shadowing may reduce α for the kinematics[3] by $\Delta\alpha \sim -0.08$.

2. Experimental Considerations

The requirements for observing the influence of color transparency were discussed in 1993[1]. The two jets should have a very small total transverse momentum. The relative transverse momentum should be greater than about 2 GeV and the mass of the diffractive state should be described by the formula

$$m_f^2 = \frac{m_q^2 + \kappa_t^2}{z(1-z)}. \tag{7}$$

Maintaining this condition is necessary to suppress diffraction into a $q\bar{q}g$ pair in which the gluon transverse momentum is not too small.

It is impossible for high energy fixed target experiments to identify the final nucleus as the target ground state, so another technique must be used here. The technique used in [3] is to isolate the dependence of the elastic diffractive peak on the momentum transfer to the target, t, as the distinctive property of the coherent processes. A strong coherent peak was observed with the slope consistent with the coherent contribution. The background was fitted as a sum of the coherent peak, inelastic diffraction with a nuclear break up and the term due to inelastic events where some hadrons were not detected.

The amplitude for the non-spin flip excitations of low-lying even-parity nuclear levels $\sim -t$, due to the orthogonality of the ground and excited state nuclear wave functions. Thus the cross section of these kinds of soft nuclear excitations integrated over t is suppressed by an additional factor of $1/R_A^4 \approx A^{-4/3}$ compared to the nuclear coherent process. For $\sqrt{-t}\, R_A \gg 1$ where q is the momentum transfer to the nucleus $q = \sqrt{-t}$ the background processes involving nuclear excitations vary as A, so an unwanted counting of such would actually weaken the signal we seek. For $q R_A \gg 1$ the diffractive peak cannot be used as signature of diffractive processes to distinguish them from non-diffractive processes whose cross section $\propto \sigma(\pi N) \propto A^{0.75}$. Thus substantial A-dependence, $\sigma \propto A^\alpha$ (with $\alpha \approx 1.5$), as predicted by QCD for large enough values of κ_t, should be distinguishable from the background processes.

The amplitude varies [1] as : $M(A) \sim \alpha_s / \kappa_t^4$

$$M(A) \sim \alpha_s x_N G_N(x_N, Q_{eff}^2)/\kappa_t^4, \tag{8}$$

where $Q_{eff}^2 \sim 2\kappa_t^2$. For the kinematics of the E791 experiment, where x_N increases $\propto \kappa_t^2$, the factor $\alpha_s x_N G_N(x_N, Q_{eff}^2)$ is a rather weak function of κ_t. For example, if we use the standard CTEQ5M parameterization we find $\sigma(A) \sim 1/\kappa_t^{8.5}$ for $1.5 \leq \kappa_t \leq 2.5$ GeV which is consistent with the data [3]. For the amplitude discussed here, $\sigma(A) \sim (z(1-z))^2$ which is in the excellent agreement with the data [3].

2.1. Extracting the coherent contribution

The experiment is discussed in Ref. [3]. The observed q_t^2 dependence for very low q_t^2 is consistent with that obtained from the previously measured radii $R(C) = 2.44$ fm, and $R_{Pt} = 5.27$ fm. The key feature is the identification of the coherent contribution from its rapid falloff with t.

We discuss the extraction of this signal in some detail. The total cross section includes terms in which the final nucleus is not the ground state. The nuclear excitation energy is small compared to the energy of the beam, so that one may use closure to treat the sum over nuclear excited states. Then the cross section is evaluated as a ground state matrix element of an operator $\sum_{i,j} e^{i\mathbf{q}\cdot(\mathbf{r}_i - \mathbf{r}_j)} = A + \sum_{i \neq j} e^{i\mathbf{q}\cdot(\mathbf{r}_i - \mathbf{r}_j)}$. The result, obtained by using $\mathbf{r}_i$ relative to the nuclear center of mass, and by neglecting correlations in the nuclear wave function is given by

$$\frac{d\sigma_A}{dt} = \left[A + A(A-1)F_A^2\left(t\,\frac{A}{A-1}\right)\right]\frac{d\sigma_N}{dt}. \tag{9}$$

The factor $\frac{A}{A-1}$ in the argument of F_A is due to accounting for nuclear recoil in the mean field approximation. This formula should be very accurate for the small values of t relevant here. The contribution of the coherent processes to the total cross section is given by

$$\frac{d\sigma^{coherent}}{dt} = A^2 \, F_A^2(t) \frac{d\sigma_N}{dt}, \tag{10}$$

and the contribution of excited nuclear states is the difference: $\frac{d\sigma_A}{dt} - \frac{d\sigma^{coherent}}{dt}$, which vanishes at $t = 0$. The experiment proceeds by removing a term $\propto A$ from $\frac{d\sigma_A}{dt}$ which has no rapid variation with t. This defines a new cross section which is actually measured experimentally.

$$\frac{d\tilde{\sigma}_A}{dt} = A(A-1)F_A^2\left(t\,\frac{A}{A-1}\right)\frac{d\sigma_N}{dt}. \tag{11}$$

The integral of this term over t can be extracted from the data:

$$\sigma_1 \equiv \int dt\,\frac{d\tilde{\sigma}_A}{dt} = \frac{3A(A-1)}{r_N^2 + R_A^2\frac{A}{A-1}}\frac{d\sigma_N}{dt}_{\,|t=0} \approx \frac{3(A-1)^2}{r_N^2 + R_A^2}\frac{d\sigma_N}{dt}_{\,|t=0}. \tag{12}$$

Here the factor r_N^2 takes into account the slope of the elementary cross section, assuming that it is determined solely by the nucleon vertex. Note that the result (12) differs by a factor of $\frac{(A-1)^2}{A^2}$ from the A-dependence predicted previously for coherent processes, recall Eq. (2). Using A=12, 195, the nuclear radii mentioned above, $r_N = 0.83fm$ and fitting the ratio of cross sections obtained from Eq.(12) with the parametrization $\sigma \propto A^\alpha$, gives then

$$\alpha = 1.54, \tag{13}$$

instead of $\alpha = 1.45$.

The result [3] of the experiment is

$$\alpha \approx 1.55 \pm 0.05 , \tag{14}$$

which is remarkably close to the theoretical value shown in Eq. (13). The sizes of our multiple scattering and nuclear shadowing corrections, which work in the opposite directions, and which were discussed in the previous Section, are of the order of the experimental error bar.

3. General properties of hard diffractive processes in QCD

Our interest in this talk which follows papers [1] is focussed on the calculation of the amplitudes of hard diffractive processes in the kinematics covered by the fixed target physics at FNAL, by the collider eRHIC, HERA physics. To find an adequate theoretical framework let us consider firstly kinematics.

The rapidity span covered by hard processes at HERA ($Q \geq 2GeV$) is ≤ 8 units only. For other accelerators the rapidity span is even less. Four units of rapidity are used by the fragmentation of projectile and nucleon fragmentation. Therefore in the practically interesting region of multiregge kinematics, the radiation of at most 1 or 2 hard gluons is possible. Thus in this kinematics the LO,NLO...DGLAP approximation should be rather effective method of calculation of cross sections of small x processes. At significantly larger energies -LHC and/or for DIS off heavy nuclear target, pQCD interactions become strong and exceed the unitarity limit [2]. In that case, the methods of pQCD may become ineffective. For a theoretical discussion of new interesting phenomena characteristic for this kinematics see [2].

QCD is nonlinear theory and Ward identities are nonlinear also. So a technical problem is how to tackle them in practice. The trick [1] is to explore that the dominant two-gluon exchange projectile-target amplitude has positive charge parity. Therefore this amplitude has positive signature i.e. it is symmetric on the transposition $s \to u$. For such an amplitude, which grows slowly with energy, the relation between real and imaginary parts of amplitude is: [8]

$$\frac{\mathrm{Re}A(\nu, t)}{\mathrm{Im}A(\nu, t)} = \frac{\pi}{2} \frac{\partial}{\partial \ln \nu} \ln \frac{\mathrm{Im}A(\nu, t)}{\nu}. \tag{15}$$

Thus it would be sufficient to calculate imaginary part of amplitude and to restore real part of amplitude by using Eq. (15). There is another enormous simplification related to the issue of gauge invariance. The pion wave function is not gauge invariant, but the s,u cut parts of the amplitude $\pi + g \to JJ + g$, for two gluons in a color singlet state, are calculable in terms of amplitudes of sub-processes where only one gluon is off mass shell. For such amplitudes the Ward identities[9]– the conservation of color current– have the same form as the conservation of electromagnetic current in QED.

Selection of the final state with the mass of di-jet: $\kappa_t^2/z(1 - z)$ leads to squeezing of the transverse size of the initial pion[1]. Really as a result of large value of l_c after hard pion collision with target gluon there should be gluon bremsstrahlung.To avoid gluon bremsstrahlung the contribution of quark-gluon configurations corresponding to the transverse size of pion r_{pion} should be suppressed by the powers of Sudakov type form factor and by the additional form factor accounting for the fact that gluons carry around half of pion momentum in the average quark-gluon configuration within light cone (lc) pion wave function.

4. Selection of diagrams for the LO DGLAP approximation

In this talk we use formulae deduced in [1] within the leading $\alpha_s \ln \frac{\kappa_t^2}{\Lambda_{QCD}^2}$ approximation when $\alpha_s \ll 1$. In LO, the dominant contribution is given by diagrams in which the pion wave function is dominated by its high κ_t tail because of gluon exchange. The NLO calculation is significantly more cumbersome, but presents no unsurmountable barriers to calculate cross sections of hard diffractive processes.

The Feynman diagrams corresponding to the interaction between q$\bar{q}$ in the final state (fsi) due to hard gluon exchange considered in [10,1,11] are a NLO effect-they are suppressed by additional power of α_s. For QCD evolution, the potentially important region of integration over quark momenta in the pion wave function is $\kappa_0^2 \ll r_t^2 \ll l_t^2 \ll \kappa_t^2$. Here l_t is quark momentum within pion wave function, r_t is transverse momentum of nuclear gluon. The contribution of the region of small $l_t^2 \ll r_t^2$ is suppressed by Sudakov form factor accounting for the requirement of lack of radiation in the hard collision off nuclear gluon. The contribtion of small κ_o^2 is killed by Sudakov type form factor and by the small probability to find a $q\bar{q}$ pair with no gluons at an average separation between constituents[1]. Thus there insufficient phase space to produce $\ln \kappa_t^2$ in the integration over l_t^2. Similarily it is easy to demonstrate that $q\bar{q}g$ intermediate state leads to NLO effect[1]. To check these technical statements, one should use renormalization invariance accounted for in Brodsky-Lepage evolution equation and decomposition over Gegenbauer polinomials.

Additional suppression of FSI and $q\bar{q}g$ intermediate state arises from the properties of skewed parton distributions. The logarithmic region of integration over l_t corresponds to $-x_2 \gg x_1$. Within the parton model approximation, such a structure function should be higher twist effect if a parton with negative x is not a wee parton.

5. Asymptotic light-cone pion wave function and z κ_t dependence of cross section

The next technical problem is to account correctly for the renormalization of pion wave function. To kill infrared divergencies, we may follow the reasoning of [14] or choose light-cone gauge $p(pion)_\mu A_\mu = 0$ [1]. The resulting expression for the amplitude has been calculated in [1].

6. Acknowledgments

We would like to thank S. Brodsky, J. Collins, Yu. Dokshitzer and A. Mueller for useful discussions. This work has been supported in part by the USDOE and GIF.

References

1. L.L. Frankfurt, G.A. Miller and M. Strikman, Phys. Lett. **B304**, 1 (1993);Found. Phys. **30**, 533 (2000), Phys.Rev. D65:094015,2002.
2. L.Frankfurt, V. Guzey, M. McDermott and M.Strikman, Phys.Rev.Lett.87:192301,2001
3. E. M. Aitala *et al.* [Fermilab E791 Collaboration], Phys. Rev. Lett. **86**, 4773 (2001); E. M. Aitala *et al.* [Fermilab E791 Collaboration], Phys. Rev. Lett. **86**, 4768 (2001). D.Ashery.Talk at this conference.
4. S. J. Brodsky, L. Frankfurt, J. F. Gunion, A. H. Mueller and M. Strikman, Phys. Rev. **D50**, 3134 (1994).
5. L. L. Frankfurt, M. I. Strikman and S. Liuti, Phys. Rev. Lett. **65**, 1725 (1990).
6. T. Gousset and H. J. Pirner, Phys. Lett. B **375**, 349 (1996) [hep-ph/9601242].
7. K. J. Eskola, V. J. Kolhinen and C. A. Salgado, Eur. Phys. J. **C9**, 61 (1999) [hep-ph/9807297].
8. V. N. Gribov and A. A. Migdal, Sov. J. Nucl. Phys. **8**, 583 (1969);
9. G. 't Hooft, Nucl. Phys. **B33**, 173 (1971); G. 't Hooft, Nucl. Phys. **B35**, 167 (1971); G. 't Hooft and M. Veltman, Nucl. Phys. **B44**, 189 (1972).
10. B.K. Jennings and G.A. Miller, Phys. Rev. C **50**, 3018 (1994).
11. V.Chernyak. Talk at this conference.
12. L. Frankfurt, A. Freund, V. Guzey, and M. Strikman, Phys. Lett. B **418**, 345 (1998) [Erratum-ibid. B **429**, 414 (1998)] [hep-ph/9703449].
13. See the review: L.L. Frankfurt, G.A. Miller and M. Strikman, Ann. Rev. Nucl. Part. Sci. **44**, 501 (1994) hep-ph/9407274.
14. S.J. Brodsky G.P. Lepage, Phys. Rev. **D22**, 2157 (1982)

DIFFRACTIVE HARD DIJETS AND NUCLEAR PARTON DISTRIBUTIONS

I.P. IVANOV[A,B], N.N. NIKOLAEV[B,C], W. SCHÄFER[D],
B.G. ZAKHAROV[C], V.R. ZOLLER[E]

[A] Novosibirsk State Univeristy, Novosibirsk, Russia
[B] Institut f. Kernphysik, Forschungszentrum Jülich, Germany
[C] L.D.Landau Institute for Theoretical Physics, Chernogolovka, Russia
[D] Nordita, Blegdamsvej 17, DK-2100 Copenhagen, Denmark
[E] Institute for Theoretical and Experimental Physics, Moscow, Russia
E-mail: N.Nikolaev@fz-juelich.de

Diffraction plays an exceptional rôle in DIS off heavy nuclei. First, diffraction into hard dijets is an unique probe of the unintegrated glue in the target. Second, because diffraction makes 50 per cent of total DIS off a heavy target, understanding diffraction in a saturation regime is crucial for a definition of saturated nuclear parton densities. After brief comments on the Nikolaev-Zakharov (NZ) pomeron-splitting mechanism for diffractive hard dijet production, I review an extension of the Nikolaev-Schäfer-Schwiete (NSS) analysis of diffractive dijet production off nuclei to the definition of nuclear partons in the saturation regime. I emphasize the importance of intranuclear distortions of the parton momentum distributions.

1. The Dominance of the Pomeron-Splitting Mechanism for Diffractive Hard Dijets

The point that diffraction excitation probes the wave function of composite systems has been made some 50 years ago by Landau, Pomeranchuk, Feinberg and Glauber [1] - it is very much relevant to QCD too!

The pQCD diagrams for production of diffractive dijets are shown in fig 1. In the Landau-Pomeranchuk diagram (b) the limited transverse momentum $\mathbf{p}$ of the quark jet comes from the intrinsic momentum of quarks and/or antiquarks in the beam particle, whereas in the Pomeron splitting diagram (a) hard jets receive the transverse momentum from gluons in the pomeron [2,3]. As shown by NSS [4] the corresponding diffractive amplitude is proportional to the unintegrated gluon structure function of the target proton, $dG(x, \mathbf{p}^2)/d \log \mathbf{p}^2$, and the so-called lightcone distribution amplitude for the beam particle.

The NSS dominance of the pomeron-splitting contribution for hard di-

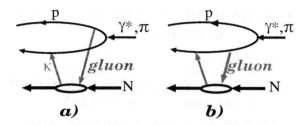

Figure 1.   The pQCD diagrams for coherent diffractive dijet excitation

jets has fully been confirmed by the NLO order analysis of Chernyak et al. [5] and Braun et al. [6]. The NLO correction to the NSS amplitude is found to be proportional to the asymptotic distribution amplitude and numerically quite substantial, so that the experimental data by E791 [7] can not distinguish between the asymptotic and double-humped distribution amplitudes. According to NSS [4] realistic model distributions do not differ much from the asymptotic one, though. To my view, the only caveat in the interpretation of the NLO results is that the issue of partial reabsorption of these corrections into the evolution/renormalization of the pion distribution amplitude has not yet been properly addressed. Anyway, there emerges a consistent pattern of diffraction of pions into hard dijets and in view of these findings the claims by Frankfurt et al. [8] that the diffractive amplitude is proportional to the integrated gluon structure function of the target must be regarded null and void. Hopefully, some day the E791 collaboration shall report on the interpretation of their results within the correct formalism.

The current status of the theory has been comprehensively reviewed at this Workshop by Chernyak [9] and Radyushkin en lieu of Dima Ivanov [10] and there is no point in repeating the same the third time - the principal conclusions by NSS have been published some years ago and are found in [4]. I would rather report new results [11] on the relevance of diffractive DIS to the hot issue of nuclear saturation of parton densities.

2. Diffractive and Truly Inelastic DIS off Free Nucleons and Heavy Nuclei

While the above cited NSS papers focused on diffraction on nuclei in the hard regime, in the rest of my talk I would like to discuss the opposite regime of nuclear saturation. Nuclear saturation is an opacity of heavy nuclei for color dipole states of the beam be it a hadron or real, and virtual, photons. The fundamental point about diffractive DIS is the counterintuitive result by Nikolaev, Zakharov and Zoller [12] that for a very heavy nucleus coherent diffractive DIS in which the target nucleus does not break and is retained in the ground state makes precisely 50 per cent of the total DIS events. Consequently, diffractive DIS is a key to an understanding of nuclear saturation. I note in passing that because of the very small fraction of DIS off free nucleons which is diffractive one, $\eta_D \lesssim$ 6-10 %, there is little room for a genuine saturation effects at HERA. Intuitively, such an importance of diffractive DIS which can not be treated in terms of parton densities in the target casts shadow on the interpretation of the saturation regime in terms of parton densities, which is one of the points from our analysis [11].

The alternative interpretation of nuclear opacity in terms of a fusion and saturation of nuclear partons goes back to the 1975 papers by Nikolaev and Valentine Zakharov [13]: the Lorentz contraction of relativistic nuclei entail a spatial overlap of partons with $x \lesssim x_A \approx 1/R_A m_N$ from different nucleons and the fusion of overlapping partons results in the saturation of parton densities per unit area in the impact parameter space. More recently this idea has been revived in the quantitative pQCD framework by McLerran et al. [14].

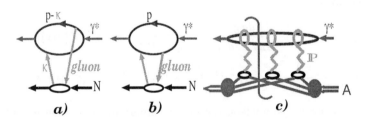

Figure 2. The pQCD diagrams for DIS off protons (a,b) and nuclei (c). Diagrams (a) and (b) show the 2-gluon tower approximation for the QCD pomeron. The diagram (c) shows the nuclear multiple scattering for virtual Compton scattering off nuclei; the diffractive unitarity cut is indicated.

We base our analysis on the color dipole formulation of DIS [15,3,16,12]. The total cross section for interaction of the color dipole $\mathbf{r}$ with the target nucleon equals

$$\sigma(r) = \alpha_S(r)\sigma_0 \int d^2\kappa f(\kappa)\left[1 - \exp(i\kappa\mathbf{r})\right], \qquad (1)$$

where $f(\kappa)$ is related to the unintegrated glue of the target by

$$f(\kappa) = \frac{4\pi}{N_c\sigma_0} \cdot \frac{1}{\kappa^4} \cdot \frac{\partial G}{\partial \log \kappa^2} \qquad (2)$$

and is normalized as $\int d^2\kappa f(\kappa) = 1$. Here σ_0 describes the saturated total cross section for very large dipoles. The total virtual photoabsorption cross section for a free nucleon target equals

$$\sigma_N(x, Q^2) = \langle \gamma^* | \sigma(r) | \gamma^* \rangle = \int_0^1 dz \int d^2 r \Psi_\gamma^*(z, r) \sigma(r) \Psi_\gamma^*(z, r) \quad (3)$$

$$\frac{d\sigma_N}{d^2 p dz} = \frac{\sigma_0}{2} \cdot \frac{\alpha_S(p^2)}{(2\pi)^2} \int d^2\kappa f(\kappa) \, |\langle \gamma^* | p \rangle - \langle \gamma^* | p - \kappa \rangle|^2 \quad (4)$$

where p is the transverse momentum, and z the Feynman variable, of the leading quark in the final state prior the hadronization, see figs. 2a,b. Notice that the target nucleon is color-excited and there is no rapidity gap in the final state. This is a starting point for a definition of the small-x sea generated from the glue. The relevant wave functions of the photon are found in [2,15,3].

Because of the smallness of the electromagnetic coupling, the diffractive DIS of fig. 1 amounts to quasielastic scattering of CD states of the photon off the target proton [15,3,16]. In this case the target nucleon is left in the color singlet state and there is a rapidity gap in the final state. For the forward diffractive DIS, $\gamma^* p \to (q\bar{q}) + p'$, with the vanishing (p, p') momentum transfer, $\Delta = 0$,

$$\left. \frac{d\sigma_D}{d\Delta^2 dz d^2 p} \right|_{\Delta^2=0} = \frac{1}{16\pi} \cdot \frac{1}{(2\pi)^2} |\langle \gamma^* | \sigma(r) | p \rangle|^2$$

$$= \frac{1}{16\pi} \cdot \frac{1}{(2\pi)^2} [\sigma_0 \alpha_S(p^2)]^2 \left| \int d^2\kappa f(\kappa) \left(\langle \gamma^* | p \rangle - \langle \gamma^* | p - \kappa \rangle \right) \right|^2 \quad (5)$$

Because η_D for a free nucleon target is so small, in the parton model interpretation of the proton structure functions one customarily neglects diffractive absorption corrections , see however warnings in [16].

Now consider DIS off nuclei at $x \sim x_A$, when interaction of the $q\bar{q}$ states dominates. The coherent diffractive cross section equals [15,12]

$$\sigma_D = \int d^2 b \langle \gamma^* | \left| 1 - \exp[-\frac{1}{2}\sigma(r)T(b)] \right|^2 |\gamma^* \rangle$$

$$= \int d^2 b \int_0^1 dz \int \frac{d^2 p}{(2\pi)^2} \left| \langle \gamma^* | \left\{ 1 - \exp[-\frac{1}{2}\sigma(r)T(b)] \right\} | p \rangle \right|^2. \quad (6)$$

Here $T(b) = \int dz n_A(z, b)$ is the optical thickness of a nucleus at an impact parameter b. The σ_D sums all the unitarity cuts between the exchanged pomerons, so that none of the nucleons of the nucleus is color-excited and there is a rapidity gap in the final state, see fig. 2c.

The inelastic DIS describes all events in which one or more nucleons of the nucleus are color-excited and there is no rapidity gap in the final state. I omit a somewhat tricky derivation [11] which is based on the technique developed in [17] and cite only the final result

$$\frac{d\sigma_{in}}{d^2pdz} = \frac{1}{(2\pi)^2} \int d^2b \int d^2r'd^2r \exp[ip(r'-r)]\Psi^*(r')\Psi(r)$$

$$\left\{ \exp[-\frac{1}{2}\sigma(r-r')T(b)] - \exp[-\frac{1}{2}[\sigma(r)+\sigma(r')]T(b)] \right\}.$$ (7)

The effect of nuclear distortions on the observed momentum distribution of quarks is obvious: the dependence of nuclear attenuation factors on r, r' shall affect strongly a computation of the Fourier transform (7).

Upon the integration over p one recovers the familiar color dipole Glauber-Gribov formulas [15,3,12] for the inelastic and total cross sections

$$\sigma_{in} = \int d^2b\langle\gamma^*|1 - \exp[-\sigma(r)T(b)|\gamma^*\rangle$$ (8)

$$\sigma_A = \sigma_D + \sigma_{in} = 2\int d^2b\langle\gamma^*|1 - \exp[-\frac{1}{2}\sigma(r)T(b)|\gamma^*\rangle$$ (9)

3. Nuclear Parton Distributions as Defined by Diffraction

The next issue is whether nuclear DIS can be given the conventional parton model interpretation or not. For the evaluation of the inclusive spectrum of quarks in inelastic DIS we resort to the NSS representation [4]

$$\Gamma_A(b,r) = 1 - \exp\left[-\frac{1}{2}\sigma(r)T(b)\right] = \int d^2\kappa\phi_{WW}(\kappa)[1 - \exp(i\kappa r)].$$ (10)

There is a close analogy to the representation (1),(2) in terms of $f(\kappa)$ and

$$\phi_{WW}(\kappa) = \sum_{j=1}^{\infty} \nu_A^j(b) \cdot \frac{1}{j!}f^{(j)}(\kappa)\exp\left[-\nu_A(b)\right]$$ (11)

can be interpreted as the unintegrated nuclear Weizsäcker-Williams (WW) glue per unit area in the impact parameter plane, normalized as

$$\int d^2\kappa\phi_{WW}(\kappa) = 1 - \exp[-\nu_A(b)].$$ (12)

Here

$$\nu_A(b) = \frac{1}{2}\alpha_S(r)\sigma_0 T(b)$$

defines the nuclear opacity and the j-fold convolutions

$$f^{(j)}(\kappa) = \int \prod_i^j d^2\kappa_i f(\kappa_i)\delta(\kappa - \sum_i^j \kappa_i)$$

describe the contribution to the diffractive amplitudes from the j split pomerons [4]. The hard asymptotics of the WW glue has been analyzed by NSS, here I only mention that broadening of convolutions compensates completely the nuclear attenuation effects obvious in the expansion (11) and, furthermore, leads to a nuclear antishadowing for hard dijets [4].

A somewhat involved analysis of properties of convolutions in the soft region shows that they develop a plateau-like behaviour with the width of the plateau which expands $\propto j$. Here I only point out that the gross features of WW glue in the soft region are well reproduced by

$$\phi_{WW}(\kappa) \approx \frac{1}{\pi} \frac{Q_A^2}{(\kappa^2 + Q_A^2)^2}, \tag{13}$$

where the saturation scale $Q_A^2 = \nu_A(b)Q_0^2 \propto A^{1/3}$. The soft parameters Q_0^2 and σ_0 are related to the integrated glue of the proton in soft region,

$$Q_0^2 \sigma_0 \sim \frac{2\pi^2}{N_c} G_{soft}, \quad G_{soft} \sim 1.$$

Making use of the NSS representation (10) and the normalization (12), after some algebra one finds for the saturation domain of $\mathbf{p}^2 \lesssim Q^2 \lesssim Q_A^2$

$$\frac{d\sigma_{in}}{d^2\mathbf{b}d^2\mathbf{p}dz} = \frac{1}{(2\pi)^2} \int d^2\kappa \phi_{WW}(\kappa) |\langle\gamma^*|\mathbf{p}+\kappa\rangle|^2 \tag{14}$$

$$\frac{d\sigma_D}{d^2\mathbf{b}d^2\mathbf{p}dz} = \frac{1}{(2\pi)^2} \left| \int d^2\kappa \phi_{WW}(\kappa^2)(\langle\gamma^*|\mathbf{p}\rangle - \langle\gamma^*|\mathbf{p}-\kappa\rangle) \right|^2$$

$$\approx \frac{1}{(2\pi)^2} \left| \int d^2\kappa \phi_{WW}(\kappa) \right|^2 |\langle\gamma^*|\mathbf{p}\rangle|^2 \approx \frac{1}{(2\pi)^2} |\langle\gamma^*|\mathbf{p}\rangle|^2 \tag{15}$$

The last result is obvious from (6) because in this case all the color dipoles in the virtual photon meet the opacity criterion $\sigma(r)T(\mathbf{b}) \gtrsim 1$, so that the nuclear attenuation terms can be neglected altogether.

4. The interpretation of the results

Following the conventional parton model wisdom, one may try defining the nuclear sea quark density per unit area in the impact parameter space

$$\frac{d\bar{q}}{d^2\mathbf{b}d^2\mathbf{p}} = \frac{1}{2} \cdot \frac{Q^2}{4\pi^2\alpha_{em}} \cdot \frac{d[\sigma_D + \sigma_{in}]}{d^2\mathbf{b}d^2\mathbf{p}} = \frac{1}{2} \cdot \frac{Q^2}{4\pi^2\alpha_{em}} \cdot \int dz$$

$$\times \left\{ \left| \int d^2\kappa \phi_{WW}(\kappa) \right|^2 |\langle\gamma^*|\mathbf{p}\rangle|^2 + \int d^2\kappa \phi_{WW}(\kappa) |\langle\gamma^*|\mathbf{p}+\kappa\rangle|^2 \right\} \tag{16}$$

It is a nonlinear functional of the NSS-defined WW glue of a nucleus. The quadratic term comes from diffractive DIS and measures the momentum distribution of quarks and antiquarks in the $q\bar{q}$ Fock state of the photon. It has no counterpart in DIS off free nucleons because diffractive DIS off free nucleons is negligible small even at HERA, $\eta_D \lesssim$ 6-10 %. The linear term comes from the truly inelastic DIS with color excitation of nucleons of the target nucleus. As such, it is a counterpart of standard DIS off free nucleons, but as a function of the photon wave function and nuclear WW gluon distribution it is completely different from (4) for free nucleons. This difference is entirely due to strong intranuclear distortions of the outgoing quark and antiquark waves in inelastic DIS off nuclei.

Up to now I specified neither the wave function of the photon nor the spin of charged partons - they could well have been scalar or spin-1 ones -, nor the color representation for charged partons. All our results would hold for any weakly interacting projectile such that elastic scattering is negligible small and diffraction excitation amounts to quasielastic scattering of Fock states of the projectile [15,3]. Now take the conventional spin-$\frac{1}{2}$ partons and the photon's virtuality $Q^2 \lesssim Q_A^2$ such that the opacity criterion is met for all color dipoles of the photon. Then upon the z-integration one finds for $\mathbf{p}^2 \lesssim Q^2$ the plateau-like distribution from diffractive DIS,

$$\left.\frac{d\bar{q}}{d^2\mathbf{b}d^2\mathbf{p}}\right|_D = \frac{N_c}{4\pi^4}. \tag{17}$$

The inclusive spectrum of sea quarks from inelastic DIS also exhibits a plateau, but very different from (17):

$$\left.\frac{d\bar{q}}{d^2\mathbf{b}d^2\mathbf{p}}\right|_{in} = \frac{1}{2}\cdot\frac{Q^2}{4\pi^2\alpha_{em}}\phi_{WW}(0)\int^{Q^2}d^2\boldsymbol{\kappa}\,|\langle\gamma^*|\boldsymbol{\kappa}\rangle|^2 = \frac{N_c}{4\pi^4}\cdot\frac{Q^2}{Q_A^2}. \tag{18}$$

The plateau for inelastic DIS extends up to $\mathbf{p}^2 \sim Q_A^2$ and this nuclear broadening of momentum distributions of outgoing quarks is an obvious indicator of strong intranuclear distortions. Its height does explicitly depend on Q^2 and for $Q^2 \ll Q_A^2$ the inelastic plateau contributes little to the transverse momentum distribution of soft quarks. Still, the inelastic plateau extends way beyond Q^2 and its integral contribution to the spectrum of quarks is exactly equal to that from diffractive DIS. The two-plateau structure of the nuclear quark momentum distributions has not been discussed before. For $Q^2 \gtrsim Q_A^2$ the inelastic plateau coincides with the diffractive one, the both extend up to $\mathbf{p}^2 \lesssim Q_A^2$. Here we agree with Mueller [18].

At this point I notice that after the formal mathematical manipulations with the NSS representation, the total nuclear cross section (9) can be cast in the form

$$\sigma_A = \int d^2\mathbf{b}\int dz\int\frac{d^2\mathbf{p}}{(2\pi)^2}\int d^2\boldsymbol{\kappa}\phi_{WW}(\boldsymbol{\kappa})\,|(\langle\gamma^*|\mathbf{p}\rangle - \langle\gamma^*|\mathbf{p}-\boldsymbol{\kappa}\rangle)|^2 \tag{19}$$

which resembles (4): the **p** distribution evolves from the WW nuclear glue in precisely the same manner as as in DIS off free nucleons, which suggests the reinterpretation of the differential form of (19) in terms of the nuclear IS parton density. Furthermore, in the saturation regime the crossing terms can be neglected, while the remaining two terms would coincide with σ_D and σ_{in}, respectively, giving some support to an extention of the parton model wisdom about the equality of the IS and FS parton densities to nuclear targets too. One should be aware of some caveats, though. First of all, the equality of IS and FS densities comes at the expense of a somewhat weird equating the diffractive FS spectrum to DIS spectrum from the spectator quark of fig. 2b and σ_{in} with the contribution from the scattered quark, the both evaluated in terms of the WW nuclear glue. Second, as pointed out above, (19) implicitly includes the diffractive interactions, which make up 50 % of the FS quark yield. Hence the thus defined parton density appears to be highly nonuniversal, recall that the diffractive final states are typical of DIS and would be quite irrelevant, e.g. in nuclear collisions. Furthermore, in sharp contrast to the situation on the proton target, in (19) in the saturation regime, the dominant contribution comes from the region of $\mathbf{p}^2 \lesssim \kappa^2$, just opposite to the at not too small x dominant strongly ordered DGLAP contribution from $\kappa^2 \ll \mathbf{p}^2$.

One can go one step further and consider interactions with the opaque nucleus of the $q\bar{q}g$ Fock states of the photon. Then the above analysis can be extended to $x \ll x_A$ and the issue of the x-dependence of the saturation scale Q_A^2 can be addressed following the discussion in [16]. I only mention here that as far as diffraction is concerned, the WW glue remains a useful concept and the close correspondence between $\phi_{WW}(\kappa)$ for the nucleus and $f(\kappa)$ for the nucleon is retained. The details of this analysis will be published elsewhere [11]. For the shortage of space I didn't report here the phenomenological consequences.

5. Summary and Conclusions

The NSS representation for nuclear profile function gives a convenient and unique definition of the WW gluon structure function of the nucleus from soft to hard region. The conclusion by NSS that diffraction into hard dijets off nucleons and nuclei is dominated by the pomeron splitting mechanism has been confirmed by NLO calculations.

Coherent diffractive DIS is shown to dominate the inclusive spectrum of leading quarks in DIS off nuclei. The observed spectrum of diffractive leading quarks measures precisely the momentum distribution of quarks in the $q\bar{q}$ Fock state of the photon, the rôle of the target nucleus is simply to provide an opacity. It exhibits a saturation property and a universal plateau but its interpretation as a saturated density of sea quarks in a nucleus is questionable. The inelastic DIS also gives the plateau-like spectrum of observed quarks, but with the height that depends on Q^2. Nuclear broadening of the inelastic plateau is a clearcut evidence for an importance of intranuclear distortions of the spectrum of a struck quark.

This work has been partly supported by the INTAS grants 97-30494 & 00-00366 and the DFG grant 436RUS17/45/00. I'm grateful to Anatoly Radyushkin and Paul Stoler for the invitation to this Workshop.

References

1. L.D. Landau and I.Ya. Pomeranchuk, *J. Exp. Theor. Phys.* **24** (1953) 505; I.Ya. Pomeranchuk and E.L. Feinberg, *Doklady Akademii Nauk SSSR* **93** (1953) 439; E.L. Feinberg, *J. Exp. Theor. Phys.* **28** (1955) 242; E.L. Feinberg and I.Ya. Pomeranchuk, *Nuovo Cim. (Suppl.)* **4** (1956) 652; R.J. Glauber, *Phys. Rev.* **99** (1955) 1515.
2. N.N. Nikolaev and B.G. Zakharov, *Phys. Lett.* **B332** (1994) 177.
3. N.N. Nikolaev and B.G. Zakharov, *Z. Phys.* **C53** (1992) 331.
4. N.N. Nikolaev, W. Schäfer and G. Schwiete, *JETP Lett.* **72** (2000) 583; *Pisma Zh. Eksp. Teor. Fiz.* **72** (2000) 583; *Phys. Rev.* **D63** (2001) 014020.
5. V.L. Chernyak, *Phys. Lett.* **B516** (2001) 116; V.L. Chernyak, A.G. Grozin, *Phys. Lett.* **B517** (2001) 119.
6. V.M. Braun, D.Yu. Ivanov, A. Schaefer, L. Szymanowski, *Phys. Lett.* **B509** (2001) 43; hep-ph/0204191.
7. E791 Collaboration, E.M. Aitala et al., *Phys. Rev. Lett.* **86** (2001) 4773, 4768; D. Ashery, these proceedings.
8. L. Frankfurt, G.A. Miller, M. Strikman, *Phys. Lett.* **B304** (1993) 1; *Phys. Rev.* **D65** (2002):094015; L. Frankfurt, these proceedings.
9. V.L. Chernyak, these proceedings
10. D. Ivanov, these proceedings
11. N.N. Nikolaev, W. Schäfer, B.G. Zakharov and V.R. Zoller,
12. N.N. Nikolaev, B.G. Zakharov and V.R. Zoller, *Z. Phys.* **A351** (1995) 435.
13. N.N. Nikolaev and V.I. Zakharov, *Sov. J. Nucl. Phys.* **21** (1975) 227; *Yad. Fiz.* **21** (1975) 434; *Phys. Lett.* **B55** (1975) 397.
14. L. McLerran and R. Venugopalan, *Phys. Rev.* **D49** (1994) 2233; **D55** (1997) 5414.
15. N.N. Nikolaev and B.G. Zakharov, *Z. Phys.* **C49**, 607 (1991)
16. N.N.Nikolaev and B.G.Zakharov, *J. Exp. Theor. Phys.* **78** (1994) 806; *Zh. Eksp. Teor. Fiz.* **105** (1994) 1498; *Z. Phys.* **C64** (1994) 631.
17. N.N. Nikolaev, *J. Exp. Theor. Phys.* **54** (1981) 434.
18. A.H. Mueller, *Nucl. Phys.* **B558** (1999) 285.

GPDS, FORM FACTORS AND COMPTON SCATTERING

P. KROLL

Fachbereich Physik, Universität Wuppertal,
D-42097 Wuppertal, Germany
Email: kroll@physik.uni-wuppertal.de

The basic theoretical ideas of the handbag factorization and its application to wide-angle scattering reactions are reviewed. With regard to the present experimental program carried out at JLab, wide-angle Compton scattering is discussed in some detail.

1. Introduction

As is well-known factorization is an important property of QCD whithout which we would not be able to calculate form factors or cross sections. Factorization into a hard parton-level subprocess to be calculated from perturbative QED and/or QCD, and soft hadronic matrix elements which are subject to non-perturbative QCD and are not calculable at present, has been shown to hold for a number of reactions provided a large scale, i.e. a large momentum transfer, is available. For other reactions factorization is a reasonable hypothesis. In the absence of a large scale we don't know how to apply QCD and, for the interpretation of scattering reactions, we have to rely upon effective theories or phenomenological models as for instance the Regge pole one.

For hard exclusive processes there are two different factorization schemes available. One of the schemes is the handbag factorization (see Fig. 1) where only one parton participates in the hard subprocess (e.g. $\gamma q \to \gamma q$ in Compton scattering) and the soft physics is encoded in generalized parton distributions (GPDs) [1,2]. The handbag approach applies to deep virtual exclusive scattering (e.g. DVCS) where one of the photons has a large virtuality, Q^2, while the squared invariant momentum transfer, $-t$, from the ingoing hadron to the outgoing one is small. It also applies to wide-angle scattering (WACS) where Q^2 is small while $-t$ (and $-u$) are large [3,4]. This class of reactions is the subject of my talk. For wide-angle scattering there is an alternative scheme, the leading-twist factorization [5]. Here all valence quarks the involved hadrons are made off participate in the hard scattering

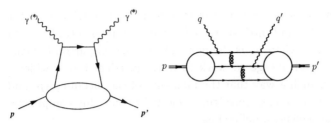

Figure 1. Handbag (left) and leading-twist (right) factorization for Compton scattering.

(e.g. $\gamma qqq \to \gamma qqq$ in Compton scattering) while the soft physics is encoded in distribution amplitudes representing the probability amplitudes for finding quarks in a hadron with a given momentum distribution (see Fig. 1). Since neither the GPDs nor the distribution amplitudes can be calculated whithin QCD at present, it is difficult to decide which of the factorization schemes provides an appropriate description of, say, wide-angle Compton scattering at $-t \simeq 10$ GeV^2. The leading-twist factorization probably requires larger $-t$ than the handbag one since more details of the hadrons have to be resolved. Recent phenomenological and theoretical developments [6] support this conjecture. In the following I will discuss the handbag contribution only, assuming that the leading-twist one is negligibly small for momentum transfers of about 10 GeV^2. The ultimate decision whether or not this assumption is correct, is to be made by experiment.

It should be noted that immediately after the discovery of the partons in the late sixties constituent scattering models had been invented [7,8] which bear resemblance to the handbag contribution. As compared to these early attempts the handbag factorization has now a sound theoretical foundation. Particularly the invention of GPDs effectuated a decisive step towards a theoretical understanding of hard exclusive reactions.

2. The handbag in wide-angle Compton scattering

In 1998 Radyushkin [3] calculated the handbag contribution to Compton scattering starting from double distributions. Somewhat later Diehl, Feldmann, Jakob and myself calculated it on the basis of parton ideas [4]. Both approaches arrived at essentially the same results. Here, I will briefly describe our approach because I am more familiar with it. Our kinematical requirements are that the three Mandelstam variables s, $-t$, $-u$ are much larger than Λ^2 where Λ is a typical hadronic scale of order 1 GeV. The bubble in the handbag is viewed as a sum over all possible parton configurations as in deep ineleastic lepton-proton scattering (DIS). The contribution we calculate is defined by the requirement of restricted parton virtualities, $k_i^2 < \Lambda^2$, and intrinsic transverse parton momenta, $\mathbf{k}_{\perp i}$, which satisfy $k_{\perp i}^2/x_i < \Lambda^2$, where x_i is the momentum fraction parton i carries.

It is of advantage to work in a symmetrical frame which is a c.m.s rotated in such a way that the momenta of the incoming (p) and outgoing (p') proton momenta have the same light-cone plus components. In this frame the skewness, defined as

$$\xi = \frac{(p-p')^+}{(p+p')^+} \, , \tag{1}$$

is zero. One can then show that the subprocess Mandelstam variables $\hat{s}$ and $\hat{u}$ are the same as the ones for the full process, Compton scattering off protons, up to corrections of order Λ^2/t:

$$\hat{s} = (k_j + q)^2 \simeq (p+q)^2 = s \, , \quad \hat{u} = (k_j - q')^2 \simeq (p - q')^2 = u \, . \tag{2}$$

The active partons, i.e. the ones to which the photons couple, are approximately on-shell, move collinear with their parent hadrons and carry a momentum fraction close to unity, $x_j, x_j' \simeq 1$. Thus, like in DVCS, the physical situation is that of a hard parton-level subprocess, $\gamma q \to \gamma q$, and a soft emission and reabsorption of quarks from the proton. The helicity amplitudes for WACS then read

$$M_{\mu'+,\,\mu+}(s,t) = 2\pi\alpha_{\text{elm}} \left[T_{\mu'+,\,\mu+}(s,t) \left(R_V(t) + R_A(t) \right) \right.$$

$$\left. + T_{\mu'-,\,\mu-}(s,t) \left(R_V(t) - R_A(t) \right) \right] , \tag{3}$$

$$M_{\mu'-,\,\mu+}(s,t) = -\pi\alpha_{\text{elm}} \frac{\sqrt{-t}}{m} \left[T_{\mu'+,\,\mu+}(s,t) + T_{\mu'-,\,\mu-}(s,t) \right] R_T(t) \, .$$

μ, μ' denote the helicities of the incoming and outgoing photons, respectively. The helicities of the protons in M and quarks in the hard scattering amplitude T are labeled by their signs. The hard scattering has been calculated to next-to-leading order perturbative QCD [9], see Fig. 2. To this order the gluonic subprocess, $\gamma g \to \gamma g$ has to be taken into account as well. The form factors R_i represent $1/x$-moments of GPDs at zero skewness. R_T controls the proton helicity flip amplitude while the combination $R_V + R_A$ is the response of the proton to the emission and reabsorption of quarks with the same helicity as it and $R_V - R_A$ that one for opposite helicities. The identification of the form factors with $1/x$-moments of GPDs is possible because the plus components of the proton matrix elements dominate as in DIS and DVCS. This is non-trivial feature given that, in contrast to DIS and DVCS, not only the plus components of the proton momenta but also their minus and transverse components are large here. A more technical

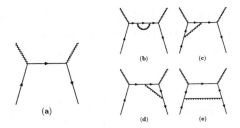

Figure 2.   Sample Feynman graphs for $\gamma q \to \gamma q$ to NLO in perturbative QCD.

aspect is the fact that the handbag approach naturally demands the use of light-cone techniques. Thus, (3) is a light-cone helicity amplitude. To facilitate comparison with experiment one may transform the amplitudes (3) to the ordinary c.m.s. helicity basis [9,10].

3. Modeling the GPDs

The structure of the handbag amplitude, namely its representation as a product of perturbatively calculable hard scattering amplitudes and t-dependent form factors

$$M(s,t) \sim T(s,t)\, R(t) \qquad (4)$$

is the essential result. Refuting the handbag approach necessitates experimental evidence against the structure (4). In oder to make actual predictions for Compton scattering however models for the soft form factors or rather for the underlying GPDs are required. A first attempt to parameterize the GPDs H and $\tilde{H}$ at zero skewness reads [3,4,9,11] (see also [12,13])

$$H^a(\bar{x},0;t) = \exp\left[a^2 t \frac{1-\bar{x}}{2\bar{x}}\right] q_a(\bar{x}),$$

$$\tilde{H}^a(\bar{x},0;t) = \exp\left[a^2 t \frac{1-\bar{x}}{2\bar{x}}\right] \Delta q_a(\bar{x}), \qquad (5)$$

where $q(x)$ and $\Delta q(x)$ are the usual unpolarized and polarized parton distributions in the proton. a, the transverse size of the proton, is the only free parameter and even it is restricted to the range of about 0.8 to 1.2 GeV^{-1} for a realistic proton. Note that a mainly refers to the lowest Fock states of the proton which, as phenomenological experience tells us, are rather compact. The model (5) is designed for large $-t$. Hence, forced by the Gaussian in (5), large x is implied, too. Despite of this the normalization of the model GPDs at $t = 0$ is correct.

The model (5) can be motivated by overlaps of light-cone wave functions. As has been shown [4,14,15] GPDs possess a representation in terms of such overlaps. Assuming a Gaussian $k_\perp$ dependence for the N-particle Fock state wave function

$$\Psi_N = \Phi_N(x_1, \cdots x_N) \exp\left[-a_N^2 \sum_{i=1}^{N} k_{\perp i}^2/x_i\right], \qquad (6)$$

which is in line with the central assumption of the handbag approach of restricted $k_{\perp i}^2/x_i$, necessary to achieve factorization of the amplitudes into soft and hard parts, and assuming further $a_N = a$ for all N in order to simplify matters, each overlap provides the Gaussian appearing in (5). The remainder of the overlaps summed over all N is just the Fock state representation of the parton distribution [5]. Thus, there is no need to specify the full x dependence of the light-cone wave function in order to arrive at (5). Note that Φ_N may depend on quark masses.

The simple model (5) may be improved in various ways. For instance, one may treat the lowest Fock states explicitly [4], take into account the evolution of the GPDs [16] or improve the parameterization in such a way that it also holds for small x [17]. One may also consider wave function with a power-law dependence on $\mathbf{k}_\perp$ instead of the Gaussian in (6) [18].

From the GPDs one can calculate the various form factors by taking appropriate moments, e.g.

$$F_1 = \sum_q e_q \int_{-1}^{1} d\bar{x} H^q(\bar{x}, 0; t), \qquad R_V = \sum_q e_q^2 \int_{-1}^{1} \frac{d\bar{x}}{\bar{x}} H^q(\bar{x}, 0; t). \qquad (7)$$

Results for the form factors are shown in Fig. 3. Obviously, as the comparison with experiment [19] reveals the model GPDs work quite well in the case of the Dirac form factor [4]. The scaled form factors $t^2 F_1$ and $t^2 R_i$ exhibit broad maxima which mimick dimensional counting in a range of $-t$ from, say, 3 to about 20 GeV^2. For very large values of $-t$, well above 100 GeV^2, the form factors turn gradually into a $\propto 1/t^4$ behaviour; this is the region where the leading-twist contribution takes the lead. The position of the maximum of a scaled form factor is approximately located at

$$t_0 \simeq -4a^{-2} \left\langle \frac{1-x}{x} \right\rangle_{F(R)}^{-1}. \qquad (8)$$

The mildly t-dependent mean value $\langle(1-x)/x\rangle$ has a value of about $1/2$.

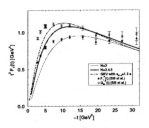 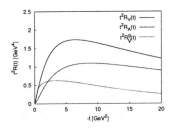

Figure 3. Predictions for the Dirac form factor of the proton (left) and for the Compton form factors (right)[4]. Data are taken from Ref. [19].

The Pauli form factor F_2 and its Compton analogue R_T contribute to proton helicity flip matrix elements and are related to the GPD E

$$F_2 = \sum_q e_q \int_{-1}^{1} d\bar{x}\, E^q(\bar{x}, 0; t)\,, \quad R_T = \sum_q e_q^2 \int_{-1}^{1} \frac{d\bar{x}}{\bar{x}} E^q(\bar{x}, 0; t)\,. \quad (9)$$

The overlap representation of E [14] involves components of the proton wave functions where the parton helicities do not sum up to the helicity of the proton. In other words, parton configurations with non-zero orbital angular momentum contribute to it. A simple ansatz for a proton valence Fock state wave function that involves orbital angular momentum is

$$\Psi_3^- \sim \sum \frac{\mathbf{k}_{\perp i}}{\sqrt{x_i}} \exp\left[-a_-^2 \sum k_{\perp i}^2 / x_i\right]\,. \quad (10)$$

Evaluating the overlap contributions to F_2 and R_T from this wave function and from (6), one finds

$$R_T/R_V\,, \quad F_2/F_1 \propto m/\sqrt{-t} \quad (11)$$

rather than $\propto m^2/t$. (11) is in agreement with the recent JLab measurement [20] while the SLAC data [21] are rather compatible with a $\propto m^2/t$ behaviour. The new experimental results on F_2/F_1 have been discussed in the same spirit as here in Ref. [22]. Clearly, more phenomenological work on E, F_2 and R_T is needed.

For an estimate of the size of R_T one may simply assume that R_T/R_V roughly behaves as its electromagnetic counter part F_2/F_1. Hence,

$$\kappa_T = \frac{\sqrt{-t}}{2m} \frac{R_T}{R_V} \simeq \frac{\sqrt{-t}}{2m} \frac{F_2}{F_1} \quad (12)$$

has a value of 0.37 [20].

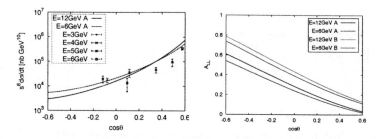

Figure 4. Predictions for the Compton cross section (left) and for the helicity correlation A_{LL} (right). NLO corrections and the tensor form factor are taken into account (scenario A) [9], in scenario B they are neglected. Data are taken from Ref. [23].

4. Results for Compton scattering

I am now ready to discuss results for Compton scattering. The cross section reads

$$\frac{d\sigma}{dt} = \frac{d\hat{\sigma}}{dt} \left\{ \frac{1}{2}[R_V^2(t)(1 + \kappa_T^2) + R_A^2(t)] \right.$$

$$\left. - \frac{us}{s^2 + u^2}[R_V^2(t)(1 + \kappa_T^2) - R_A^2(t)] \right\} + O(\alpha_s), \qquad (13)$$

where $d\hat{\sigma}/dt$ is the Klein-Nishina cross section for Compton scattering of point-like spin-1/2 particles. This cross section is multiplied by a factor that describes the structure of the proton in terms of the three form factors. The predictions from the handbag are in fair agreement with experiment [23], see Fig. 4. The approximative s^6-scaling behaviour is related to the broad maximum at about 8 GeV^2 the form factors exhibit, see (8). Clearly, more accurate date are needed for a detailed comparison. The JLab will provide such data soon.

Another interesting observable for Compton scattering is the helicity correlation, A_{LL}, between the initial state photon and proton or, equivalently, the helicity transfer, K_{LL}, from the incoming photon to the outgoing proton. From the handbag approach one obtains [9,24]

$$A_{LL} = K_{LL} \simeq \hat{A}_{LL} \frac{R_A}{R_V} + O(\kappa_T, \alpha_s, \beta), \qquad (14)$$

where $\hat{A}_{LL}$ is the corresponding observable for $\gamma q \to \gamma q$

$$\hat{A}_{LL} = \frac{s^2 - u^2}{s^2 + u^2}. \qquad (15)$$

The subprocess observable is diluted by the ratio of the form factors R_A and R_V as well as by other corrections but its shape essentially remains unchanged. The predictions for A_{LL} from the leading-twist approach drastically differ from the ones shown in Fig. 4. For $\theta \lesssim 110°$ negative values for A_{LL} are obtained for all but one examples of distribution amplitudes. The diquark model[26], a variant of the leading-twist approach, also leads to a negative value for A_{LL}. The JLab E99-114 collaboration [27] has presented a first measurement of A_{LL} at a c.m.s. scattering angle of 120° and a photon energy of 4.3 GeV. This still preliminary data point is in agreement with the prediction from the handbag, the leading-twist calculations fails badly. A measurement of the angular dependence of A_{LL} would be highly welcome for establishing the handbag approach [a]. For predictions of other polarization observables for Compton scattering I refer to Refs. [9,24].

5. Other applications of the handbag mechanism

The handbag approach has been applied to several other high-energy wide-angle reactions. Thus, as shown in Ref. [24], the calculation of real Compton scattering can be straightforwardly extended to *virtual Compton scattering* provided $Q^2/-t \ll 1$. *Elastic hadron-hadron scattering* can be treated as well [24]. Details have not yet been worked out but it has been shown that form factors of the type discussed in Sect. 3 control elastic scattering, too. The experimentally observed scaling behaviour of these cross sections can be attributed to the broad maximum the scaled form factors show, see Fig. 3.

The time-like processes *two-photon annihilations into pairs of mesons or baryons* can also be calculated, the arguments for handbag factorization hold here as well as has recently been shown in Refs. [29,30], see also the talk by Weiss [31]. The cross section for the production of baryon pairs read

$$\frac{d\sigma}{dt}(\gamma\gamma \to B\overline{B}) = \frac{4\pi\alpha_{\mathrm{elm}}^2}{s^2 \sin^2\theta}\Big\{\big|R_A^B(s) + R_P^B(s)\big|^2 + \cos^2\theta \big|R_V^B(s)\big|^2 + \frac{s}{4m^2}\big|R_P^B(s)\big|^2\Big\}. \quad (16)$$

The form factors represent integrated $B\overline{B}$ distribution amplitudes $\Phi_{B\overline{B}\,i}$ which are time-like versions of GPDs at a time-like skewness of 1/2. They read $(i = V, A, P)$

$$R_i^B(s) = \sum_q e_q^2 F_i^{Bq}(s), \qquad F_i^{Bq}(s) = \int_0^1 dz\, \Phi_{B\overline{B}\,i}^q(z, \zeta = 1/2, s). \quad (17)$$

---

[a]Note, however, that, over a wide range of scattering angles, a Regge model leads to very similar predictions for A_{LL} as the handbag [28].

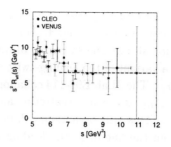

Figure 5. The scaled annihilation form factor $s^2|R_A^p|$ as extracted from the data of Refs. [32,33]. The dashed line represents a fit to the data above 6.5 GeV [2].

The form factors have not been modeled by us, they are extracted from the measured intergrated cross sections. The result for the effective form factor for $\gamma\gamma \to p\bar{p}$, being a combination of the dominant axial vector form factor and the pseudoscalar one, is shown in Fig. 5. The form factors behave similar to the magnetic one, $G_M(s)$, in the time-like region and have the same size as it within about a factor of two [34]. A characterisic feature of the handbag is the $q\bar{q}$ intermediate state implying the absence of isospin-two components in the final state. A consequence of this property is

$$\frac{d\sigma}{dt}(\gamma\gamma \to \pi^0\pi^0) = \frac{d\sigma}{dt}(\gamma\gamma \to \pi^+\pi^-)\,, \qquad (18)$$

which is independent of the soft physics input and is, in so far, a hard prediction of the handbag approach. The absence of the isospin-two components combined with flavor symmetry allows one to calculate the cross section for other $B\overline{B}$ channels using the form factors for $p\bar{p}$ as the only soft physics input. It is important to note that the leading-twist mechanism has difficulties to account for the size of the cross sections [35] while the diquark model [36] is in fair agreement with experiment for $\gamma\gamma \to B\overline{B}$.

Photo- and electroproduction of mesons have also been discussed within the handbag approach [11] using, as in deep virtual electroproduction [37], a leading-twist mechanism for the generation of the meson. It turns out, however, that the photoproduction cross section is way below experiment. The reason for this failure is not yet understood. Either the vector meson dominance contribution is still large or the leading-twist generation of the meson underestimates the handbag contribution. Despite of this the handbag contribution to photo-and electroproduction has several interesting properties which perhaps survive an improvement of the approach. For instance, the helicity correlation $\hat{A}_{LL}$ for the subprocess $\gamma q \to \pi q$ is the same as for $\gamma q \to \gamma q$, see (15). A_{LL} for the full process is diluted by form factors similar to the case of Compton scattering. Another result is the ratio of the production of π^+ and π^- which is approximately given by

$$\frac{d\sigma(\gamma n \to \pi^- p)}{d\sigma(\gamma p \to \pi^+ n)} \simeq \left[\frac{e_d u + e_u s}{e_u u + e_d s}\right]^2 . \tag{19}$$

6. Summary

I have reviewed the theoretical activties on applications of the handbag mechanism to wide-angle scattering. There are many interesting predictions still awaiting their experimental examination. At this workshop many new, mainly preliminary data for wide-angle scattering from JLab have been presented, more data will come soon. There are first hints that the handbag mechanism plays an important role. However, before we can draw firm conclusions we have to wait till the data have been finalized. For the kinematical situation available at JLab substantial corrections to the handbag contribution are to be expected. This may render a detailed quantitative comparison between theory and experiment difficult.

Acknowledgments. It is a pleasure to thank the organizers of this workshop Carl Carlson, Jean-Marc Laget, Anatoly Radyushkin, Paul Stoler and Bogdan Wojtsekhowski for inviting me to attend this interesting and well organized workshop at Jefferson Lab.

References

1. D. Müller, D. Robaschik, B. Geyer, F. M. Dittes and J. Hořejši, Fortsch. Phys. **42**, 101 (1994) [hep-ph/9812448].
2. A. V. Radyushkin, Phys. Rev. **D56** 5524 (1997) [hep-ph/9704207]; X. Ji, Phys. Rev. **D55** 7114 (1997) [hep-ph/9609381].
3. A. V. Radyushkin, Phys. Rev. D **58**, 114008 (1998) [hep-ph/9803316].
4. M. Diehl, T. Feldmann, R. Jakob and P. Kroll, Eur. Phys. J. C **8**, 409 (1999) [hep-ph/9811253].
5. G. P. Lepage and S. J. Brodsky, Phys. Rev. D **22**, 2157 (1980).
6. J. Bolz and P. Kroll, Z. Phys. A **356**, 327 (1996) [hep-ph/9603289]; V. M. Braun, A. Lenz, N. Mahnke and E. Stein, Phys. Rev. D **65**, 074011 (2002) [hep-ph/0112085]; D. Diakonov and V. Y. Petrov, hep-ph/0009006.
7. J.D. Bjorken and E.A. Paschos, Phys. Rev. **185**, 1975 (1969); D.M. Scott, Phys. Lett. **53B**, 185 (1974).
8. D. Sivers, S.J. Brodsky and R. Blankenbecler, Phys. Rep. **23 C**, 1 (1976) and references therein.
9. H. W. Huang, P. Kroll and T. Morii, Eur. Phys. J. C **23**, 301 (2002) [hep-ph/0110208].
10. M. Diehl, Eur. Phys. J. C **19**, 485 (2001) [hep-ph/0101335].

11. H. W. Huang and P. Kroll, Eur. Phys. J. C **17**, 423 (2000) [hep-ph/0005318].
12. A. V. Afanasev, hep-ph/9910565.
13. V. Barone *et al.*, Z. Phys. C **58**, 541 (1993).
14. M. Diehl, T. Feldmann, R. Jakob and P. Kroll, Nucl. Phys. B **596**, 33 (2001) [Erratum-ibid. B **605**, 647 (2001)] [hep-ph/0009255].
15. S. J. Brodsky, M. Diehl and D. S. Hwang, Nucl. Phys. B **596**, 99 (2001) [hep-ph/0009254].
16. C. Vogt, Phys. Rev. D **63**, 034013 (2001) [hep-ph/0007277].
17. M. Vanderhaeghen, these proceedings; K. Goeke, . V. Polyakov and M. Vanderhaeghen, Prog. Part. Nucl. Phys. **47**, 401 (2001) [hep-ph/0106012].
18. A. Mukherjee, I.V. Musatov, H.C. Pauli and A.V. Radyushkin, hep-ph/0205315.
19. A.F. Sill *et al.*, Phys. Rev. D**48**, 29 (1993); A. Lung *et al.*, Phys. Rev. Lett. **70**, 718 (1993).
20. O. Gayou *et al.* [Jefferson Lab Hall A Collaboration], Phys. Rev. Lett. **88**, 092301 (2002) [nucl-ex/0111010].
21. L. Andivahis *et al.*, *Phys. Rev.* D **50**, 5491 (1994).
22. J.P. Ralston, P. Jain and R. Buniy, these proceedings and in those of the Conf. of Intersections of Particle and Nuclear Physics, Quebec (2000).
23. M.A. Shupe *et al.*, Phys. Rev. **D19**, 1921 (1979).
24. M. Diehl, T. Feldmann, R. Jakob and P. Kroll, Phys. Lett. B **460**, 204 (1999) [hep-ph/9903268].
25. T. C. Brooks and L. Dixon, *Phys. Rev.* D **62**, 114021 (2000) [hep-ph/0004143].
26. P. Kroll, M. Schürmann and W. Schweiger, Intern. J. Mod. Phys. A **6**, 4107 (1991).
27. A. Nathan, for the E99-114 JLab collaboration, these proceedings
28. F. Cano and J. M. Laget, Phys. Rev. D **65**, 074022 (2002) [hep-ph/0111146].
29. M. Diehl, P. Kroll and C. Vogt, Phys. Lett. B **532**, 99 (2002) [hep-ph/0112274; C. Vogt, these proceedings.
30. M. Diehl, P. Kroll and C. Vogt, hep-ph/0206288.
31. C. Weiss, these proceedings; A. Freund, A. Radyushkin, A. Schäfer and C. Weiss, work in progress.
32. M. Artuso *et al.* [CLEO Collaboration], Phys. Rev. D **50**, 5484 (1994).
33. H. Hamasaki *et al.* [VENUS Collaboration], Phys. Lett. B **407**, 185 (1997).
34. T. A. Armstrong *et al.* [E760 Collaboration], Phys. Rev. Lett. **70**, 1212 (1993).
35. G. R. Farrar, E. Maina and F. Neri, Nucl. Phys. B **259**, 702 (1985) [Erratum-ibid. B **263**, 746 (1985)].
36. P. Kroll *et al.*, Phys. Lett. B **316**, 546 (1993) [hep-ph/9305251]; C.F. Berger, B. Lechner and W. Schweiger, Fizika B **8**, 371 (1999).
37. J.C. Collins, L. Frankfurt and M. Strikman, Phys. Rev. **D56**, 2982 (1997) [hep-ph/9611433]; A. V. Radyushkin, Phys. Lett. B **385**, 333 (1996) hep-ph/9605431].

REAL COMPTON SCATTERING FROM THE PROTON

ALAN M. NATHAN

(FOR THE RCS COLLABORATION)

University of Illinois,
Department of Physics
1110 W. Green Street
Urbana, IL 61801, USA
E-mail: a-nathan@uiuc.edu

Real Compton Scattering on the proton in the hard scattering regime is discussed. Recent theoretical developments are reviewed. Initial results from JLab E99-114 are presented.

1. Introduction

Real Compton Scattering (RCS) in the hard scattering limit is a potentially powerful but seldom used probe of the short-distance structure of the nucleon. It is a natural complement to other hard exclusive reactions that are currently being pursued at JLab and elsewhere, such as high Q^2 elastic form factors, deeply virtual pion electroproduction, and deeply virtual Compton Scattering (DVCS). The common feature of these reactions is a hard energy scale, leading to the factorization of the scattering amplitude into a part involving a hard perturbative scattering amplitude, which describes the coupling of the external particles to the active quarks, and the overlap of soft nonperturbative wave functions, which describes the coupling of the active quarks to the proton. For RCS, the hard scale is achieved when the Mandelstam variables s, $-t$, and $-u$ are all large, or equivalently when $p_\perp$ is large, on the hadronic scale.

There has been a considerable theoretical effort in recent years in calculating RCS cross sections and polarization observables in the hard scattering limit, which I review in Section 2). Despite this renewed theoretical interest in RCS, the only Compton scattering data available in this kinematic regime are the 20-year old Cornell data,[1] which are sparse and of limited statistical precision in the theoretically interesting range of high $p_\perp$. In order to provide the high quality data necessary to discriminate among reaction mechanisms and gain new insight into the structure of the proton, a new experiment has recently been performed at Jefferson Laboratory (E99-114) to measure RCS cross sections over a broad range of s and t and to have an initial look at polarization observables. The experiment and some very preliminary results will be discussed in Section 3. A summary is presented in Section 4.

2. Theoretical Overview

Various theoretical approaches have been applied to RCS in the hard scattering regime, and these can be distinguished by the number of active quarks participating in the hard scattering subprocess, or equivalently, by the mechanism for sharing the transferred momentum among the constituents. Two extreme pictures have emerged. In the perturbative QCD (pQCD) approach,[2,3,4,5] there are three active quarks which share t by the exchange of two hard gluons. In the soft overlap approach,[6,7,8] the handbag diagram dominates in which there is one active quark and t is shared by the overlap of the high momentum components of the soft wave function. In any given kinematic regime, both mechanisms will contribute, in principle, to the cross section. It is generally believed that at sufficiently high energies, the pQCD mechanism dominates. However, the question of how high is "sufficiently high" is still an open question to be answered by more precise experiments, and it is not known with any certainty what is the dominant mechanism in the kinematic regime appropriate to JLab ($p_\perp \sim 1-2$ GeV).

I next examine the two reaction mechanisms in somewhat more detail. In the pQCD approach, the exchange of two hard gluons leads naturally to the constituent counting rule and scaling[9]

$$\frac{d\sigma}{dt} = \frac{f(\theta_{cm})}{s^n}, \tag{1}$$

where n=6 for RCS. The soft physics enters through the valence quark distribution amplitude (DA), with higher Fock states suppressed by factors of $1/s$. Experimentally, the existing data support scaling with $n \approx 6$,[1] albeit with modest statistical precision. Nevertheless, it has been shown [2,4,5] that the cross sections calculated using the asympotic DA badly underpredict the data. This has led to the suggestion[6,7] that the dominant mechanism at experimentally accessible energies is the soft overlap mechanism.

In the soft overlap approach to RCS, the hard physics is contained in the scattering from a single active quark, which is calculable in pQCD, whereas the soft physics is contained in the wave function describing how the active quark couples to the proton. This coupling is described in terms of Generalized Parton Distributions (GPD's)[10,11] which are superstructures of the nucleon from which are derived the normal parton distribution functions (PDF's), elastic form factors, and other quantities that have yet to be measured, including new form factors accessible through Compton scattering. The GPDs provide links among diverse physical processes, including both inclusive and exclusive reactions. For example, the dominant GPD in RCS is $\mathcal{F}^a(x,t)$, which is related to both the Dirac form factor F_1 and the RCS vector form factor R_V through the expressions

$$F_1(t) = \Sigma_a e_a \int_0^1 \mathcal{F}^a(x;t)dx \qquad R_V(t) = \Sigma_a e_a^2 \int_0^1 \mathcal{F}^a(x;t)\frac{dx}{x}, \qquad (2)$$

whereas the $t = 0$ limit of $\mathcal{F}^a(x,t)$ is the PDF $q_a(x)$. Additional links are shown in Table 1. Despite the similarity between the (e,e) and RCS form factors (Eq. 2), an important distinction is the weighting by the quark charge which is linear for (e,e) and quadratic for RCS, reflecting the one- and two-photon nature of the interaction, respectively.

Table 1. GPD's and their associated electroweak scattering (EWS) and RCS form factors, where a labels the quark flavor. The last column shows the relationship with the PDF's.

GPD	EWS Form Factor	RCS Form Factor	$t = 0$ limit
$\mathcal{F}^a(x;t)$	$F_1(t)$	$R_V(t)$	$q^a(x)$
$\mathcal{K}^a(x;t)$	$F_2(t)$	$R_T(t)$	$2J^a(x)/x - q^a(x)$
$\mathcal{G}^a(x;t)$	$G_A(t)$	$R_A(t)$	$\Delta q^a(x)$

The relationship between the RCS cross section and form factors was initially worked out with several simplfying approximations, including the neglect of terms such as R_T which involve hadron helicity flip. This leads to the factorization of the RCS cross sections into a simple product of the Klein-Nishina (KN) cross section describing the hard scattering from a single quark and a sum of form factors depending only on t [6,7]:

$$\frac{d\sigma}{d\sigma_{\mathrm{KN}}} = f_V R_V^2(t) + (1 - f_V)R_A^2(t), \qquad (3)$$

For the the interesting region of large $p_\perp$, the kinematic factor f_V is always close to 1. Consequently the unpolarized cross sections are largely insensitive to R_A, and the left-hand-side of Eq. 3 is nearly s-independent at fixed t. Since no such simple scaling behavior is predicted by the pQCD mechanism, measurements of $d\sigma/d\sigma_{\mathrm{KN}}$ allow a powerful test of the reaction mechanism as well as a determination of R_V. Recent calculations of the soft overlap approach beyond the leading order, taking into account both photon and proton helicity-flip amplitudes, do not change this prediction in any appreciable way.[8]

Polarization observables provides further tests of the reaction mechanism as well as access to additional form factors. The longitudinal and sideways polarization transfer observables, K_{LL} and K_{LS}, respectively, are defined by

$$K_{LL}\frac{d\sigma}{dt} \equiv \frac{d\sigma(\uparrow\downarrow)}{dt} - \frac{d\sigma(\uparrow\downarrow)}{dt} \qquad K_{LS}\frac{d\sigma}{dt} \equiv \frac{d\sigma(\uparrow\rightarrow)}{dt} - \frac{d\sigma(\downarrow\rightarrow)}{dt} \qquad (4)$$

where the first arrow refers to the incident photon helicity and the second to the recoil proton helicity ($\uparrow$) or sideways polarization ($\rightarrow$). For the handbag mechanism, these are related to the form factors by the expressions [7,8]

$$K_{LL} \approx K_{LL}^{KN}\frac{R_A(t)}{R_V(t)} \qquad K_{LS} \approx -K_{LL}^{KN}\frac{\sqrt{-t}}{2M}\frac{R_A(t)R_T(t)}{R_V^2(t)}, \qquad (5)$$

where K_{LL}^{KN} is the longitudinal asymmetry for a structureless Dirac particle. The predictions for K_{LL} are plotted in Fig. 1, where one sees that the handbag calculation essentially follows the point result, whereas the pQCD prediction looks very different, thereby providing another stringent test of the reaction mechanism in addition to a measurement of the axial R_A form factor. Of particular interest is K_{LS}/K_{LL} which is proportional to R_T/R_V, which Kroll has suggested is approximately equal to F_2/F_1.[8] Recent JLab experiments[12] have shown that the latter quantity falls as $1/\sqrt{-t}$ rather than as $1/t$, as predicted by pQCD.

3. A New Experiment: JLab E99-114

Experiment E99-114, which was recently run in Hall A at JLab, measured differential cross sections for Compton scattering from the proton at incident photon energies between 3 and 6 GeV (s=6-12 GeV^2) and over a wide range of CM scattering angles ($-t$=2-7 GeV^2). The goal was to achieve a statistical precision of order 5%, with systematic errors less than 6%. In addition a measurement of K_{LL} and K_{LS} was measured at the kinematic point $s = 7$, $-t = 4$ GeV^2. The measurements utilize the technique shown schematically in Fig. 2.

A high duty factor electron beam with current ≥ 10 μA is incident on a 6% copper radiator located just upstream of the scattering target. The mixed beam of electrons and bremsstrahlung photons is incident on a 15-cm LH$_2$ target. For incident photons near the bremsstrahlung endpoint, the scattered photon is detected in a calorimeter consisting of an array of 704 blocks of lead-glass, while the associated recoil proton is detected in one

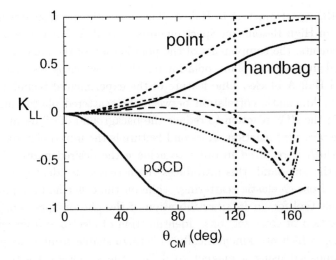

Figure 1. Theoretical predictions for the longitudinal polarization transfer observable for RCS for $s = 7$ GeV^2. The vertical dashed line is the angle at which E99-114 was done.

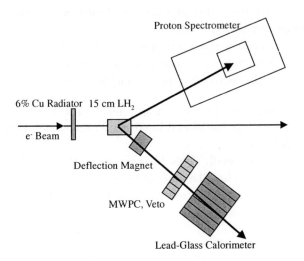

Figure 2. Schematic layout of JLab E99-114.

of the HRS spectrometers in Hall A. The magnetic spectrometer is one of the pair of High Resolution Spectrometers in Hall A. For the polarization measurements, the components of the polarization of the recoil proton are measured in the focal plane polarimeter (FPP) which is also part of the standard Hall A blocks. One feature of the experimental technique is the use of the kinematic correlation between the scattered photon and recoil proton in the RCS reaction to reduce the background of π^0 decay photons from the $p(\gamma, \pi^0 p)$ reaction. A second feature is the mixed electron-photon beam, which is required in order to achieve the desired photon luminosity. On the one hand, this introduces the necessity to identify and reject electrons from ep elastic scattering, while on the other hand it provides a convenient tool for an *in situ* calibration of the photon spectrometer and normalization of cross sections. Identification of electrons from ep elastic scattering, which are kinematically indistinguishable from RCS photons, is accomplished using a magnet to deflect the ep elastic electrons by a sufficient amount on the front face of the calorimeter to allow identification by the altered kinematic correlation with the recoil proton relative to undeflected RCS photons. In addition, electrons are identified in a plexiglass Čerenkov veto detector that is segmented to allow for a veto that is spatially correlated with an event in the calorimeter. The data in Fig. 3 demonstrate that these techniques allow us to distinguish the RCS events from the competing backgrounds.

As of this writing, there is a preliminary analysis of only the polarization transfer data. Although these results are subject to change after a more detailed analysis, the following general conclusions have been reached:

- K_{LL} is large and positive, in agreement with the overall prediction of the handbag model and in strong disagreement with the pQCD calculations.
- K_{LS}/K_{LL} is small. This is expected in the pQCD calculations, whereas in the handbag model the result suggests that $R_T \ll R_V$.
- K_{LL} for the $p(\gamma, \pi^0 p)$ reaction is comparable to that of the RCS reaction.

It is expected that the analysis of both the polarization results and the unpolarized cross sections will be largely completed in about one year.

4. Summary

In this contribution, I have summarized our present theoretical understanding of the RCS process in the hard scattering regime, discussed the setup for

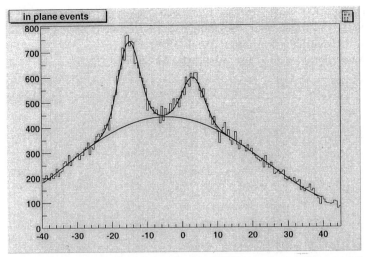

Figure 3. In-plane angular correlation plot from the RCS experiment. The events in the peak near x=0 are the RCS events. The peak at negative x is due to deflected electrons from elastic *ep* scattering. The continuum background is due to $p(\gamma, \pi^0 p)$ events

the recent experiment E99-114, and summarized some of the preliminary results.

Acknowledgments

It is a pleasure to acknowledge the RCS collaboration for their significant efforts in preparing and running the experiment and in doing a preliminary analysis of the data. This work was supported in part by the National Science Foundation under Grant No. PHY-00-72044.

References

1. M.A. Shupe *et al.*, Phys. Rev. **D19** (1979) 1929.
2. A. S. Kronfeld and B. Nizic, Phys. Rev. **D44** (1991) 3445.
3. G. R. Farrar and H. Zhang, Phys. Rev. Lett. **65** (1990) 1721;
   Glennys R. Farrar and H. Zhang, Phys. Rev. **D41** (1990) 3348; and
   E. Maina and G. R. Farrar, Phys. Lett. **B206** (1988) 120.
4. M. Vanderhaeghen, *et al.*, Nucl. Phys. **A622** (1997) 144c.
5. T. C. Brooks and L. Dixon, Phys. Rev. D **62** (2000) 114021.
6. A. Radyushkin, Phys. Rev. D **58** (1998) 114008.
7. M. Diehl, Th. Feldman, R. Jakob, and P. Kroll, hep-ph/9811253, hep-ph/9903268.
8. H. W. Huang, P. Kroll, and T. Morii, Eur. Phys. J. C **23** (2002) 301; P. Kroll, these proceedings.

9. S.J. Brodsky and G. Farrar, Phys. Rev. Lett. **31** (1973) 1953.
10. X. Ji, Phys. Rev. Lett. **78** (1997) 610; Phys. Rev. D **55** (1997) 7114.
11. A. Radyushkin, Phys. Rev. D **56** (1997) 5524.
12. M. K. Jones, *et al.*, Phys. Rev. Lett. **84** (2000) 1398.

RESONANCE EXCHANGE CONTRIBUTIONS TO WIDE-ANGLE COMPTON SCATTERING: THE D–TERM

T. OPPERMANN

Institute of Theoretical Physics,
University of Regensburg
D-93040 Regensburg, Germany
E-mail: tim.oppermann@physik.uni-regensburg.de

We present results for the differential cross section of Wide Angle Compton Scattering (WACS) at moderately large momentum transfer, including resonance–exchange type contribution parametrized by the the so-called D-term. We derive the tensor structure of these contributions and model their dependence on the momentum transfer. We find that the resonance exchange contribtions are of the same order of magnitude as the usual wave function overlap contributions and improve the agreement with the data.

1. Introduction

Compton scattering off nucleons provides us with valuable information on the nucleon structure. While in general it is very difficult to describe the response of a relativistic bound state to electromagnetic scattering, there exist certain special kinematic situations, where the description of the Compton process becomes simpler. One example is wide angle Compton scattering of real photons off protons (large momentum transfer). In the case of asymptotically large momentum transfer, this process is dominated by the lowest Fock state of the proton, and the hard scattering off partons can be described perturbatively. The amplitude is described by the convolution of nonperturbative distribution amplitudes and a hard scattering kernel.[1] Among others, a recent analysis[2] suggests that at moderately large momentum transfer the hard mechnism does not describe the data from the Cornell experiment.[3] In this region WACS seems to be dominated by the handbag contributions (soft mechanism),[4,5] in which the Compton form factors are expressed in terms of generalized parton distributions (GPD's). These GPD's are modeled from the overlap of soft wave functions.[4,5] Investigations of the structure of GPD's in connection with deeply-virtual

Compton scattering have shown the existence of contributions to GPD's from resonance exchange $(D\text{-term})$,[6] which cannot be included in the overlap picture. The D-Term affects e.g. the spin asymmetry of the DVCS cross section measured recently by the HERMES experiment.[7] We calculate the contributions of the D-term to the WACS amplitude and compare our results with the previous calculations in the soft mechanism and the experimental data.

2. Soft Mechanism and Generalized Parton Distributions

In the soft mechanism, the photon scatters off a single quark, which is emitted and absorbed by the nucleon through soft interactions.[8] The proton matrix elements describing these soft interactions are parametrized through GPD's,[9,10] which can be visualized as the sum over all Fock states that contain one active parton. There exist different representations of GPD's, namely the skewed parton distributions[9] and the double distributions.[10] In our approach we incorporate the double distributions, which are a spectral representation with respect to the forward momentum and the momentum transfer. Resonance exchange effects are included through the so-called D-term, which is required by the polynomiality condition of the GPD's, a direct consequence of the Lorentz invariance of the proton matrix elements[9].

3. Resonance Exchange in the soft mechanism: The D–term

The contribution from the D-term is distinct from the other contributions. It can be visualized as a resonance exchange of even-spin resonances and only depends on the momentum transfer. The contribution of the D-term to the WACS tensor amplitude is

$$T_{\mu\nu}^D = \frac{4(r^2 g_{\mu\nu} - 2r_\mu r_\nu)}{t m_p} R_D(t), \qquad (1)$$

where we have introduced a new WACS form factor $R_D(t)$ defined by

$$R_D(t) \equiv \left(\frac{1}{N_f} \sum_q e_q^2\right) \int d\alpha D(\alpha; t) \frac{\alpha}{\alpha^2 - 1}. \qquad (2)$$

Note that for the D-term one is not able to distinguish the contributions of the direct and crossed handbag diagrams. This is a specific feature of the WACS kinematics: The propagators of the diagrams are identical, unlike the propagators of the overlap contributions.

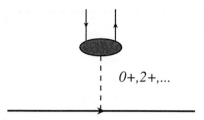

Figure 1. Visualization of the D-term as a resonance exchange

4. Modeling the t–dependence of the D–term

The t-dependence of the D-term at moderate $|t| \approx 1\text{GeV}^2$ is not well known at present. Contrary to the double distribution, it is not possible to model the t-dependence of this term from the overlap of soft nucleon wave functions. However, we can use the double distribution model of Ref.[4] as a guideline and write the t-dependent D-term as

$$D(\alpha; t) \equiv D(\alpha)S(\alpha; t) \qquad (3)$$

Here $D(\alpha)$ is the D-term at $t = 0$, for which we use the parametrization derived in Ref.[11] from model calculations within the chiral quark-soliton model. Furhtermore, $S(\alpha, t)$ is a cutoff function, satisfying $S(\alpha, t = 0) = 1$, which we model as

$$S(\alpha; t) = \exp\left[\frac{t}{(1 - \alpha^2)\lambda_D^2}\right] \qquad (4)$$

To fix the parameter λ we make use of the fact that the second moment of the D-term is related to the form factor of the nucleon matrix element of the energy momentum tensor.[6] Lacking data for this form factor, we assume here that it can be described by a simple 'monopole form'

$$F^\sigma(t) = \frac{1}{1 - t/m_\sigma^2} \qquad (5)$$

where for m_σ^2 we use the mass of the "scalar meson" introduced in meson-exchange parametrizations of the nucleon-nucleon interaction, $m_\sigma \approx 550$ MeV.[12] We then fix the parameter such that

$$F^\sigma(t) \approx \frac{\int d\alpha\, \alpha D(\alpha)S(\alpha; t)}{\int d\alpha\, \alpha D(\alpha)} \qquad (6)$$

The best agreement is found at $\lambda = 1.76$.

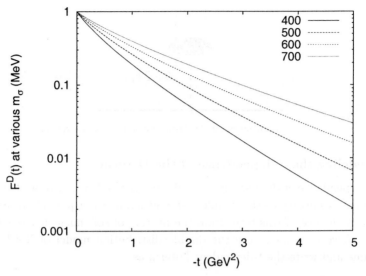

Figure 2. Effect of the parameter λ_D^2 on the D-term form factor $F^D(t)$. The form factor increases in value for growing λ_D. The legend shows the masses m_σ in MeV corresponding to λ_D^2

5. Results

We have computed the WACS differential cross section including the resonance exchange (D–term) contribution. For the overlap (double distribution) contribution we quote the expression given by Radyushkin.[4]

$$\frac{d\sigma}{dt} = \frac{2\pi\alpha^2}{\tilde{s}^2} \left(\frac{\tilde{s}+t}{\tilde{s}} + \frac{\tilde{s}}{\tilde{s}+t} \right) R_1^2(t)$$
$$-8\frac{\pi\alpha^2}{\tilde{s}^2} \left(\frac{t}{m_p^2} - 4 \right) R_D(t)^2 \tag{7}$$

From Eqn.(7) we expect the D-term contributions to be quite large: The momentum transfer dependence of the D-term wave function is similar to the dependence of the wave function used to model the previously known WACS form factors and the numerical prefactor of the D-term WACS form factor $R_D(t)$ is 4 times larger than the prefactor of the overlap. Looking at the numerical results Fig.(3), we see that the resonance exchange contributions can add contributions not only of a few percent, but of the same magnitude. In general, the agreement of the experimental data and our theoretical prediction is good in the region of small photon energies. For large photon energies, the overlap contribution alone is sufficient to describe the data, i.e. the contributions from resonance exchange is too large. While

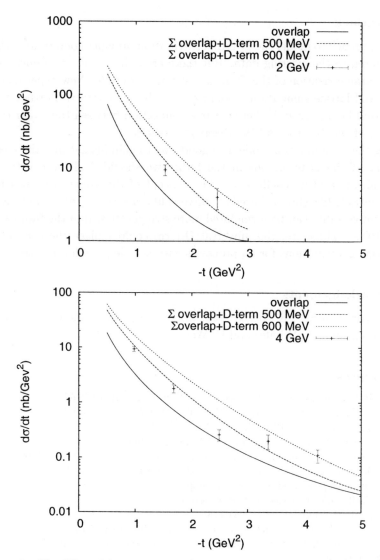

Figure 3.   The differential cross section $d\sigma/dt$, at incident photon energies of 2 GeV (top) and 4 GeV (bottom), as a function of t. Shown are the overlap (double distribution) contribution, and the sum of the overlap and the resonance exchange contributions at $m_\sigma = 500$ MeV and 600 MeV. The crosses denote the experimental data from the 1979 Cornell experiment.[3]

the Klein-Nishina formula enhances the region of large momentum transfer, the resonance exchange contributions drop exponentially.

6. Outlook

Our approach to model the D-term contribution provides good results when compared to the existing data. But there are still various open questions: The parametrization of the D-term is still rather crude. New experimental data and lattice simulations should reduce the uncertainty. Furthermore, it should be analyzed, if there exists a connection between the scalar resonances of the D-term and the Regge resonances.

Using the σ-meson as a model is a good way to gain insight into the mechanisms of the D-term, but in the future we should deal more with the normalization of its coefficients. The concept of the σ-meson as an effective, particle-like state is based on universality of certain scalar amplitudes. This universality can be formulated more stringently within the framework of GPD's. Therefore, the universal D-term could replace the concept of σ-meson contributions for all processes with an adequate hard scale.

Acknowledgements

This contribution is based on work done in collaboration with A. Radyushkin, A. Schäfer and C. Weiss.

References

1. G. P. Lepage and S. J. Brodsky, Phys. Rev. D **22** (1980) 2157.
2. T. C. Brooks and L. J. Dixon, Phys. Rev. D **62** (2000) 114021
3. M. A. Shupe et al., Phys. Rev. D **19** (1979) 1921.
4. A. V. Radyushkin, Phys. Rev. D **58** (1998) 114008
5. M. Diehl, T. Feldmann, R. Jakob and P. Kroll, Eur. Phys. J. C **8** (1999) 409
6. M. V. Polyakov and C. Weiss, Phys. Rev. D **60** (1999) 114017
7. J. Ely, talk presented at this workshop
8. P. Kroll, talk presented at this workshop
9. X. D. Ji, J. Phys. G **24** (1998) 1181
10. A. V. Radyushkin, Nucl. Phys. Proc. Suppl. **90** (2000) 113.
11. K. Goeke, M. V. Polyakov and M. Vanderhaeghen, Prog. Part. Nucl. Phys. **47** (2001) 401
12. R. Machleidt, K. Holinde and C. Elster, Phys. Rept. **149** (1987) 1.

THE HANDBAG CONTRIBUTION TO TWO-PHOTON ANNIHILATION INTO MESON PAIRS

C. VOGT

Nordita
Blegdamsvej 17
2100 Copenhagen, Denmark
Email: cvogt@nordita.dk

We report on the handbag contribution to two-photon annihilation into pion and kaon pairs at large energy and momentum transfer. The underlying physics of the mechanism is outlined and characteristic features and predictions are presented.

1. Introduction

Meson pair production in the collision of two real photons at asymptotically large energies can be described in the hard scattering approach [1], where to leading-twist the transition amplitude factorizes into a perturbatively calculable $\gamma\gamma \to q\bar{q}\,q\bar{q}$ subprocess and single-meson distribution amplitudes for the hadronization of each of the $q\bar{q}$ pairs. The perturbative contribution to the cross section for $\gamma\gamma \to \pi^+\pi^-$, however, turns out to be well below the experimental data if single-pion distribution amplitudes consistent with other data are employed [2].

In the following, we discuss a complementary approach [3] for large values of s, t, u, where the process amplitude factorizes into a hard subprocess for the production of a single $q\bar{q}$ pair, and a subsequent soft transition $q\bar{q} \to \pi\pi$ (see Fig.1). The latter is described in terms of a new annihilation form factor which is given by a moment of the two-pion distribution amplitude [4,5]. This mechanism is analogous to the handbag contribution to wide-angle Compton scattering [6,7].

2. The Physics of the Handbag Mechanism

We consider the process $\gamma\gamma \to \pi^+\pi^-$ in the kinematical region where $s \sim -t \sim -u$. The condition for the transition $q\bar{q} \to \pi^+\pi^-$ to be soft is that there be no large invariants at the parton-hadron vertices, cf. Fig. 1b. In particular, all virtualities occurring at these vertices are to be of $O(\Lambda^2)$,

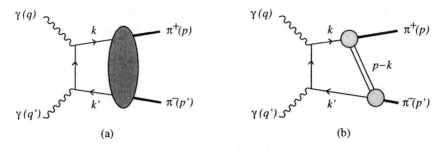

Figure 1. (a) Handbag factorization of the process $\gamma\gamma \to \pi\pi$ for large s and t. The hard scattering subprocess is shown at leading order in α_s, and the blob represents the two-pion distribution amplitude. The second contributing graph is obtained by interchanging the photon vertices. (b) The handbag resolved into two pion-parton vertices connected by soft partons. There is another diagram with the π^+ and π^- interchanged.

where Λ is a typical hadronic scale. Moreover, the momenta of the additional $q\bar{q}$ pair and possibly other partons are required to be soft. This implies that up to corrections of $O(\Lambda^2/s)$ the initial quark and antiquark approximately carry the momenta of their respective parent pions, and we have the condition $k \simeq p$ or $k' \simeq p$.

Note that configurations where the blob in Fig. 1a contains hard gluon exchange are part of the leading-twist contribution and not included in the the soft handbag amplitude. There are also diagrams of the cat's-ears topology, where the photons couple to different quark lines. At large s, t, u, these diagrams however require the presence of a large virtuality in one of the quark lines or hard gluon exchange.

In order to display the factorization it is advantageous to choose a symmetrical c.m. frame where the pions carry the same light-cone plus momentum, i.e., the skewness, defined by $\zeta = p^+/(p + p')^+$, has the value $1/2$. From the collinearity condition it follows that the initial quark and antiquark also have approximately equal light-cone plus momenta. Thus, we have $z = k^+/(p+p')^+ = 1/2 + O(\Lambda^2/s)$. Furthermore, corrections from partonic off-shell effects and partonic transverse momenta are of $O(\Lambda^2/s)$. Exploiting the on-shell and collinearity conditions one can show that the helicity amplitude for the process $\gamma\gamma \to \pi^+\pi^-$ can be written in the simple form

$$A_{\mu\mu'} = -4\pi\alpha_0\,\delta_{\mu,-\mu'}\,\frac{s^2}{tu}\,R_{2\pi}(s)\,, \tag{1}$$

where μ, μ' are the photon helicities and the soft part is encoded in the annihilation form factor defined by

$$R_{2\pi}(s) = \sum_q e_q^2 R_{2\pi}^q(s), \quad R_{2\pi}^q(s) = \frac{1}{2} \int_0^1 dz \, (2z - 1) \, \Phi_{2\pi}^q(z, 1/2, s). \quad (2)$$

Here the summation is over u, d, s quarks and $\Phi_{2\pi}^q(z, 1/2, s)$ is the two-pion distribution amplitude in light-cone gauge [5] at $\zeta = 1/2$. We remark that the operator corresponding to this form factor is the quark part of the energy-momentum tensor, and that the form factor is C even due to the weight $(2z - 1)$, as one could expect for a pion pair produced in two-photon annihilation. The differential cross section of the process is given by

$$\frac{d\sigma}{dt}(\gamma\gamma \to \pi^+\pi^-) = \frac{8\pi\alpha_0^2}{s^2} \frac{1}{\sin^4\theta} \left| R_{2\pi}(s) \right|^2. \quad (3)$$

3. Properties and Predictions

Considerations of isospin of the intermediate $q\bar{q}$ pair together with charge conjugation of the final state imply that charged and neutral pion pairs are only produced in isospin zero states. Hence, one can relate their corresponding form factors and finds the key result

$$\frac{d\sigma}{dt}(\gamma\gamma \to \pi^0\pi^0) = \frac{d\sigma}{dt}(\gamma\gamma \to \pi^+\pi^-), \quad (4)$$

which is a parameter-free prediction of the handbag approach and in striking contrast to the leading-twist approach, where the differential cross sections for $\pi^0\pi^0$ production is found about an order of magnitude smaller than that for $\pi^+\pi^-$ pairs. By using U-spin symmetry, i.e., the symmetry under the exchange $d \leftrightarrow s$, one can further relate the form factor for K^+K^- to that of $\pi^+\pi^-$ production, which leads to the relation

$$\frac{d\sigma}{dt}(\gamma\gamma \to K^+K^-) \simeq \frac{d\sigma}{dt}(\gamma\gamma \to \pi^+\pi^-). \quad (5)$$

The approximate symbol indicates that, in general, flavour symmetry breaking effects have to be expected. Note that (5) holds in any dynamical approach respecting SU(3) flavour symmetry. Isospin also provides a link between the form factors for charged and neutral kaon pairs, resulting in a further relation characteristic for the handbag mechanism:

$$\frac{d\sigma}{dt}(\gamma\gamma \to K^0\overline{K^0}) \simeq \frac{4}{25} \frac{d\sigma}{dt}(\gamma\gamma \to K^+K^-), \quad (6)$$

where we have neglected non-valence contributions and the numerical factor stems from the ratio of the corresponding charge factors.

The annihilation form factors and the two-pion distribution amplitude can as yet not be calculated within QCD. They also do not allow for an

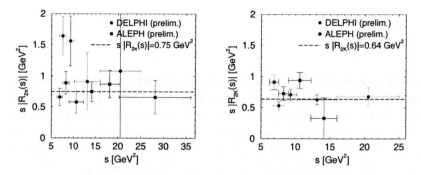

Figure 2. The scaled annihilation form factors $s|R_{2\pi}|$ (left) and $s|R_{2K}|$ (right) versus s. The preliminary ALEPH and DELPHI data is taken from [9,10]. Dashed lines represent our fitted values (7).

overlap representation of light-cone wave functions [8]. Therefore, we use the preliminary data of ALEPH [9] and DELPHI [10] on the new measurements of $\gamma\gamma \to \pi^+\pi^-$, K^+K^- in order to extract the form factors from experiment. We find that the scaling of the form factors is compatible with dimensional counting rule behaviour and a fit for $\sqrt{s} \gtrsim 2.5$ GeV provides the values (cf. Fig. 2)

$$s|R_{2\pi}(s)| = 0.75 \pm 0.07 \text{ GeV}^2 \quad \text{and} \quad s|R_{2K}(s)| = 0.64 \pm 0.04 \text{ GeV}^2 . \quad (7)$$

The pion annihilation form factor is comparable in magnitude with the timelike electromagnetic pion form factor. Note that although the handbag contribution formally provides a power correction to the leading-twist contribution, it appears to dominate at experimentally accessible energies.

Another characteristic result of our approach is the angular dependence (see Eq. (3)), which is in good agreement with the preliminary ALEPH data, as can be seen from Fig. 3.

Having determined the normalization of the pion and kaon annihilation form factors from experiment, we can also compare with the CLEO data for the integrated cross section [11], where pions and kaons have not been separated. The result in displayed in Fig. 4 and again we find good agreement.

As already mentioned in the introduction, the perturbative contribution is way below the experimental data [2]. In order to facilitate comparison of the handbag and the leading-twist approach, we make a rather conservative estimate of the latter and employ a fixed coupling $\alpha_s = 0.5$. We further use the asymptotic form for both the pion and kaon distribution amplitudes. The leading-twist prediction thus obtained amounts to about 15% of our

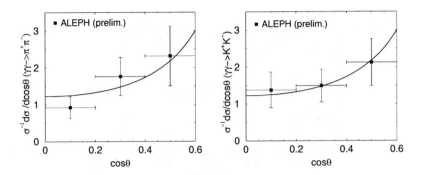

Figure 3.   The normalized angular distribution for $\gamma\gamma \to \pi^+\pi^-$ (left) and $\gamma\gamma \to K^+K^-$ (right), compared to the preliminary ALEPH data [9] for 4 GeV$^2 < s <$ 36 GeV^2.

fitted handbag result as shown in Fig. 4.

4. Conclusions

We have presented a brief discussion of the handbag contribution to $\gamma\gamma \to \pi\pi, KK$. In this approach the process amplitude factorizes into a hard $\gamma\gamma \to q\bar{q}$ subamplitude and an annihilation form factor for the soft transition to a meson pair. The form factor is a moment of the two-meson distribution amplitude at skewness $\zeta = 1/2$. In lack of a model for the form factors and the two-meson distribution amplitude in the kinematical range we are interested, we fit the form factors for $\pi^+\pi^-$ and K^+K^- to the data and we find that their scaling is compatible with dimensional counting rule behaviour. Note that in the spacelike case of wide-angle Compton scattering off protons the overlap representation in terms of light-cone wave functions, together with a plausible model for the latter, explicitly shows how the soft part of the Compton form factors can mimic counting rule behaviour.

Key results of the approach are the prediction that the differential cross sections for $\pi^+\pi^-$ and $\pi^0\pi^0$ production are the same, and the angular distribution $d\sigma/(d\cos\theta) \propto 1/\sin^4\theta$, which agrees well with data.

The handbag mechanism has recently also been applied the production of baryon-antibaryon pairs [12]. In Ref. [13] $p\bar{p}$ annihilation into photon pairs is investigated in an approach based on double distributions.

244

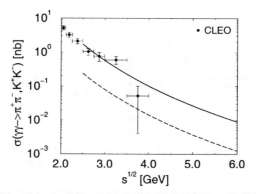

Figure 4. The CLEO data [11] for the cross section $\sigma(\gamma\gamma \to \pi^+\pi^-) + \sigma(\gamma\gamma \to K^+K^-)$ integrated with $|\cos\theta| < 0.6$. The solid line is the result of the handbag approach with our fitted annihilation form factors (7). The dashed line is the estimate of the leading-twist contribution described in the text.

Acknowledgments

This work is partially funded by the European Commission IHP program under contract HPRN-CT-2000-00130.

References

1. G. P. Lepage and S. J. Brodsky, Phys. Rev. D **22**, 2157 (1980).
2. C. Vogt, hep-ph/0010040.
3. M. Diehl, P. Kroll and C. Vogt, Phys. Lett. B **532**, 99 (2002).
4. D. Müller, D. Robaschik, B. Geyer, F. M. Dittes and J. Hořejši, Fortsch. Phys. **42**, 101 (1994).
5. M. Diehl, T. Gousset, B. Pire and O. Teryaev, Phys. Rev. Lett. **81**, 1782 (1998).
6. A. V. Radyushkin, Phys. Rev. D **58**, 114008 (1998).
7. M. Diehl, T. Feldmann, R. Jakob and P. Kroll, Eur. Phys. J. C **8**, 409 (1999).
8. M. Diehl, T. Feldmann, R. Jakob and P. Kroll, Nucl. Phys. B **596**, 33 (2001), Erratum-ibid. B **605**, 647 (2001);
   S. J. Brodsky, M. Diehl and D. S. Hwang, Nucl. Phys. B **596**, 99 (2001).
9. A. Finch [for the ALEPH collaboration], to appear in the Proceedings of the *International Conference on The Structure and Interactions of the Photon (PHOTON 2001)*, Ascona, Switzerland, September 2–7, 2001 (World Scientific, Singapore).
10. K. Grzelak [for the DELPHI collaboration], *ibid.*
11. J. Dominick *et al.* [CLEO Collaboration], Phys. Rev. D **50**, 3027 (1994).
12. M. Diehl, P. Kroll and C. Vogt, hep-ph/0206288.
13. C. Weiss, these proceedings;
    A. Freund, A. V. Radyushkin, A. Schäfer and C. Weiss, in preparation.

PROTON-ANTIPROTON ANNIHILATION INTO TWO PHOTONS AT LARGE S

C. WEISS

Institut für Theoretische Physik
Universität Regensburg
D-93053 Regensburg, Germany
E-mail: christian.weiss@physik.uni-regensburg.de

Exclusive proton–antiproton annihilation into two photons can be viewed as the Compton process in the crossed channel. At large s ($\approx 10\,\mathrm{GeV}^2$) and $|t|, |u| \sim s$ this process can be described by a generalized partonic picture, analogous to the "soft mechanism" in wide–angle real Compton scattering. The two photons are emitted in the annihilation of a single fast quark and antiquark ("handbag graph"). The transition of the $p\bar{p}$ system to a $\bar{q}q$ pair through soft interactions is described by double distributions, which can be related to the timelike proton elastic form factors as well as, by crossing symmetry, to the usual quark–antiquark distributions in the nucleon. We estimate that this reaction should be observable with reasonable statistics at the proposed $1.5 \ldots 15\,\mathrm{GeV}$ high–luminosity antiproton storage ring (HESR) at GSI. (Talk presented at the Workshop on Exclusive Processes at High Momentum Transfer, Jefferson Lab, Newport News, VA, May 15–18, 2002)

Compton scattering, *i.e.*, the scattering of real or virtual photons off hadrons, has proven to be a useful tool for investigating the structure of the nucleon. Of particular interest are certain extreme kinematical situations where the reaction mechanism simplifies. In the limit of large virtuality of the incoming photon and fixed t ("deeply virtual Compton scattering") QCD implies that the amplitude factorizes into a hard photon–quark amplitude and a generalized parton distribution, representing the information about the structure of the nucleon. Similarly, it has been argued that real Compton scattering at large s ($\approx 10\,\mathrm{GeV}^2$) and $|t|, |u| \sim s$ (wide–angle scattering) is dominated by contributions in which the photon scatters off a single quark/antiquark in the nucleon ("handbag graph").[1,2,3] The soft interactions responsible for the emission and absorption of the active quark/antiquark by the nucleon are parametrized by double distributions, which can be related to both the usual quark/antiquark distributions measured in inclusive deep–inelastic scattering and the elastic form factors of the proton. This so-called "soft mechanism" describes well the existing

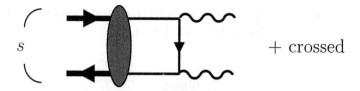

Figure 1.   The "handbag" contribution to $p\bar{p} \to \gamma\gamma$ annihilation

data for the total cross section,[4] and also recent results for the spin asymmetry from the JLAB Hall A experiment.[5] The hard scattering mechanism, in which the struck quarks rescatters via gluon exchange of virtuality $\sim t$, is relevant only at asymptotically large t and falls short of the measured cross section at JLAB energies.[1]

Exclusive proton–antiproton annihilation into two photons, $p\bar{p} \to \gamma\gamma$, can be regarded as the Compton process in the crossed channel. This reaction could be studied with the proposed high–luminosity $1.5 \dots 15 \,\mathrm{GeV}$ antiproton storage ring (HESR) at GSI.[6] In this talk I would like to argue that at large s and $|t|, |u| \sim s$ (wide–angle scattering) this process can be described by a generalized partonic picture analogous to the "soft mechanism" in wide–angle real Compton scattering. The results reported here have been obtained in collaboration with A. Freund (Regensburg U.), A. V. Radyushkin (Jefferson Lab and Old Dominion U.), and A. Schäfer (Regensburg U.). A detailed account will be published shortly.[7] Similar ideas have been presented by P. Kroll at this meeting.[8,9]

To motivate the partonic picture of proton–antiproton annihilation let us recall e^+e^- annihilation into two photons in QED. This process proceeds via the t (or u) channel exchange of a virtual electron/positron with spacelike momentum. In $p\bar{p} \to \gamma\gamma$ annihilation in QCD, the exchanged system consists of at least three quarks. At large momentum transfer such an exchange should be strongly suppressed by the proton and antiproton wave functions. In this situation the most efficient way of accommodating a large momentum transfer is the handbag diagram shown in Fig. 1. In the first part the proton–antiproton system makes a transition to a quark–antiquark pair by exchanging a virtual qq ("diquark") type system, denoted by the blob in the diagram, whose spacelike virtuality is limited by the bound–state wave functions. In the second part, the quark–antiquark pair annihilates into two photons by exchanging a highly virtual quark/antiquark, exactly as in e^+e^- annihilation in QED. This picture is consistent: Because of the limit on the virtuality of the exchanged diquark–type system the active quark (antiquark) carries a significant fraction of the proton (antiproton)

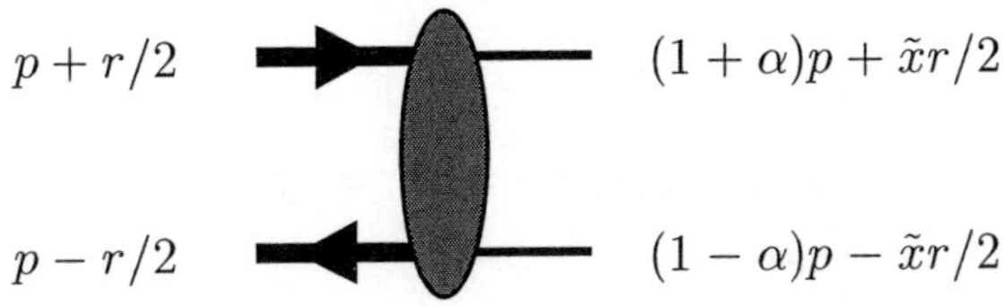

$p + r/2$ $(1+\alpha)p + \tilde{x}r/2$

$p - r/2$ $(1-\alpha)p - \tilde{x}r/2$

Figure 2. Graphical representation of the double distribution describing the transition of the $p\bar{p}$ system to a $q\bar{q}$ pair.

momentum, which in turn makes for a large virtuality in the quark propagator connecting the photon vertices. In short, our picture states that the dominant contribution to exclusive $p\bar{p} \to \gamma\gamma$ at large s, $|t|$ and u comes from the annihilation of "fast" quarks and antiquarks in the proton and antiproton, respectively.

The matrix element describing the transition of the $p\bar{p}$ system to a $q\bar{q}$ pair through "soft" interactions can be parametrized by double distributions, which are the timelike analogue of the functions parametrizing the matrix element in wide–angle Compton scattering. These distributions measure the momenta of the active quark and antiquark in terms of the proton and antiproton momenta, or, equivalently, their average and difference, p and r, see Fig. 2. Transverse momenta are neglected here, and the variables $\tilde{x}$ and α obey the "partonic" restrictions $|\tilde{x}| + |\alpha| < 1$. There are two types of matrix elements, corresponding to Dirac structures $(\gamma_\mu)_{ij}$ and $(\gamma_\mu\gamma_5)_{ij}$ in the quark spinor indices. Following Refs.[10], we parametrize them by two double distributions $F_a(\tilde{x}, \alpha; s)$ and $G_a(\tilde{x}, \alpha; s)$, depending on the partonic variables α and $\tilde{x}$ as well as on s (the subscript $a = u, d$ represents the quark flavor); details will be given elsewhere.[7] When integrating over the spectral variables and summing over the contributions of the different quark flavors, the new double distribution $F_a(\tilde{x}, \alpha; s)$ reduces to the timelike proton Dirac form factor,

$$\sum_a e_a^2 \iint_{|\tilde{x}|+|\alpha|<1} d\tilde{x}\, d\alpha\, F_a(\tilde{x}, \alpha; s) = F_1(s); \qquad (1)$$

a similar relation holds for $G_a(\tilde{x}, \alpha; s)$ and the axial nucleon form factor. (For simplicity we neglect here components of the matrix element corresponding to the Pauli and pseudoscalar form factors; these components should be included in a more complete treatment.) Another useful relation is found in the limit $s \to 0$, where one can use crossing symmetry to relate the annihilation–type double distribution of Fig. 2 to the usual scattering–type double distributions of wide–angle Compton scattering. In particular, this implies

$$\int_{-1+|\tilde{x}|}^{1-|\tilde{x}|} d\alpha \; F_a(\tilde{x}, \alpha; s = 0) \;\; = \;\; f_a(\tilde{x}), \tag{2}$$

where $f_a(\tilde{x}) = \theta(\tilde{x})q_a(\tilde{x}) - \theta(-\tilde{x})\bar{q}_a(-\tilde{x})$ is the usual unpolarized quark/antiquark distribution of flavor a in the proton, as measured in deep–inelastic scattering. In a similar way the function $G_a(\tilde{x}, \alpha; s)$ reduces to the polarized distribution. These "reduction relations" provide constraints for models of the double distributions.

To construct an explicit model for the double distributions we factorize them into an s–independent double distribution and a "cutoff function" containing the s–dependence,

$$F_a(\tilde{x}, \alpha; s) \;\; = \;\; f_a(\tilde{x}, \alpha) \; S(\tilde{x}, \alpha; s). \tag{3}$$

The s–independent double distribution we model as $f_a(\tilde{x}, \alpha) = f_a(\tilde{x})h(\tilde{x}, \alpha)$, where $f_a(\tilde{x})$ is the usual quark/antiquark distribution, and $h(\tilde{x}, \alpha)$ a normalized profile function; a particularly simple choice is $h(\tilde{x}, \alpha) = \delta(\alpha)$.[10] The distributions refer to a scale of the order $|t| \sim 1\,\text{GeV}^2$. The cutoff function $S(\tilde{x}, \alpha; s)$, which is defined to be unity at $s = 0$, implements the restriction on the virtuality of the exchanged "diquark"–type system in the handbag diagram of Fig. 1:

$$\frac{[(1 - \tilde{x})^2 - \alpha^2]s}{4\tilde{x}(1 - \tilde{x})} < \lambda^2, \tag{4}$$

where λ^2 is a parameter. For the scattering–type double distributions appearing in wide–angle Compton scattering this cutoff could be derived from the overlap of light–cone wave functions;[1,2] such an interpretation is no longer possible in the annihilation channel. In practice, we choose a Gaussian cutoff and fix the parameter λ^2 by fitting the proton form factor, cf. Eq.(1); for details we refer to the forthcoming publication.[7]

The helicity–averaged differential cross section obtained from our simple model is

$$\frac{d\sigma}{d\cos\theta} \;\; = \;\; \frac{2\pi\alpha_{\text{em}}^2}{s} \; \frac{R_V^2(s)\cos^2\theta + R_A^2(s)}{\sin^2\theta}, \tag{5}$$

where θ is the scattering angle in the center–of–mass system. The information about the structure of the proton is contained in generalized form factors, defined as

$$R_V(s) \;\; \equiv \;\; \sum_a e_a^2 \iint_{|\tilde{x}|+|\alpha|<1} d\tilde{x} \, d\alpha \; \frac{F_a(\tilde{x}, \alpha; s)}{\tilde{x}}; \tag{6}$$

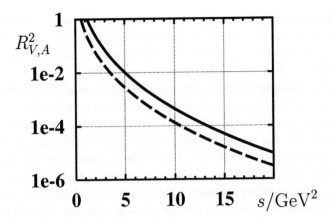

Figure 3. The form factors $R_V^2(s)$ (solid line) and $R_A^2(s)$ (dashed line) calculated from the double distribution model, *cf.* Eq.(6).

$R_A(s)$ is given by the corresponding integral over the double distribution $G_f(\tilde{x}, \alpha; s)$. The integral here differs from the one in the proton elastic form factor, Eq.(1), by an additional factor $1/\tilde{x}$ in the integrand; this is just the "remnant" of the quark propagator in the hard $q\bar{q} \to \gamma\gamma$ scattering amplitude in the handbag graph of Fig. 1. For $R_V(s) \equiv R_A(s) \equiv 1$ the expression (5) would reproduce the Klein–Nishina formula for the $e^+e^- \to \gamma\gamma$ cross section in QED. Note that our partonic picture is applicable only for $|t|, |u| \sim s$, which implies that θ should be sufficiently far from 0 or π (wide–angle scattering). The numerical results for the form factors are shown in Fig. 3.

It is interesting to make a quick estimate of the counting rate for $p\bar{p} \to \gamma\gamma$ annihilation expected for the proposed $1.5 \dots 15\,\text{GeV}$ antiproton storage ring (HESR) at GSI.[6] With a fixed solid target the luminosity could be as high as $L = 2 \times 10^{32}\,\text{cm}^{-2}\,\text{s}^{-1} = 2 \times 10^6\,\text{fm}^{-2}\,\text{s}^{-1}$. Since our partonic picture applies only for $|t|, |u| \sim s$ we have to exclude the region of θ close to 0 or π. For $s = 10\,\text{GeV}^2$, the range $45° < \theta < 135°$ would correspond to $|t|, |u| > 1.5\,\text{GeV}^2$, which is hopefully sufficient for our approximations to work. The integrated cross section over this region is $5.2 \times 10^{-10}\,\text{fm}^2$, which corresponds to a counting rate of $1.0 \times 10^{-3}\,\text{sec}^{-1}$, that is ~ 2800 events per month. Note that already slightly lower values of s would increase the counting rate dramatically, *cf.* Fig. 3. The process should thus be measurable with reasonable statistics at the proposed facility.

To summarize, we have discussed exclusive annihilation $p\bar{p} \to \gamma\gamma$ in a generalized partonic picture. Our approach represents an attempt to extend the "soft mechanism", which successfully describes wide–angle Compton scattering at JLAB energies, to the annihilation channel. The timelike double distributions describing the transition of the $p\bar{p}$ system to a $q\bar{q}$ pair are strongly constrained by data for the timelike proton form factor and the quark/antiquark distributions in the proton, which have been obtained from independent measurements. Our estimate of the cross section, based on a simple model for the double distributions, suggests that this reaction could be observed with reasonable statistics at the proposed GSI HESR facility. This would offer the exciting possibility of studying the hadronic Compton process in the annihilation channel. The results could also be compared to data on hadron production in photon–photon collisions observed in e^+e^- experiments.

C.W. is supported by a Heisenberg Fellowship from Deutsche Forschungsgemeinschaft (DFG).

References

1. A. V. Radyushkin, Phys. Rev. D **58**, 114008 (1998).
2. M. Diehl, T. Feldmann, R. Jakob and P. Kroll, Eur. Phys. J. C **8**, 409 (1999).
3. H. W. Huang, P. Kroll and T. Morii, Eur. Phys. J. C **23**, 301 (2002).
4. M. A. Shupe et al., Phys. Rev. D **19**, 1921 (1979).
5. A. Nathan, Talk presented at the Workshop on Exclusive Processes at High Momentum Transfer, Jefferson Lab, Newport News, VA, May 15–18, 2002
6. "An International Accelerator Facility for Beams of Ions and Antiprotons", GSI Conceptual Design Report, November 2001
7. A. Freund, A. V. Radyushkin, A. Schäfer, and C. Weiss, to be published.
8. P. Kroll, Talk presented at the Workshop on Exclusive Processes at High Momentum Transfer, Jefferson Lab, Newport News, VA, May 15–18, 2002
9. See also: M. Diehl, P. Kroll and C. Vogt, Phys. Lett. B **532**, 99 (2002).
10. A. V. Radyushkin, Phys. Rev. **D59** (1999) 014030; Phys. Lett. **B449** (1999) 81.

THE ELECTROMAGNETIC FORM FACTORS OF THE PROTON AND NEUTRON: FUNDAMENTAL INDICATORS OF NUCLEON STRUCTURE

GERALD A. MILLER

Department of Physics, University of Washington
Seattle, WA 98195-1560
E-mail: miller@phys.washington.edu

We present a relativistic interpretation for why the proton's G_E/G_M falls and QF_2/F_1 is approximately constant. Reproducing the observed G_E^n then mandates the inclusion of the effects of the pion cloud.

1. Introduction

An alternate title would be "Surprises in the Proton". This talk owes its existence to the brilliant, precise, stunning and exciting recent experimental work on measuring G_E/G_M (or QF_2/F_1) for the proton and G_E for the neutron. My goal here is to interpret the data. Symmetries including Poincaré invariance and chiral symmetry will be the principal tool I'll use.

If, a few years ago, one had asked participants at a meeting like this about the Q^2 dependence of the proton's G_E/G_M or QF_2/F_1. Almost everyone one have answered that for large enough values of Q^2, G_E/G_M would be flat and QF_2/F_1 would fall with increasing Q^2. The reason for the latter fall being conservation of hadron helicity. Indeed, the shapes of the curves have been obtained in the new measurements, except for the mis-labeling of the ordinate axes. The expected flatness of G_E/G_M holds for QF_2/F_1, and the quantity G_E/G_M falls rapidly and linearly with Q^2. This revolutionary behavior needs to be understood!

2. Outline

I will begin with a brief discussion of Light Front Physics. Stan Brodsky has long been advocating this technique, I have become a convert. Then I will discuss a particular relativistic model of the nucleon, and proceed to apply it. The proton calculations will be discussed first, but recent high accuracy experiments make it necessary for us to compute observables for the neutron as well.

3. Light Front

Light-front dynamics is a relativistic many-body dynamics in which fields are quantized at a "time" $=\tau = x^0 + x^3 \equiv x^+$. The τ-development operator is then given by $P^0 - P^3 \equiv P^-$. These equations show the notation that a four-vector A^μ is expressed as $A^\pm \equiv A^0 \pm A^3$. One quantizes at $x^+ = 0$ which is a light-front, hence the name "light front dynamics". The canonical spatial variable must be orthogonal to the time variable, and this is given by $x^- = x^0 - x^3$. The canonical momentum is then $P^+ = P^0 + P^3$. The other coordinates are $\mathbf{x}_\perp$ and $\mathbf{P}_\perp$.

The most important consequence of this is that the relation between energy and momentum of a free particle is given by: $p_\mu p^\mu = m^2 = p^+ p^- - p_\perp^2 \to p^- = \frac{p_\perp^2 + m^2}{p^+}$, a relativistic kinetic energy which does not contain a square root operator. This allows the separation of center of mass and relative coordinates, so that the computed wave functions are frame independent.

The use of the light front is particularly relevant for calculating form factors, which are probability amplitudes for an nucleon to absorb a four momentum q and remain a nucleon. The initial and final nucleons have different total momenta. This means that the final nucleon is a boosted nucleon, with different wave function than the initial nucleon. In general, performing the boost is difficult for large values of $Q^2 = -q^2$. However the light front technique allows one to set up the calculation so that the boosts are independent of interactions. Indeed, the wave functions are functions of relative variables and are independent of frame.

3.1. Definitions

Let us define the basic quantities concerning us here. These are the independent form factors defined by

$$\langle N, \lambda' p' | J^\mu | N, \lambda p \rangle = \bar{u}_{\lambda'}(p') \left[F_1(Q^2)\gamma^\mu + \frac{\kappa F_2(Q^2)}{2M_N} i\sigma^{\mu\nu}(p' - p)_\nu \right] u_\lambda(p). \quad (1)$$

The Sachs form factors are defined by the equations:

$$G_E = F_1 - \frac{Q^2}{4M_N^2}\kappa F_2, \ G_M = F_1 + \kappa F_2. \quad (2)$$

There is an alternate light front interpretation, based on field theory, in which one uses the "good" component of the current, J^+, to suppress the effects of quark-pair terms. Then, using nucleon light-cone spinors:

$$F_1(Q^2) = \frac{1}{2P^+}N, \langle \uparrow | J^+ | N, \uparrow \rangle, Q\kappa F_2(Q^2) = \frac{-2M_N}{2P^+}\langle N, \uparrow | J^+ | N, \downarrow \rangle. \quad (3)$$

4. Why I am Giving This Talk

In 1996, Frank, Jennings & I [1] examined the point-like-configuration idea of Frankfurt & Strikman[2]. We needed to start with a relativistic model of the free nucleon. The resulting form factors are shown in Figs. 10 and 11 of our early paper. The function G_M was constrained[3] by experimental data to define the parameters of the model, but we predicted a very strong decrease of G_E/G_M as a function of Q^2. This decrease has now been measured as a real effect, but the task of explaining its meaning remained relevant. That was the purpose of our second paper[4] in which imposing Poincaré invariance was shown to lead to substantial violation of the helicity conservation rule as well as an analytic result that the ratio QF_2/F_1 is constant for the Q^2 range of the Jefferson Laboratory experiments. Although the second paper is new, the model is the same. Ralston *et al.*[5] have been talking about non-conservation of helicity for a long time.

5. Three-Body Variables and Boost

We use light front coordinates for the momentum of each of the i quarks, such that $\mathbf{p}_i = (p_i^+, \mathbf{p}_{i\perp}), p^- = (p_\perp^2 + m^2)/p^+$. The total (perp)-momentum is $\mathbf{P} = \mathbf{p_1} + \mathbf{p_2} + \mathbf{p_3}$, the plus components of the momenta are denoted as

$$\xi = \frac{p_1^+}{p_1^+ + p_2^+}, \qquad \eta = \frac{p_1^+ + p_2^+}{P^+}, \quad (4)$$

and the perpendicular relative coordinates are given by

$$\mathbf{k}_\perp = (1 - \xi)\mathbf{p}_{1\perp} - \xi\mathbf{p}_{2\perp}, \quad \mathbf{K}_\perp = (1 - \eta)(\mathbf{p}_{1\perp} + \mathbf{p}_{2\perp}) - \eta\mathbf{p}_{3\perp}. \quad (5)$$

In the center of mass frame we find:

$$\mathbf{p}_{1\perp} = \mathbf{k}_\perp + \xi\mathbf{K}_\perp, \quad \mathbf{p}_{2\perp} = -\mathbf{k}_\perp + (1 - \xi)\mathbf{K}_\perp, \quad \mathbf{p}_{3\perp} = -\mathbf{K}_\perp. \quad (6)$$

The coordinates $\xi, \eta, \mathbf{k}, \mathbf{K}$ are all relative coordinates so that one obtains a frame independent wave function $\Psi(\mathbf{k}_\perp, \mathbf{K}_\perp, \xi, \eta)$.

Now consider the computation of a form factor, taking quark 3 to be the one struck by the photon. One works in a special set of frames with $q^+ = 0$ and $Q^2 = \mathbf{q}_\perp^2$, so that the value of $1 - \eta$ is not changed by the photon. The coordinate $\mathbf{p}_{3\perp}$ is changed to $\mathbf{p}_{3\perp} + \mathbf{q}_\perp$, so only one relative momentum, $\mathbf{K}_\perp$ is changed:

$$\mathbf{K}'_\perp = (1 - \eta)(\mathbf{p}_{1\perp} + \mathbf{p}_{2\perp}) - \eta(\mathbf{p}_{3\perp} + \mathbf{q}_\perp) = \mathbf{K}_\perp - \eta\mathbf{q}_\perp, \quad \mathbf{k}'_\perp = \mathbf{k}_\perp, \quad (7)$$

The arguments of the spatial wave function are taken as the mass-squared operator for a non-interacting system:

$$M_0^2 \equiv \sum_{i=1,3} p_i^- P^+ - P_\perp^2 = \frac{K_\perp^2}{\eta(1-\eta)} + \frac{k_\perp^2 + m^2}{\eta\xi(1-\xi)} + \frac{m^2}{1-\eta}. \tag{8}$$

This is a relativistic version of the square of a the relative three-momentum. Note that the absorption of a photon changes the value to:

$$M_0'^2 = \frac{(K_\perp - \eta q_\perp)^2}{\eta(1-\eta)} + \frac{k_\perp^2 + m^2}{\eta\xi(1-\xi)} + \frac{m^2}{1-\eta}. \tag{9}$$

6. Wave function

Our wave function is based on symmetries. The wave function is anti-symmetric, a function of relative momenta, independent of reference frame, an eigenstate of the spin operator and rotationally invariant (in a specific well-defined sense). The use of symmetries is manifested in the construction of such wave functions, as originally described by Terent'ev [6], Coester[7] and their collaborators. A schematic form of the wave functions is

$$\Psi(p_i) = \Phi(M_0^2)u(p_1)u(p_2)u(p_3)\psi(p_1, p_2, p_3), \quad p_i = \mathbf{p}_i s_i, \tau_i \tag{10}$$

where ψ is a spin-isospin color amplitude factor, the p_i are expressed in terms of relative coordinates, the $u(p_i)$ are ordinary Dirac spinors and Φ is a spatial wave function.

We take the the spatial wave function from Schlumpf[3]:

$$\Phi(M_0) = \frac{N}{(M_0^2 + \beta^2)^\gamma}, \beta = 0.607 \text{ GeV}, \gamma = 3.5, m = 0.267 \text{ GeV}. \tag{11}$$

The value of γ is chosen that $Q^4 G_M(Q^2)$ is approximately constant for $Q^2 > 4 \text{ GeV}^2$ in accord with experimental data. The parameter β helps govern the values of the perp-momenta allowed by the wave function Φ and is closely related to the rms charge radius, and m is mainly determined by the magnetic moment of the proton.

At this point the wave function and the calculation are completely defined. One could evaluate the form factors as $\langle \Psi | J^+ | \Psi \rangle$ and obtain the previous numerical results[1].

7. Simplify Calculation- Light Cone Spinors

The operator $J^+ \sim \gamma^+$ acts its evaluation is simplified by using light cone spinors. These solutions of the free Dirac equation, related to ordinary Dirac spinors by a unitary transformation, conveniently satisfy:

$$\bar{u}_L(p^+, \mathbf{p}', \lambda')\gamma^+ u_L(p^+, \mathbf{p}, \lambda) = 2\delta_{\lambda\lambda'}p^+. \tag{12}$$

To take advantage of this, re-express the wave function in terms of light-front spinors using the completeness relation: $1 = \sum_\lambda u_L(p, \lambda)\bar{u}_L(p, \lambda)$. We then find

$$\Psi(p_i) = u_L(p_1, \lambda_1)u_L(p_2, \lambda_2)u_L(p_3, \lambda_3)\psi_L(p_i, \lambda_i), \tag{13}$$

$$\psi_L(p_i, \lambda_i) \equiv [\bar{u}_L(\mathbf{p}_1, \lambda_1)u(\mathbf{p}_1, s_1)][\bar{u}_L(\mathbf{p}_2, \lambda_2)u(\mathbf{p}_2, s_2)]$$
$$\times [\bar{u}_L(\mathbf{p}_3, \lambda_3)u(\mathbf{p}_3, s_3)] \psi(p_1, p_2, p_3). \tag{14}$$

This is the very same Ψ as before, it is just that now it is easy to compute the matrix elements of the γ^+ operator.

The unitary transformation is also known as the Melosh rotation. The basic point is that one may evaluate the coefficients in terms of Pauli spinors: $|\lambda_i\rangle, |s_i\rangle$, with $\langle\lambda_i|R_M^\dagger(\mathbf{p}_i)|s_i\rangle \equiv \bar{u}_L(\mathbf{p}_i, \lambda_i)u(\mathbf{p}_i, s_i)$. It is easy to show that

$$\langle\lambda_3|R_M^\dagger(\mathbf{p}_3)|s_3\rangle = \langle\lambda_3|\left[\frac{m + (1-\eta)M_0 + i\boldsymbol{\sigma}\cdot(\mathbf{n}\times\mathbf{p}_3)}{\sqrt{(m+(1-\eta)M_0)^2 + p_{3\perp}^2}}\right]|s_3\rangle. \tag{15}$$

The important effect resides in the term $(\mathbf{n}\times\mathbf{p}_3)$ which originates from the lower components of the Dirac spinors. This large relativistic spin effect can be summarized: the effects of relativity are to replace Pauli spinors by Melosh rotation operators acting on Pauli spinors. Thus

$$|\uparrow \mathbf{p}_i\rangle \equiv R_M^\dagger(\mathbf{p}_i)\begin{pmatrix}1\\0\end{pmatrix}, \quad |\uparrow\mathbf{p}_3\rangle \neq |\uparrow\rangle. \tag{16}$$

8. Proton F_1, F_2-Analytic Insight

The analytic insight is based on Eq. (15). Consider high momentum transfer such that $Q = \sqrt{\mathbf{q}_\perp^2} \gg \beta = 560$ MeV. *Each* of the quantities: M_0, M_0', $\mathbf{p}_{3\perp}$, $\mathbf{p}_{3\perp}$ can be of order $q_\perp$, so the spin-flip term is as large as the non-spin flip term. In particular, $(s_3 = +1/2)$ may correspond to $(\lambda_3 = -1/2)$, so the spin of the struck quark $\neq$ proton spin. This means that there is no hadron helicity selection rule[5,8].

The effects of the lower components of Dirac spinors, which cause the spin flip term $\boldsymbol{\sigma}\times\mathbf{p}_3$, are the same as having a non-zero L_z, if the wave functions are expressed in the light-front basis. See Sect. 9.

We may now qualitatively understand the numerical results, since

$$F_1(Q^2) = \int\frac{d^2q_\perp d\xi}{\xi(1-\xi)}\frac{d^2K_\perp d\eta}{\eta(1-\eta)}\cdots\langle\uparrow\mathbf{p}_3'|\uparrow(\mathbf{p}_3)\rangle \tag{17}$$

$$Q\kappa F_2(Q^2) = 2M_N\int\frac{d^2q_\perp d\xi}{\xi(1-\xi)}\frac{d^2K_\perp d\eta}{\eta(1-\eta)}\cdots\langle\uparrow\mathbf{p}_3'|\downarrow(\mathbf{p}_3)\rangle, \tag{18}$$

where the $\cdots$ represents common factors. The term $F_1 \sim \langle \uparrow\mathbf{p}_3'|\uparrow\mathbf{p}_3 \rangle$ is a spin-non-flip term and $F_2 \sim \langle \uparrow\mathbf{p}_3'|\uparrow\mathbf{p}_3 \rangle$ depends on the spin-flip term. In doing the integral each of the momenta, and M_0, M_0' can take the large value Q for some regions of the integration. Thus in the integral

$$\langle \uparrow\mathbf{p}_3'|\uparrow\mathbf{p}_3 \rangle \sim \frac{Q}{Q}, \quad \langle \uparrow\mathbf{p}_3'|\downarrow\mathbf{p}_3 \rangle \sim \frac{Q}{Q}, \tag{19}$$

so that F_1 and QF_2 have the same Q^2 dependence. This is shown in Fig. 1.

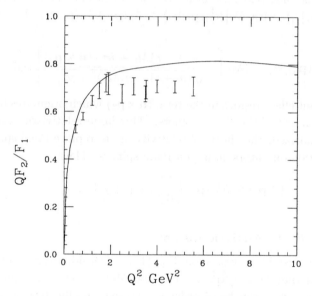

Figure 1.   Calculation of Refs. [2,5], data are from Jones *et al.*[10] and from Gayou *et al.*.

9. Relation between ordinary Dirac Spinors and L_z

Our use of ordinary Dirac spinors corresponds to the use of a non-zero L_z in the light front basis. We may represent Dirac spinors as Melosh rotated Pauli spinors, and this is sufficient to show $L_z \neq 0$.

It is worthwhile to consider the pion as an explicit example. Then our version of the light-front wave function χ_π would be[11]:

$$\chi_\pi(k^+, \mathbf{k}_\perp, \lambda, \lambda') \propto \langle \lambda | i\sigma_2 m - (k_1 - i\sigma_3 k_2)|\lambda' \rangle, \tag{20}$$

while the Gousset-Pire-Ralston[5] pion Bethe-Salpeter amplitude Φ is

$$\Phi = P_{0\pi} \rlap{/}{p}_\pi + P_{1\pi}[\rlap{/}{p}_\pi, \rlap{/}{k}_\perp], \qquad (21)$$

where p_π is the pion total momentum, $P_{i\pi}$ are scalar functions of relative momentum, and the term with $P_{1\pi}$ is the one which carries orbital angular momentum. The relation[12] between the Bethe-Salpeter amplitude and the light-front wave function ϕ_π is

$$\phi_\pi(k^+, \mathbf{k}_\perp, \lambda, \lambda') = \bar{u}_L(k^+, \mathbf{k}_\perp, \lambda)\gamma^+ \Phi\gamma^+ v_L(P_\pi^+ - k^+, -\mathbf{k}_\perp, \lambda'). \qquad (22)$$

Doing the Dirac algebra and choosing suitable functions $P_{i\pi}$, leads to $\chi_\pi = \phi_\pi$. The Melosh transformed Pauli spinors, which account for the lower components of the ordinary Dirac spinors, contain the non-zero angular momentum of the wave function Φ.

10. Neutron Charge Form Factor

The neutron has no charge, $G_{En}(Q^2 = 0) = 0$, and the square of its charge radius is determined from the low Q^2 limit as $G_{En}(Q^2) \to -Q^2 R^2/6$. The quantity R^2 is well-measured[13] as $R^2 = -0.113 \pm 0.005$ fm^2. The Galster parameterization[14] has been used to represent the data for $Q^2 < 0.7$ GeV^2.

Our proton respects charge symmetry, the interchange of u and d quarks, so it contains a prediction for neutron form factors. This is shown in Fig. 2. The resulting curve labeled relativistic quarks is both large and small. It is very small at low values of Q^2. Its slope at $Q^2 = 0$ is too small by a factor of five, if one compares with the straight line. But at larger values of Q^2 the prediction is relatively large.

Our model gives $R^2_{\text{model}} = -0.025$ fm^2, about five times smaller than the data. The small value can be understood in terms of $F_{1,2}$. Taking the definition (2) for small values of Q^2 gives

$$-Q^2 R^2/6 = -Q^2 R_1^2/6 - \kappa_n Q^2/4M^2 = -Q^2 R_1^2/6 - Q^2 R_F^2/6, \qquad (23)$$

where the Foldy contribution, $R_F^2 = 6\kappa_n/4M^2 = -0.111$ fm^2, is in good agreement with the experimental data. That a point particle with a magnetic moment can explain the charge radius has led some to state that G_E is not a measure of the structure of the neutron. However, one must include the Q^2 dependence of F_1 which gives R_1^2. In our model $R_1^2 = +0.086$ fm^2 which nearly cancels the effects of R_F^2. Isgur[17] showed that this cancellation is a natural consequence of including the relativistic effects of the lower components of the Dirac spinors. Thus our relativistic effects are standard. We need another source of R^2. This is the pion cloud.

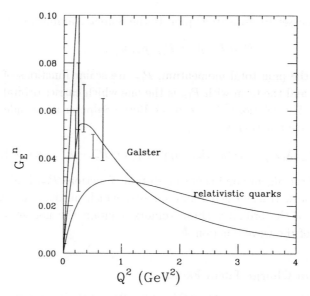

Figure 2.   Calculation of G_E^n. The data are from Ref. [15], with more expected soon[16].

11. Pion Cloud and the Light Front Cloudy Bag Model

The effects of chiral symmetry require that sometimes a physical nucleon can be a bare nucleon emersed in a pion cloud. An incident photon can interact electromagnetically with a bar nucleon, a pion in flight or with a nucleon while a pion is present. These effects were included in the cloudy bag model, and are especially pronounced for the neutron. Sometimes the neutron can be a proton plus a negatively charged pion. The tail of the pion distribution extends far out into space (see Figs. 10 and 11) of Ref.[18], so that the square of the charge radius is negative.

It is necessary to modernize the cloudy bag model, so as to make it relativistic. This involves using photon-nucleon form factors from our model, using a relativistic π-nucleon form factor, and treating the pionic contributions relativistically by doing a light front calculation. This has been done. The result is the light front cloudy bag model, and the preliminary results are shown in Fig. 3. We see that the pion cloud effects are important for small values of Q^2 and, when combined with those of the relativistic quarks coming from the bare nucleon, leads to a good description of the low Q^2 data. The total value of G_E is substantial for large values of Q^2.

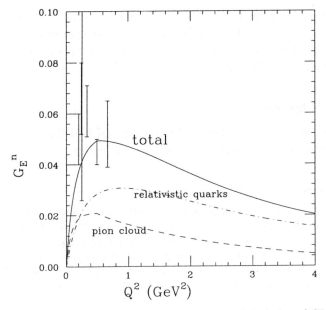

Figure 3. Light front cloudy bag model LFCBM Calculation of G_E^n.

12. Summary

Poincaré invariance is needed to describe the new exciting experimental results. Ordinary Dirac spinors carry light front orbital angular momentum. Including the effects of these spinors, in a way such that the proton is an eigenstate of spin leads naturally to the result that QF_2/F_1 is constant for values of Q^2 between 2 and about 20 GeV^2.

The prediction of hadron helicity conservation is that Q^2F_2/F_1 is constant, so we see that this is not respected in present data and there is no need to expect it to hold for a variety of exclusive reactions occurring at high $Q^2 \leq 5.5$ GeV^2. Examples include the anomalies seen in pp elastic scattering and the large spin effects seen in the reactions $\gamma d \rightarrow np$ and $\gamma p \rightarrow \pi^0 p$.

The results for the neutron G_E can be concisely stated. At small values of Q^2 the effects of a pion cloud is needed to counteract the relativistic effects which cancel the effects of the Foldy term. At large values of Q^2 relativistic effects give a "large" value of G_E; large in the sense that this form factor is predicted to be larger than that of the Galster parameterization.

At the time of this workshop, I had not yet used the light front cloudy bag model to compute proton form factors or the neutron's G_M. Including the effects of the pion cloud (with a parameter to describe the pion-nucleon form factor) allows the use of different quark-model parameters. The result is an excellent description of all four nucleon electromagnetic form factors, and I plan to publish that soon.

Acknowledgments

This work is partially supported by the U.S. DOE. I thank R. Madey for encouraging me to compute the neutron form factors.

References

1. M.R. Frank, B.K. Jennings and G.A. Miller, Phys. Rev. C **54**, 920 (1996).
2. L. L. Frankfurt and M. I. Strikman, Nucl. Phys. B **250**, 143 (1985).
3. F. Schlumpf, U. Zurich Ph. D. Thesis, hep-ph/9211255.
4. G. A. Miller and M. R. Frank, nucl-th/0201021 to appear Phys. Rev. C.
5. P. Jain, B. Pire and J. P. Ralston, Phys. Rept. **271**, 67 (1996) ; T. Gousset, B. Pire and J. P. Ralston, Phys. Rev. D **53**, 1202 (1996)
6. V. B. Berestetskii and M. V. Terent'ev. Sov. J. Nucl. Phys. **25**, 347 (1977).
7. P. L. Chung and F. Coester. *Phys. Rev. D* **44**, 229, (1991).
8. V. M. Braun, A. Lenz, N. Mahnke and E. Stein, Phys. Rev. D **65**, 074011 (2002).
9. M. K. Jones *et al.*, Phys. Rev. Lett. **84**, 1398 (2000)
10. O. Gayou *et al.*, Phys. Rev. Lett. **88**, 092301 (2002).
11. P.L. Chung, F. Coester and W.N. Polyzou, Phys. Lett 205B,545 (1988).
12. H. H. Liu and D. E. Soper, Phys. Rev. D **48**, 1841 (1993).
13. S. Kopecky *et al.*, Phys. Rev. Lett. **74**, 2427 (1995)
14. S. Galster *et al.*, Nucl. Phys. B **32**, 221 (1971).
15. T. Eden *et al.*, Phys. Rev. C **50**, 1749 (1994); M. Meyerhoff *et al.*, Phys. Lett. B **327**, 201 (1994); M. Ostrick *et al.*, Phys. Rev. Lett. **83**, 276 (1999). J. Becker *et al.*, Eur. Phys. J. A **6**, 329 (1999). I. Passchier *et al.*, Phys. Rev. Lett. **82**, 4988 (1999) D. Rohe *et al.*, Phys. Rev. Lett. **83**, 4257 (1999). H. Zhu *et al.* Phys. Rev. Lett. **87**, 081801 (2001)
16. Jefferson Laboratory Experiment 93-038, R. Madey Spokesperson; R.Madey, for the Jlab E93-038 collaboration, "Neutron Electric Form Factor Via Recoil Polarimetry", contribution to Baryons 2002.
17. N. Isgur, Phys. Rev. Lett. **83**, 272 (1999)
18. S. Théberge, A. W. Thomas and G. A. Miller, Phys. Rev. **D22** (1980) 2838; (1981) 2106, A. W. Thomas, S. Théberge, and G. A. Miller, Phys. Rev. **D24** (1981) 216; S. Théberge, G. A. Miller and A. W. Thomas, Can. J. Phys. **60**, 59 (1982). G. A. Miller, A. W. Thomas and S. Théberge, Phys. Lett. B **91**, 192 (1980).

PROTON ELASTIC FORM FACTOR RATIO

C.F. PERDRISAT

College of William and Mary, Williamsburg, VA 23187
E-mail: perdrisa@jlab.org

FOR THE JEFFERSON LAB HALL A COLLABORATION

The ratio of the electric and magnetic proton form factors, G_{Ep}/G_{Mp}, has been obtained in two Hall A experiments, from measurements of the longitudinal and transverse polarization of the recoil proton, P_ℓ and P_t, respectively, in the elastic scattering of polarized electrons, $\vec{e}p \rightarrow e\vec{p}$. Together these experiments cover the Q^2- range 0.5 to 5.6 GeV^2.

1. Introduction

The nucleon elastic form factors describe the internal structure of the nucleon; in the non-relativistic limit, for small four-momentum transfer squared, Q^2, they are Fourier transforms of the charge and magnetization distributions in the nucleon. At high Q^2 values, the nucleon must be treated as a three valence quarks system; perturbative QCD predicts the Q^2-dependence[1] of the form factors. At Q^2 between 1 and 10 GeV^2, relativistic constituent quark models [2,3] currently give the best understanding of the nucleon form factors, with the strongest dynamical input; Vector Meson Dominance (VMD) (see e.g. Refs. [4,5]) also describe them factors well.

The unpolarized elastic ep cross section is given by:

$$\frac{d\sigma}{d\Omega} = \frac{\alpha^2 E_e' \cos^2 \frac{\theta_e}{2}}{4E_e^3 \sin^4 \frac{\theta_e}{2}} \left[G_{Ep}^2 + \frac{\tau}{\epsilon} G_{Mp}^2 \right] \left(\frac{1}{1+\tau} \right), \tag{1}$$

where G_{Ep} and G_{Mp} are the electric and magnetic form factors, $\epsilon = [1 + 2(1 + \tau) \tan^2(\frac{\theta_e}{2})]^{-1}$, θ_e is the scattering angle of the electron in the laboratory and $\tau = Q^2/4M_p^2$, with M_p the proton mass; E_e and E_e' are the energies of the in- and outgoing electrons, respectively. For a given Q^2, G_{Ep} and G_{Mp} can be extracted from cross section measurements made at fixed Q^2, over a range of ϵ values with the Rosenbluth method. At Q^2 below 1 GeV^2, G_{Ep} and G_{Mp} have been determined by this method and

$\mu_p G_{Ep}/G_{Mp}$ has been found to be ≈ 1. At larger Q^2, the cross section becomes dominated by the G_{Mp} contribution; G_{Mp} is known up to $Q^2 = 31$ GeV^2 [11]. In Fig. 1a, the error bars on $\mu_p G_{Ep}/G_{Mp}$ from the World cross section data are seen to grow with Q^2. Above $Q^2 \approx 1$ GeV^2, systematic differences between different experiments are evident.

The JLab results have been obtained by measuring the recoil proton polarization in $\vec{e}p \rightarrow e\vec{p}$ [12,13]. In one-photon exchange, the scattering of longitudinally polarized electrons on unpolarized hydrogen results in a transfer of polarization to the recoil proton with two components, P_t perpendicular to, and P_ℓ parallel to the proton momentum in the scattering plane [14]:

$$I_0 P_t = -2\sqrt{\tau\,(1+\tau)}G_{E_p}G_{M_p}\tan\frac{\theta_e}{2} \tag{2}$$

$$I_0 P_\ell = \frac{1}{M_p}\left(E_e + E_{e'}\right)\sqrt{\tau\,(1+\tau)}G_{M_p}^2\tan^2\frac{\theta_e}{2} \tag{3}$$

where $I_0 \propto G_{E_p}^2 + \frac{\tau}{\epsilon}G_{M_p}^2$. Measuring simultaneously these two components and taking their ratio gives the ratio of the form factors:

$$\frac{G_{Ep}}{G_{Mp}} = -\frac{P_t}{P_\ell}\frac{(E_e + E_{e'})}{2M_p}\tan(\frac{\theta_e}{2}) \tag{4}$$

Neither the beam polarization nor the analyzing power of the polarimeter , used to measure P_t and P_ℓ, appear in Eqn. 4.

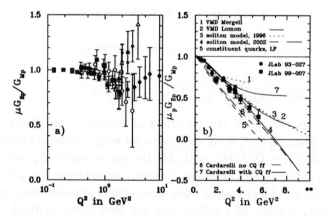

Figure 1.   (a) World data for $\mu_p G_{Ep}/G_{Mp}$ versus Q^2 from refs. 6,7,8,9 as open triangles and from ref. 10 as solid triangles; (b) Comparison between $\mu_p G_{Ep}/G_{Mp}$ measured at JLab and several theoretical models; the JLab data are from refs. 12,13.

2. Experiments

In 1998 G_{Ep}/G_{Mp} was measured for Q^2 from 0.5 to 3.5 GeV^2 [12]. Protons and electrons were detected in coincidence in the two high-resolution spectrometers (HRS) of Hall A. The polarization of the recoiling proton was measured in a graphite analyzer focal plane polarimeter (FPP).

In 2000 new measurements at Q^2 = 4.0, 4.8 and 5.6 GeV^2 with overlap points at Q^2 = 3.0 and 3.5 GeV^2 were made [13]. To extend the measurement to these higher Q^2, two changes were made. First, to increase the figure-of-merit of the FPP, a CH$_2$ analyzer was used; the thickness was increased from 50 cm of graphite to 100 cm of CH$_2$ (60 cm for Q^2 = 3.5 GeV^2). Second, the electrons were detected in a lead-glass calorimeter with 9 columns and 17 rows of 15 × 15 × 35 cm^3 blocks placed so as to achieve complete solid angle matching with the HRS detecting the proton. At the largest Q^2 the solid angle of the calorimeter was 6 times that of the HRS.

3. Results and Discussion

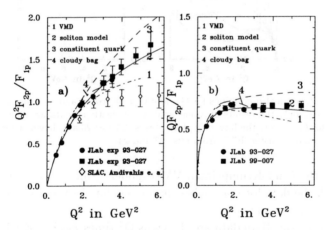

Figure 2. (a) The ratio $Q^2 F_{2p}/F_{1p}$ from the JLab experiments, compared with the data of ref. (10), (b) The ratio $Q F_{2p}/F_{1p}$ discussed in the text.

The combined results from both experiments are plotted in Fig. 1b as the ratio $\mu_p G_{Ep}/G_{Mp}$. If the $\mu_p G_{Ep}/G_{Mp}$-ratio continues its linear decrease with the same slope, it will cross zero at $Q^2 \approx 7.5$ GeV^2. In Fig. 1b, calculations based on VMD [4], a relativistic constituent quark (CQ) [2], and a soliton model[15] are shown. Also shown are results with another relativistic CQ model (rCQM) [3], with and without CQ form factor. Lomon [5] has

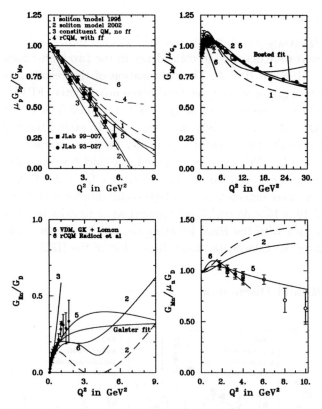

Figure 3. Theoretical predictions for G_E and G_M of proton and neutron, and selected data. For G_{En} only the results of recent analysis of elastic ed data are shown from ref. 22 shown; for G_{Mn} only the larger Q^2 data of refs. 25 and 26 are shown. Curves labeled "Bosted fit" and "Galster fit" is from refs. 23 and 24.

reworked the Gary-Krumpelman VMD model [16] and obtains good agreement with the data for reasonable parameters for the vector-meson masses and coupling constants.

In Fig. 2a the JLab data are shown as Q^2 times F_2/F_1; pQCD predicts quenching of the spin flip form factor F_2, or equivalently helicity conservation; higher order contributions should make $Q^2 F_2/F_1$ asymptotically constant. The data clearly contradict this prediction. Shown in Fig. 2b is Q times F_2/F_1, which reaches a constant value at $Q^2 \sim 2$ GeV^2. Ralston et al. [19] have proposed that this scaling is due to the non-zero orbital angular momentum part of the proton quark wave function. Miller and Frank [20] have shown that imposing Poincare invariance leads to violation of the helicity conservation rule, and reproduces the QF_2/F_1 behavior.

More demanding for models are predictions for all four form factors of

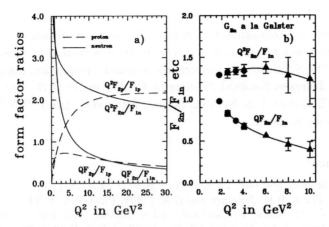

Figure 4. (a) The ratios $Q\times$ and $Q^2 \times F_2/F_1$ for proton and neutron, from ref. 3. Note the prediction that both ratios tend toward the same value for proton and neutron above 15 GeV^2; neither one of these ratio becomes constant below this same Q^2, a marked difference to the proton ratio, (b) The ratio $Q\times$ and $Q^2 \times F_2/F_1$ for the neutron from Galster values for G_{En} and the G_{Mn} data from refs. 25 and 26.

the nucleon. The VMD fits are done in terms of the isoscalar and isovector form factors and thus naturally include all four form factors. In Fig. 3 predictions from the rCQM with SU(6) of Pace et al. [3], the soliton model [15], the point form model [26], and the VMD model of Ref. [5] are shown. The soliton model does well for the proton only. The recent VMD analysis [5] reproduces G_{Ep}, G_{Mp} and G_{Mn} well, and predicts larger values for G_{En} than the fit of Ref. [25].

Isospin invariance at the quark level requires that F_2/F_1 become the same for proton and neutron starting at some large Q^2-value. In Fig. 4a we show the prediction for QF_2/F_1 and Q^2F_2/F_1 for proton and neutron from Ref. [5]; F_2/F_1 may become equal for the proton and the neutron for $Q^2 > 10$ GeV^2. The evolution of QF_2/F_1 at small Q^2 is dominated by the charge neutrality of the neutron, which results in $F_{1n} = 0$ at $Q^2 = 0$.

In Fig. 4b we show QF_2/F_1 and Q^2F2/F_1 for the neutron assuming Galster [25] fit values for G_{En} and using G_{Mn} data from Refs.[22,23]. QF_{2n}/F_{1n} will not have the scaling behavior of the proton data unless G_{En} becomes larger than the Galster fit value at Q^2 larger than 1.5 GeV^2, beyond which it is not constrained by data. New G_{En} measurement to 3.4 GeV^2 [27] will be crucial in the testing of the $\frac{1}{Q}$ scaling hypothesis of F_2/F_1.

The precise new JLab data on $\mu_p G_{Ep}/G_{Mp}$ show that this ratio continues to drop off linearly with increasing Q^2 up to 5.6 GeV^2. Comparison of model calculations to the JLab data provides a stringent test of models of the nucleon.

I thank my colleagues V. Punjabi, M. Jones, E. Brash, L. Pentchev and O. Gayou for their essential roles in the completion of these experiments. The Southeastern Universities Research Association manages the Thomas Jefferson National Accelerator Facility under DOE contract DE-AC05-84ER40150. U.S. National Science Foundation grant PHY 99 01182 supports my research.

References

1. S.J. Brodsky and G.P. Lepage, Phys. Rev. D **22**, (1981) 2157.
2. M.R. Frank, B.K. Jennings and G.A. Miller, Phys. Rev. C **54**, 920 (1996).
3. E. Pace, G. Salme, F. Cardarelli and S. Simula, Nucl. Phys. A **666&667**, 33c (2000). F. Cardarelli and S. Simula, Phys. Rev. C **62**, 65201 (2000).
4. P. Mergell, U.G. Meissner, D. Drechsler Nucl. Phys. B **A596**, 367 (1996) ; and A.W. Hammer, U.G. Meissner and D. Drechsel, Phys. Lett. B **385**, 343 (1996).
5. E. Lomon, Phys. Rev. C **64** 035204 (2001) and nucl-th/0203081
6. J. Litt *et al.*, Phys. Lett. B **31**, 40 (1970).
7. Ch. Berger *et al.*, Phys. Lett. B **35**, 87 (1971).
8. L.E. Price *et al.*, Phys. Rev. D **4**, 45 (1971).
9. W. Bartel *et al.*, Nucl. Phys. B **58**, 429 (1973).
10. L. Andivahis *et al.*, Phys. Rev. D **50**, 5491 (1994).
11. A.F. Sill *et al.*, Phys. Rev. D **48**, 29 (1993).
12. M. K. Jones *et al.*, Phys. Rev. Lett. **84**, 1398 (2000).
13. O. Gayou et al., Phys. Rev. Lett. **88** 092301 (2002)
14. A.I. Akhiezer and M.P. Rekalo, Sov. J. Part. Nucl. **3**, 277 (1974); R. Arnold, C. Carlson and F. Gross, Phys. Rev. C **23**, 363 (1981).
15. G. Holzwarth, Z. Phys. A **356**, 339 (1996).
16. M.F. Gari and W. Krumpelmann, Phys. Lett. B **274**, 159 (1992)
17. P. Kroll, M. Schurmann and W. Schweiger, Z. Phys. A **338**, 339 (1991).
18. D.H. Lu, S.N. Yang, A.W. Thomas, J. Phys. **G26**, L75 (2000). D.H. Lu, A.W. Thomas and A.G. Williams, Phys. Rev. C **57**, 2628 (1998).
19. J. Ralston *et al.*, in Proc. of 7th International Conference on Intersection of Particle and Nuclear Physics, Quebec City (2000), p. 302, and private communication (2001).
20. G.A. Miller and M.R. Frank, nucl-th 0201021 (2002)
21. R. Schiavilla and I. Sick, nucl-ex/0107004 (2001)
22. A. Lung et al., Phys. Rev.Lett. **70**, 718 (1993)
23. S. Rock et al. Phys. Rev. Lett. **49**, 1139 (1982)
24. P. E. Bosted, Phys. Rev. **51**, 409 (1995)
25. S. Galster et al. Nucl. Phys. **B32**, 221 (1971)
26. R.F. Wagenbrunn, S. Boffi, W.H. Klink, W. Plessas and M. Radici, Phys. Lett. B, **511**:33 (2001)
27. G. Catews, K. McCormick, B. Reitz and B. Wojtsekhowski et al., JLab approved proposal 02-013 (2002).

THE NEUTRON MAGNETIC FORM FACTOR AT HIGH Q^2: EXPERIMENTAL STATUS, FUTURE MEASUREMENTS

W. K. BROOKS

Thomas Jefferson National Accelerator Facility
12000 Jefferson Ave, MS 12H
Newport News, VA, 23601, USA
E-mail: brooksw@jlab.org

Recent progress in improving our knowledge of the four nucleon form factors G_M^p, G_E^p, G_M^n, G_E^n at high momentum transfer is stimulating a new wave of theoretical efforts to describe these fundamental quantities. Both model calculations and lattice QCD can predict the elastic form factors; a definitive, stringent test of these efforts is to predict all of them simultaneously. However, the limited range and quality of the data for the neutron magnetic form factor G_M^n presently reduce the discriminating power of such a test. The present status of our knowledge of G_M^n is discussed, and prospects for future improvements are presented.

1. Introduction

The quest to understand the structure of protons and neutrons has been a major theme of hadronic physics over the last half century. Elastic scattering from the proton gave early indication of its composite nature[1], confirming the much earlier indications that its magnetic moment was not consistent with that of a point particle[2]. Over the intervening decades, knowledge of the *proton* elastic form factors has improved both in precision and in the range of momentum transfer (Q^2) spanned, including very recent progress[3]. The experimental determination of the *neutron* elastic form factors is more challenging for a variety of reasons, however, this problem is now being surmounted due to substantial technical advances. These include the development of continuous-wave electron accelerators that permit high-quality coincidence measurements, and the availability of high beam and target polarizations. The following discussion focuses on the present status and future prospects for measuring the neutron magnetic form factor G_M^n at $Q^2 > 1\ GeV^2$.

268

2. Significance of High-Q^2 Elastic Form Factors

Nucleon elastic form factors reflect the charge and magnetic moment distributions that arise from the intrinsic properties and motion of the nucleonic constituents. Therefore, they can typically be predicted by nucleon models and by lattice QCD calculations. While this is true at any momentum transfer, at higher momentum transfer a greater degree of sensitivity to the nucleonic constituents is achieved. In Figure 1 is shown the wavelength of the virtual photon as a function of Q^2 for the range $Q^2 = 1 - 15 \; GeV^2$ under discussion. As may be seen, the wavelength itself ranges from approximately the nucleon radius to much smaller sizes, and one expects to be able to resolve structure to some fraction of a wavelength. Therefore, the elastic form factors in this range are very sensitive to the details of the sub-nucleonic constituents (classically, an average over their effective sizes and trajectories). The requirement to predict all four elastic nucleon form factors over a wide range in Q^2 is a powerful constraint for models of the nucleon.

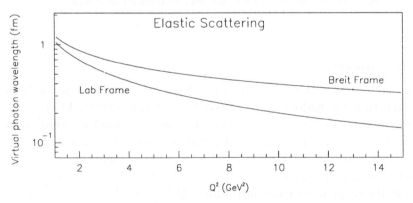

Figure 1. Wavelength of the virtual photon for elastic scattering in the Breit frame (upper curve) and the lab frame (lower curve).

Unfortunately, the quality of the world data set for these four fundamental functions varies widely. In Figure 2 is shown a comparison of representative data for the proton magnetic form factor and the neutron magnetic form factor, normalized to the dipole form $G_D(Q^2) \propto 1/(1+Q^2/Q_0^2)^2$, with $Q_0^2 = 0.71$ GeV^2. Both the extent of the Q^2 coverage and the size of the errors are poorer for the neutron. The lack of high-quality data for G_M^n clearly limits the degree to which these data can test our understanding in this range of momentum transfer.

Nucleon elastic form factors can be calculated in a wide variety of models; a number have attempted to describe the recent proton data[3]. In addition, there are predictions from pQCD, and lattice QCD efforts are underway[10]. It has been suggested that the form factors can be related to high-x structure functions using local duality[6]. Finally, these form factors are closely related to the generalized parton distributions[9,7,8], which are an important focus of the 12 GeV physics program at Jefferson Lab.

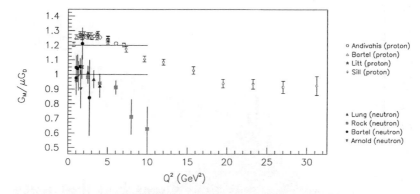

Figure 2. The normalized proton and neutron magnetic form factors ($G_M^n/G_D\mu_n$, $G_M^p/G_D\mu_p$) out to high Q^2. The proton data has been shifted upwards by 0.2 for clarity. Note the lack of high-quality data for the neutron at large Q^2. See text for further discussion.

3. Measurement Methods for the Neutron

The previous data for the neutron at large Q^2 are dominated by inclusive electron scattering measurements. This method is straightforward for measurement of G_M^p in this kinematic regime, however, to measure G_M^n there are further complications associated with the necessity of using a bound neutron as a target. The state-of-the-art for this technique may be illustrated by the NE11 experiment at SLAC [4]. In this experiment, a Rosenbluth separation of quasielastic scattering on deuterium was performed; the response functions obtained were fitted with model curves for the quasielastic process and for the inelastic background. Assuming PWIA expressions for the response functions, the contribution from the proton form factors (which were also measured in the experiment) were then subtracted from the deuteron response functions, yielding the neutron form factors G_M^n and G_E^n. Extensive studies of the model dependence of the results were performed. This method has the advantage that it only requires the detection of the electron, however, it relies heavily on modelling of the deuteron wave function and the inelastic background. A limitation in extending this method to higher Q^2 is that the peak-to-background ratio in the W^2 spectrum decreases as Q^2 increases, primarily due to kinematic broadening. In the NE11 experiment, it was estimated that the background was 15% of the signal at the quasielastic peak for the $Q^2 = 4\ GeV^2$ point.

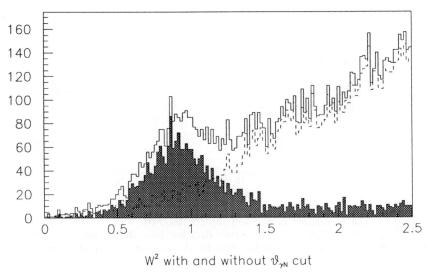

W² with and without $\vartheta_{\gamma N}$ cut

Figure 3. A plot from the CLAS E5 data of the W^2 spectrum with a cut on $\theta_{\gamma^* N} < 2.6^o$ for $Q^2 > 3\ GeV^2$. Approximately 7% of the data sample is shown. The uppermost solid line shows the W^2 spectrum for events in which a nucleon candidate was detected; the shaded peak shows the quasielastic events identified using the cut on $\theta_{\gamma^* N}$; the dashed line shows the inelastic background contribution that was removed using this cut.

An alternative approach, often referred to as the 'ratio method,' has been used at lower Q^2 for precision determinations[5] of G^n_M. Recent measurements[11] with CLAS at Jefferson Lab are extending this method to Q^2 of nearly 5 GeV^2. In the ratio method, protons and neutrons are detected in coincidence with scattered electrons from a deuterium target. The ratio of e-n scattering to e-p scattering is measured for quasielastic kinematics. Using the known values of the other three nucleon form factors, one can then use the ratio to calculate G^n_M. For larger Q^2, the method is largely insensitive to the precision of both electric form factors. In this method it is possible to suppress the inelastic background by using the additional information from the detected protons and neutrons. For instance, the angle between the emitted nucleon and the virtual photon, $\theta_{\gamma^* N}$, is small for quasielastic events and larger for the inelastic events. In Figure 3 is shown the effect of a cut in $\theta_{\gamma^* N}$ on the W^2 spectrum for $Q^2 > 3\,GeV^2$. The inelastic background is well-separated from the quasielastic peak with this simple kinematic cut, which can be applied to both proton and neutron. Since the distribution in $\theta_{\gamma^* N}$ narrows with increasing Q^2, this cut remains effective at high momentum transfer. Additional suppression of inelastic final states can be obtained by vetoing events with additional charged tracks, and this technique also continues to be effective at high momentum transfer, particularly for spectrometers with substantial hermiticity. This method has the advantage that the reaction of interest is identified via experimental information with little model input, and numerous sources of error cancel or are reduced due to the use of the ratio. The experimental challenges include the requirement of a precise measurement of the neutron detection efficiency, and the equivalency of the neutron and proton solid angles. The primary technique in this measurement that addresses those challenges was to have a hydrogen target in the beam simultaneously with the deuterium target. This allows measurement of tagged neutrons in exclusive reactions on the hydrogen target as well as many cross-checks of neutral and charged particle angles and reconstruction efficiencies.

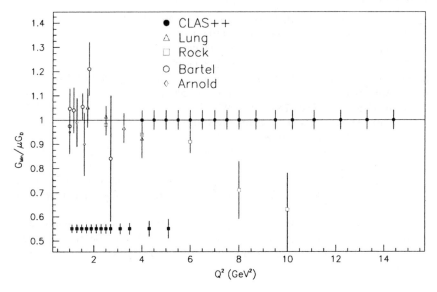

Figure 4. Measurements of G_M^n that will be feasible with an upgraded CLAS, with the 12 GeV upgrade of Jefferson Lab. See text for further discussion. An indication of the coverage of the CLAS E5 data that is currently under analysis is plotted at 0.55 in solid squares.

4. Future Prospects

New, high-precision data for the neutron magnetic form factor for $Q^2 <$ 5 GeV^2 will be forthcoming in the near future[11], derived using the ratio method as described in the previous section. The approximate coverage for these data is indicated in solid squares plotted at 0.55 in Figure 4. To go to higher Q^2 using the same technique requires an upgraded CLAS and Jefferson Lab. After the 12 GeV upgrade of CLAS and Jefferson Lab, CLAS will be capable of measuring G_M^n out to 14 GeV^2; at a luminosity of $10^{35} cm^{-2} s^{-1}$, the uncertainties are expected to be due entirely to systematic errors for reasonable run times. Figure 4 shows the expected quality and coverage of the data accessible after the 12 GeV upgrade.

Acknowledgments

This work was supported by the U. S. Department of Energy. The Southeastern Universities Research Association (SURA) operates the Thomas Jefferson National Accelerator Facility for the United States Department of Energy under contract number DE-AC05-84ER40150.

References

1. R. Hofstadter, R. W. McAllister, *Phys. Rev.* **98**, 217 (1955).
2. R. Frisch, O. Stern, *Z. Phys.* **85**, 4 (1933).
3. O. Gayou *et al.*, *Phys. Rev. Lett.* **88**, 092301 (2002).
4. A. Lung *et al.*, *Phys. Rev. Lett.* **70**, 718 (1993).
5. See the discussion in W. Xu *et al.*, *Phys. Rev. Lett.* **85**, 2900 (2000).
6. W. Melnitchouk, *Phys. Rev. Lett.* **86**, 36 (2001).
7. X. Ji, *Phys. Rev. Lett.* **78**, 610 (1997).
8. A. Radyushkin, *Phys. Rev.* **D 56**, 5524 (1997).
9. P. Stoler, *Phys. Rev. Lett.* **86**, 36 (2001).
10. The Lattice Hadron Physics Collaboration; see, e.g., "National Computational Infrastructure for Lattice Gauge Theory," proposal to the U.S. Department of Energy SciDAC (Scientific Discovery through Advanced Computing) program, J. Negele, W. Watson, *et al.* (2001), and documents referenced therein.
11. "The Neutron Magnetic Form Factor from Precision Measurements of the Ratio of Quasielastic Electron-Neutron to Electron-Proton Scattering in Deuterium," Jefferson Lab Experiment 94-017, W.K. Brooks, M.F. Vineyard, spokespersons.

PROSPECT FOR MEASURING G_E^N AT HIGH MOMENTUM TRANSFERS

B. WOJTSEKHOWSKI

Physics Division, Thomas Jefferson National Accelerator Facility,
12000 Jefferson Avenue, Newport News, VA 23606, USA
E-mail: bogdanw@jlab.org

Experiment E02-013, approved by PAC21, will measure the neutron electric form factor at Q^2 up to 3.4 $(GeV/c)^2$, which is twice that achieved to date. The main features of the new experiment will be the use of the electron spectrometer BigBite, a large array of neutron detectors, and a polarized $^3\vec{H}e$ target. We present the parameters and optimization of the experimental setup. A concept of an experiment for G_E^n where precision G_E^p data is used for calibration of the systematics of a Rosenbluth type measurement is also discussed.

1. Introduction

Elastic electron scattering, which in the one-photon approximation is characterized by two form factors, is the simplest exclusive reaction on the nucleon. It provides important ingredients to our knowledge of nucleon structure. There are well-founded predictions of pQCD for the Q^2 dependence of the form factors and their ratio in the limit of large momentum transfer [1]. Predictions of a fundamental theory always attract substantial attention from experimentalists. Recent surprising results on G_E^p show that the ratio G_E^p/G_M^p declines sharply as Q^2 increases, and therefore pQCD is not applicable up to 10 $(GeV/c)^2$. According to [2,3] the electric and magnetic form factors behave differently, starting at $Q^2 \approx 1$ $(GeV/c)^2$. The same mechanisms causing this deviation should also be present in the neutron. It is an intriguing question, how the ratio G_E^n/G_M^n develops in this Q^2 regime, where confinement plays an important role.

2. World data on G_E^n

The study of G_E^n has been a priority in electromagnetic labs for the last 15 years. Figure 1 presents recent data [4,5,6,7,8,9] along with points representing the accuracy of JLab experiments [10,11] which have already collected data, and the expected statistical accuracy of experiment E02-013. Presently

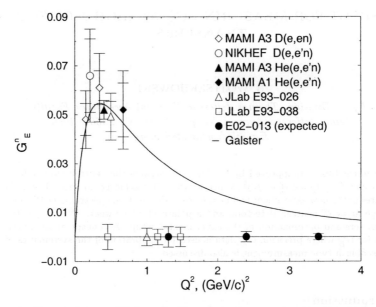

Figure 1. Recent data on G_E^n from double polarization experiments, the obtained accuracy and Q^2 of JLab experiments, and the expected accuracy of E02-013.

published results can be fitted by the Galster approximation [12]. The double polarization technique used in these experiments was introduced more than 20 years ago [14,15,16]. The experiments used a polarized electron beam and three different targets: unpolarized deuterium (together with a neutron polarimeter), polarized ND_3, and polarized ^3He.

3. Experiment E02-013

The steady progress of the E93-028 [10] and the E93-026 [11] experiments has made possible the accurate determination of G_E^n up to 1.47 $(GeV/c)^2$. The next step in Q^2 requires an experimental approach with much higher Figure-of-Merit (FOM).

In E02-013 [13] we optimized the setup in several respects:

- the solid angle of the electron spectrometer,
- the neutron detector efficiency and the trigger logic,
- the type of polarized target.

A recent addition in Hall A at JLab, the BigBite spectrometer developed by NIKHEF [17], has a 76 msr solid angle for a 40 cm long target. We found that for the identification of quasi-elastic scattering, the momentum

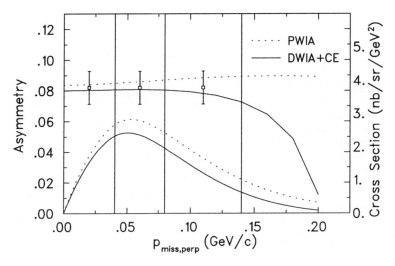

Figure 2. The GEA predictions for cross section and asymmetry in E02-013. The error bars show the expected accuracy. For each Q^2 the asymmetry will be measured for three values of $p_{miss,perp}$.

resolution of BigBite ($\approx 1\%$) is sufficient for electron momenta up to 1.5 GeV/c. The luminosity available with the $^3\vec{H}e$ target is about 10^{36} Hz/cm^2. According to our calculations it can be used with BigBite in spite of the direct view of the target by the detectors. Neutrons with kinetic energy above 1 GeV with which we have to deal at the proposed momentum transfers, can be efficiently detected with a relatively high detector threshold, which allows to suppress background and is crucial for the operation at the expected luminosity, which is about a factor of 10 higher than used in a recent JLab experiment [10] with a polarized ND$_3$ target.

In the last several years the theoretical development of the Generalized Eikonal Approximation (GEA) [18] has provided a framework for taking into account nuclear effects in the extraction of G_E^n from the experimental asymmetry. The GEA prediction for the asymmetry as a function of the missing transverse momenta $p_{miss,perp}$ is shown in Fig. 2. The GEA calculations and experimental data from JLab Hall B for the unpolarized reaction ^{3}He(e,e'p) have demonstrated the dominance of quasi-elastic scattering at $p_{miss,perp}$ below 0.15 GeV/c, when a modest cut of 0.5 GeV/c is applied on $p_{miss,parallel}$.

Table 1 summarizes the contributions to the error budget for the highest Q^2 point. For each Q^2 the measurement will be done with $\sim 14\%$ statistical accuracy for three intervals of $p_{miss,perp}$. As a result the systematics will be evaluated by comparison of an experimental asymmetry and the GEA prediction vs $p_{miss,perp}$.

Table 1. The contributions to the error budget in G_E^n for the data point at $Q^2=3.4$ $(GeV/c)^2$.

quantity	expected value	rel. uncertainty
statistical error in raw asymmetry A_{exp}	-0.0233	13.4%
beam polarization P_e	0.75	3%
target polarization P_{He}	0.40	4%
neutron polarization P_n	$0.86 \cdot P_{He}$	2%
dilution factor D	0.94	3%
dilution factor V	0.91	4%
correction factor for $A_{parallel}$	0.94	1%
G_M^n	0.057	5%
nuclear correction factor	1.0 – 0.85	5%
statistical error in G_E^n		13.8%
systematic error in G_E^n		10.4%

4. Future considerations

Experiment E02-013 is based on presently achieved parameters of the $^3\vec{H}e$ target and the existing electron spectrometer. With additional developments the FOM of the experiment can be increased by a factor of 5 and a measurement of G_E^n will be feasible at Q^2 up to 5 $(GeV/c)^2$.

4.1. Luminosity with the $^3\vec{H}e$ target

The present configuration of the $^3\vec{H}e$ target has the highest FOM at a beam current of 12-15 μA, when the beam-induced depolarization time is on the order of 30 hours. The use of the higher beam current requires a higher rate of polarizing and faster delivery of the polarized gas to the target cell. Advances in solid-state laser technology have made available 100 and even 200 W light power suitable for polarizing Rb atoms. Fig. 3 shows the target cell where the polarized gas flows through two tubes connecting the pumping and target cells. The flow will dramatically reduce the time for exchange of the polarized atoms between the pumping cell and the target cell.

4.2. High momentum spectrometer

The FOM of the experiment is approximately proportional to $E_f^2/E_i^2 = (E_i - Q^2/2M)^2/E_i^2$, where the $E_{i(f)}$ is the initial(final) electron energy. By using a beam energy of 7.8 GeV it is possible to increase the FOM by a

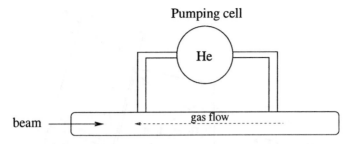

Figure 3. The target cell with two attachments to the pumping cell which allow the gas flow.

factor of 2.7 in comparison to the plan in E02-013 [13] for Q^2=3.4 (GeV/c)^2. It requires a new spectrometer for scattered electrons with a momentum 6 GeV/c and a solid angle of 75 msr. For Q^2= 5 (GeV/c)^2 the gain of FOM is 3.4. The relative momentum resolution should be of 0.5% to keep a W resolution sufficient for identification of the quasi-elastic events. The base component of the spectrometer is a dipole magnet with a 4.5 T·m field integral and a 35 cm open gap. The scheme of a spectrometer based on such a dipole magnet is shown in Fig. 4 [19]. We call it Super BigBite. Its characteristics are similar to BigBite, but with the momentum range extended by a factor of 5-8. As in the case of BigBite, the detector will be open to the target, so it can be used mainly with a polarized target luminosity.

5. Rosenbluth approach

In the Rosenbluth method the form factors ratio $g = G_E/G_M$ is obtained from two (or more) measurements at different beam energies at a fixed value of Q^2. The following equation is used to find g:

$$g^2 = \tau \cdot \frac{F_{\epsilon_1}^2 \epsilon_2^{-1} - F_{\epsilon_2}^2 \epsilon_1^{-1}}{F_{\epsilon_2}^2 - F_{\epsilon_1}^2} \tag{1}$$

where F is the total form factor measured experimentally, $F^2 = \left(G_E^2 + \frac{\tau}{\epsilon}G_M^2\right)/(1+\tau)$, $\tau = \frac{Q^2}{4M_N^2}$ and ϵ is virtual photon polarization. The uncertainty of g, which is growing with Q^2, can be estimated from the equation

$$\sigma(g^2) \approx \frac{\sigma(F_\epsilon^2)}{F_\epsilon^2} \frac{\sqrt{2} \cdot \tau}{\epsilon_1 - \epsilon_2} \tag{2}$$

278

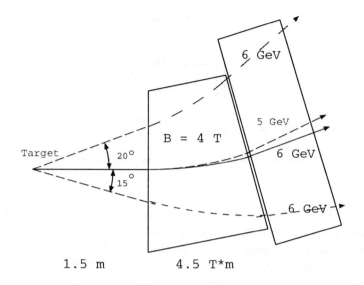

Figure 4. A side view of the Super BigBite.

where we neglect uncertainties in ϵ and τ. The total form factor is calculated from the event rates and other parameters of the experiment as

$$F^2 = N_{events}/[I_{beam} \cdot d_{target} \cdot t_{DAQ} \cdot \sigma_{Mott} \cdot \Omega_e \cdot \eta_e] \qquad (3)$$

Each of these experimental parameters - the beam current I_{beam}, the target density d_{target}, the data taking time t_{DAQ}, the Mott cross section σ_{Mott}, the detector solid angle Ω_e, and the detection efficiency η_e - is known with limited accuracy, which contributes to the systematics of the measurement. Some of them cancel in the calculation of the ratio g, because of the good stability of the target and detectors. A sufficiently accurate determination of the beam energy, the detector solid angle, and the scattering angle present a big challenge for the experiment. In the best case the overall systematic error is on the level of a few percent. By detecting the recoiling proton, as was suggested in LOI99-103 [20], the acceptance of the detector can be excluded from the list of problems, because at a given value of the proton momentum the solid angle of the detector is fixed. Experiment E01-001[21], which used such an approach, recently took data in JLab Hall A.

Quasi-elastic electron scattering from the deuteron $D(e,e'n)p$, with the ratio method suggested by Durand [22], has been used for determination of the neutron magnetic form factor in recent experiments at Bonn [23], Mainz [24], and JLab [25]. The same reaction can be used for measurement of the ratio G_E^n/G_M^n even with less stringent requirements on the knowledge of the absolute neutron detection efficiency. The small value of G_E^n made such measurement quite difficult; however, as we are proposing here, the problem can be solved by using the complementary $D(e, e'p)n$ reaction for calibration of the experiment. We will use the fact that in the Q^2 region of 5 $(\text{GeV/c})^2$ the ratio of the proton form factors G_E^p/G_M^p is already well known from JLab experiments [2,3]. In a dedicated experiment the accuracy of g_p can be improved to the level of 2-3%.

The proposed scheme will use the magnetic spectrometer as an electron arm and a non-magnetic detector as a hadron arm. The last one will consist of a large array of plastic scintillators and veto detectors. At a few $(\text{GeV/c})^2$ momentum transfer the kinetic energy of the recoiling nucleon is above 1 GeV and proton and neutron interactions with the detector are similar (nuclear interaction dominates). The neutron detection efficiency of different measurements will be similar to each other because of equal kinetic energy of the neutron in both measurements. Most of the remaining variations of the detector efficiency and solid angle will affect the same way the complementary reaction $D(e, e'p)n$.

The ratio $F_{\epsilon_2}^2/F_{\epsilon_1}^2$, which defines as the value of g_n, can be expressed in the proposed experiment as

$$\left(\frac{F_{\epsilon_2}^n}{F_{\epsilon_1}^n}\right)^2 = \left(\frac{F_{\epsilon_2}^p}{F_{\epsilon_1}^p}\right)^2 \cdot \frac{N_2^{e,e'n}}{N_1^{e,e'n}} \cdot \frac{N_1^{e,e'p}}{N_2^{e,e'p}} \cdot \frac{\Omega_{\epsilon_2}^n}{\Omega_{\epsilon_1}^n} \cdot \frac{\Omega_{\epsilon_1}^p}{\Omega_{\epsilon_2}^p} \cdot \frac{\eta_{\epsilon_2}^n}{\eta_{\epsilon_1}^n} \cdot \frac{\eta_{\epsilon_1}^p}{\eta_{\epsilon_2}^p} \qquad (4)$$

Several parameters such as the beam current, the electron-arm solid angle and efficiency, the Mott cross section, the data taking time, and the target parameters all cancel out from the final ratio of the form factors at two different values of ϵ. The remaining parameters are the neutron/proton detector solid angle Ω and efficiency η, whose variations for different ϵ need to be controlled.

For the proposed non-magnetic detector at large nucleon energy the neutron and proton detector efficiency will be almost equally affected by any change of rates and drifts of detector parameters, so it will be compensated. The detector solid angle is defined by the detector size. It can be well controlled and small changes will be the same for both proton and neutron channels.

The prospect of the Rosenbluth approach for a measurement of G_E^n depends on the high rate capability of the neutron detector. The potential FOM is higher than that possible in the double polarization approach by a factor 10-20, when it operates at a luminosity of 10^{38} Hz/cm^2. Experiment E93-038[11], which was done at a similar luminosity, developed the appropriate techniques for background reduction.

Conclusion

The experimental field of neutron electromagnetic form factors made very good progress in recent years. The present frontier for G_E^n is Q^2 above 2 $(\text{GeV/c})^2$. JLab experiment E02-013 will do the measurement of G_E^n up to $Q^2 = 3.4$ $(\text{GeV/c})^2$. There are possibilities of the further enhancements of the luminosity and polarization of the polarized $^3\vec{H}e$ target. The Rosenbluth approach may also be revived by using calibration on the proton G_E^p/G_M^p ratio.

Acknowledgments

It is pleasure to thank G. Cates, B. Reitz, and K. McCormick for collaboration in developing the E02-013 experiment. I would like also to acknowledge the crucial contributions made by P. Degterenko, V. Nelyubin, M. Sargsyan, and by members of the Hall A collaboration. Discussions with W. Brooks, K. de Jager, and B. Mecking on the possibility of the Rosenbluth approach for the G_E^n measurement are greatly appreciated. The Southeastern Universities Research Association operates the Thomas Jefferson National Accelerator Facility under Department of Energy contract DE-AC05-84ER40150.

References

1. S.J. Brodsky and G.P. Lepage, *Phys. Rev.* **D24**, 2848 (1981).
2. M. Jones *et al.*, *Phys. Rev. Lett.* **84**, 1398 (2000).
3. O. Gayou *et al.*, *Phys. Rev. Lett.* **88**, 092301 (2002).
4. I. Passchier *et al.*, *Phys. Rev. Lett.* **82**, 4988 (1999).
5. M. Ostrick *et al.*, *Phys. Rev. Lett.* **83**, 276 (1999).
6. D. Rohe *et al.*, *Phys. Rev. Lett.* **83**, 4257 (1999).
7. C. Herberg em et al., *Eur. Phys. J.* **A5**, 131 (1999).
8. H. Zhu *et al.*, *Phys. Rev. Lett.* **87**, 081801 (2001).
9. J. Golak *et al.*, *Phys. Rev.* **C63**, 034006 (2001).
10. D. Day, J. Mitchell, and G. Warren, spokespersons, JLab experiment E93-026.
11. R. Madey and S. Kowalski, spokespersons, JLab experiment E93-038.
12. S. Galster *et al.*, *Nucl. Phys.* **B32**, 221 (1971).
13. G. Cates, B. Reitz, K. McCormick, and B. Wojtsekhowski, spokespersons, JLab experiment E02-013.
14. N. Dombey, *Rev. Mod. Phys.* **41**, 236 (1969).
15. A.I. Akhiezer and M.P. Rekalo, *Sov. Phys. Doklady* **13/6**, 572 (1968), *Sov. J. Part. Nucl.* **3**, 277 (1974).

16. R. Arnold, C. Carlson, and F. Gross, *Phys. Rev.* **C23**, 363 (1981).
17. D.J.J. de Lange *et al.*, *Nucl. Instrum. Methods* **A406**, 182 (1998).
18. M.M. Sargsian, *Int. J. Mod. Phys.* **E10** 405, (2001); e-Print Archive: nucl-th/0110053.
19. V. Nelyubin, private communication.
20. W. Bertozzi, K. Fissum, D. Rowntree, and B. Wojtsekhowski, spokespersons, Letter-of-Intent 99-103 for JLab PAC18.
21. J. Arrington and R. Segal, spokespersons, JLab experiment 01-001.
22. L. Durand, *Phys. Rev.* **115**, 1020 (1959).
23. E.E.W. Bruins *et al.*, *Phys. Rev. Lett.* **75**, 21 (1995).
24. G. Kubon *et al.*, *Phys. Lett.* **B524**, 26 (2002).
25. W. Brooks, spokesperson, JLab experiment E94-017, also in this proceedings.

BARYON FORM FACTORS AT HIGH MOMENTUM
TRANSFER AND GPD'S

PAUL STOLER

Physics Department, Rensselaer Polytechnic Institute, Troy NY12180

Nucleon elastic and transition form factors at high momentum transfer $-t$ are treated in terms of generalized parton distributions in a two-body framework. In this framework the high $-t$ dependence of the form factors gives information about the high $k_\perp$, or short distance $b_\perp$ correlations of nucleon model wave functions. Applications are made to elastic and resonance nucleon form factors, and real Compton Scattering.

During the past several years there has been considerable discussion of how to describe exclusive reactions at momentum transfers which are experimentally attainable. While pQCD is an interesting mechanism which probes the simplest Fock state component of the hadron, most theoretical studies agree that even at the highest attainable momentum transfers, there is a large *soft* contribution which involves more complex components of the hadronic wave functions. The so-called handbag [1] mechanism has evolved to describe such soft processes, and achieves its full power at high momentum transfer where a process can be factorized into a fully perturbative hard amplitude and a *generalized parton distribution* (GPD) [2,3,4], which represents the off-diagonal probability of the interacting quark being placed back into the remaining hadron, keeping it intact at a different transferred longitudinal momentum. The power of the mechanism is that the same soft GPD, which contains the information about the hadronic structure is accessed in a variety of different reactions, while the hard perturbative part is reaction specific. The GPD's give us unique information about the longitudinal (x) and transverse ($k_\perp$) parton momentum distributions, and importantly, about the interference between the initial parton wave function and the phase shifted final parton wave function.

The GPD approach manifests itself in two kinematical regimes, corresponding to the t dependent *form factor type* reactions, and the $t \to t_{min}$ *off-forward* production of mesons or photons. Here we focus on the former. In such a reaction the incident real or virtual photon interacts perturbatively with one of the quarks within the hadron, which is re-absorbed into the hadron leaving it intact or in a higher resonant state. This is a Feynman type reaction which involves the full complexity of the non-perturbative nucleon structure, as opposed to the leading order pQCD mechanism, which involves only the valence quark Fock state. Form factors are the x moments of the GPD's, and as such constrain the longitudinal dependence of the nucleon structure. As a function of t they uniquely constrain the $k_\perp$ dependence of the nucleon's wave functions. Fourier transforms of the GPD's, $\mathcal{F}_b(x, \vec{b}_\perp) \propto \int d\vec{q}_\perp exp(i\vec{b}_\perp \cdot \vec{q}_\perp)\mathcal{F}(x,t)$, directly give the transverse spatial impact parameter distribution of the quarks for each longitudinal momentum fraction [5]. Thus, together with x distributions obtained in DIS, the $k_\perp$ accessed in form factor measurements give us a unique 3 dimensional picture of the quark distributions in the nucleon. Examples of reactions accessible via GPDs include the nucleon elastic Dirac and Pauli form factors F_1 and F_2 (or equivalently G_{Ep} and G_{Mp}), resonance transition amplitudes such as $A_{1/2}$ for $N \rightarrow S_{11}(1535)$, or G_M^* for $N \rightarrow \Delta$, and Compton scattering form factors, R_V and R_A, and their polarization asymmetries. The relationship of the GPD's to these various form factors is given as follows:

For elastic scattering

$$F_1(t) = \int_0^1 \sum_q \mathcal{F}^q(x,t)dx, \qquad F_2(t) = \int_0^1 \sum_q \mathcal{K}^q(x,t)dx, \qquad (1)$$

where q signifies both quark and anti-quark flavors. We work in a reference frame in which the total momentum transfer is transverse so that $\zeta=0$, and denote $\mathcal{F}^q(x,t) \equiv \mathcal{F}_0^q(x,t)$, $\mathcal{K}^q(x,t) \equiv \mathcal{K}_0^q(x,t)$.

For Compton scattering [6]

$$R_1(t) = \int_0^1 \sum_q \frac{1}{x}\mathcal{F}^q(x,t)dx, \qquad R_2(t) = \int_0^1 \sum_q \frac{1}{x}\mathcal{K}^q(x,t)dx. \qquad (2)$$

Resonance transition form factors access components of the GPD's which are not accessed in elastic scattering or Compton scattering. The $N \rightarrow \Delta$ form factors are related to isovector components of the GPD's [7,8].

$$G_M^* = \int_0^1 \sum_q \mathcal{F}_M^q(x,t)dx \qquad (3)$$

$$G_E^* = \int_0^1 \sum_q \mathcal{F}_E^q(x,t)dx \qquad (4)$$

$$G_C^* = \int_0^1 \sum_q \mathcal{F}_C^q(x,t)dx \qquad (5)$$

where G_M^*, G_E^* and G_C^* are magnetic, electric and Coulomb transition form factors [9], and $\mathcal{F}_M^q$, $\mathcal{F}_E^q$, and $\mathcal{F}_C^q$ are axial (isovector) GPD's, which can be related to elastic GPD's in the large N_c limit through isospin rotations [8]. The $N \to S_{11}$ transition form factor is also important, as it probes fundamental aspects of dynamical chiral symmetry breaking in QCD. If chiral symmetry were not broken, the S_{11} would be the nucleon's parity partner and the N and S_{11} masses would be degenerate.

As a basis for constructing the GPD's we use the two-body model [6] whose connection with the handbag is illustrated in Fig. 1.

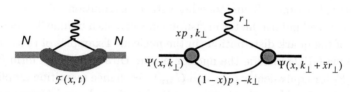

Figure 1.   Schematic relation between the two-body and handbag mechanisms discussed in the text.

In this framework the GPD is written as

$$\mathcal{F}(x,t) = \int \Psi^*(x, k_\perp + \bar{x}r_\perp)\Psi(x, k_\perp)\frac{d^2 k_\perp}{16\pi^3}, \tag{6}$$

where $\bar{x} \equiv 1 - x$.

An example of a specific model wave function [10] is

$$\Psi(x, k_\perp) = \Phi(x)\left(A_s e^{-k_\perp^2/2x\bar{x}\lambda^2} + A_h \frac{x\bar{x}\Lambda^2}{k_\perp^2 + \Lambda^2}\right) \equiv \Psi_{soft} + \Psi_{hard}. \tag{7}$$

The function $\Phi(x)$ is constrained so that $\mathcal{F}(x,0)$ reduces to the valence quark distribution $f(x)$. It was shown in Ref. 10 that although a Gaussian form of the $k_\perp$ dependence in Ψ_{soft} accounts for the magnitude and shape of the elastic F_1 for Q^2 below several GeV^2, it is inadequate at higher Q^2. However, the addition of a small Ψ_{hard} component in Eq. (7) can dramatically improve the agreement at high Q^2. As an example of a power law dependence, we choose an ad-hoc $1/k_\perp^2$ behavior with lower cutoff parameter Λ. A similar parametrization is chosen for F_2 with $\mathcal{K}^q(x,0) = \sqrt{(1-x)}\mathcal{F}^q(x,0)$. In order to constrain the parameters of Eq. (7) the available data on both G_{Mp} and G_{Ep}/G_{Mp} were simultaneously reproduced, giving $A_s = \sqrt{1 - A_h^2} = 0.97$, $A_H = 0.24$, $\lambda_1^2 = 0.6$ GeV^2 and $\lambda_2^2 = 0.45$ GeV^2. The function $\Psi(k_\perp) = \int \Psi(k_\perp, x)dx$ is shown in Fig. 2. Only at $k_\perp$ greater than about 1 GeV does the hard tail become important.

The fits to the data using respectively $\Psi = \Psi_{soft} + \Psi_{hard}$, and $\Psi = \Psi_{soft}$ are shown in Figs. 3,4, and 6.

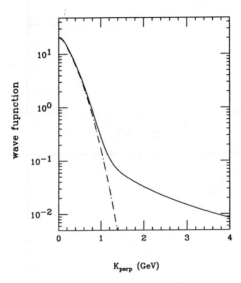

Figure 2. The function $\Psi(k_\perp) \equiv \int \Psi(x, k_\perp) dx$ vs. $k_\perp$. The dashed curve is due to the soft Gaussian component Ψ_{soft}, with $\lambda^2 = 0.6$ GeV^2. The solid curve is $\Psi_{soft} + \Psi_{hard}$, with $A_h = 0.24$, $k_{\perp,max} = 4$ GeV, and cutoff parameter $\Lambda = 0.45$ GeV.

As seen in the top panel of Fig. 3, this rather small addition of high momentum components can account for the high, as well as the low Q^2 magnetic form factor. Interestingly, Ref. 11 found that even in a pQCD calculation a power law tail is useful in reproducing the high Q^2 data.

Taking the Fourier transforms of the GPD's gives the spatial impact parameter distribution of the struck quarks. The bottom panel in Fig. 3 shows

$$\mathcal{F}_b(x, b_\perp) = \int dq_\perp e^{i\vec{b}_\perp \cdot \vec{q}_\perp} \mathcal{F}(x, t).$$

and the effect of Ψ_{hard}. Only a modest addition of small impact parameter components to the wave function accounts for most of the form factor at high Q^2.

In Fig. 4 the obtained values of G_{Ep}/G_{Mp} for $\Psi_{soft} + \Psi_{hard}$ and Ψ_{hard} alone are compared with the recent JLab data [15].

The obtained GPD's as a function of x and t are shown in Fig. 5.

One may apply the constraints of the elastic form factors to investigate properties of inelastic resonance transitions. For example, in the large N_c

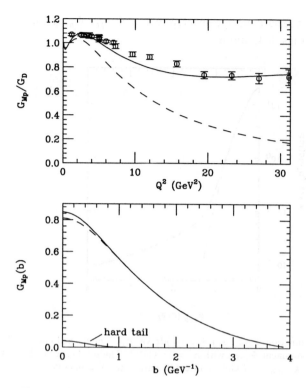

Figure 3. Upper: Proton magnetic form factor G_{Mp}/G_D, where $G_D = 1/(1 + Q^2/0.71)^2$. Data are from SLAC[9,10] with low energy data reevaluated[11]. The dashed curve uses only Ψ_{soft}, while the solid curve uses $\Psi_{soft} + \Psi_{hard}$. Lower: The impact parameter dependence of the curves in the upper figure, $G_{Mp}(b_\perp) = \int dx \mathcal{F}_b(x, b_\perp)$. The curve at the bottom left labelled "hard tail" is the difference between the solid and dashed curves, which is responsible for most of the form factor at high Q^2.

limit the GPDs for the $N \to \Delta(1232)$ transition are expected to be isovector components of the elastic GPD, which is approximately given by

$$\mathcal{F}_M^{(IV)} = \frac{2}{\sqrt{3}}\mathcal{K}_M^{(IV)} = \frac{2}{\sqrt{3}}\left(\mathcal{K}^u - \mathcal{K}^d\right),$$

where $\mathcal{K}^u$ and $\mathcal{K}^d$ are the GPD's for the up and down quarks respectively. Figure 6 shows the result of applying the GPD's from elastic scattering to the $N \to \Delta$ transition. The data was renormalized by the ratio $3/2.14$, to bring into line the nucleon isovector form factor at $Q^2=0$ with the experimental value for the $N \to \Delta$.

In summary, it is seen that complete knowledge of the various types of baryon form factors provides very strong constraints for model wave

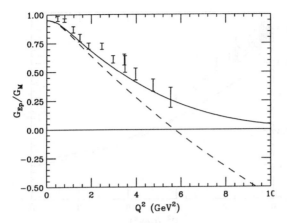

Figure 4. G_{Ep}/G_{Mp} for $\Psi_{soft} + \Psi_{hard}$ and Ψ_{hard} alone are compared with the recent JLab data[12]. The curves are as in Fig. 3.

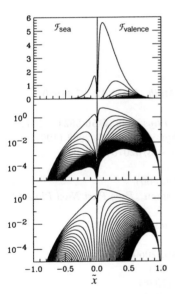 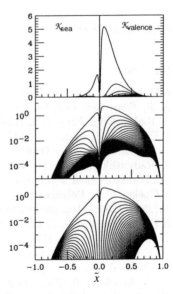

Figure 5. GPD's as a function of $\tilde{x}$ for various values of t, where $\tilde{x} = x$ for valence quarks, and $\tilde{x} = -x$ for the sea quarks. The figures on the left and right are for $\mathcal{F}$ and $\mathcal{K}$ respectively. The graphs for positive $\tilde{x}$ represent the *valence* quark contribution, while the graphs for negative $\tilde{x}$ represent the *sea* quark contributions. The individual curves range from $|t| \sim 0$ GeV^2 (highest curve in each panel) to $|t| = 35$ GeV^2 (lowest curve in each panel). The upper and middle panels are the GPD's for the full wave function $\Psi_{soft} + \Psi_{hard}$, while those in the lowest panels are obtained using the Ψ_{soft} soft only. Note that the addition of the Ψ_{hard} mainly affects the GPD's at higher $|t|$ and $\tilde{x} < 0.5$

288

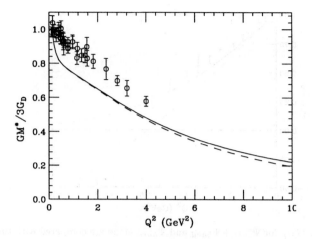

Figure 6. The $N \to \Delta$ magnetic form factor $G_M^*(Q^2)$ relative to the dipole $G_D = 3/(1 + Q^2/.71)^2$

functions and GPD's.

References

1. A.V. Efremov and A.V. Radyushkin, *Theor. Math. Phys.* **42**, 97 (1980)
2. X. Ji, *Phys. Rev. Lett.* **78**, 610 (1997);
3. A. Radyushkin, *Phys. Lett.* **B380**,417 (1996); *Phys. Rev.* **D56**,5524 (1997).
4. J. Collins, L. Frankfurt, and M. Strikman, *Phys. Rev.*, **D56**, 2982 (1997).
5. M. Burkardt,*Phys.Rev.***D62**,071503,(2000); hep-ph/0207047 (2002).
6. A. Radyushkin, *Phys. Rev.* **D58**,114008 (1998).
7. L.L.Frankfurt et al.,*Phys. Rev. Lett.* **84** ,2589 (2000).
8. K.Goeke, M.V. Polyakov, and M. Vanderhaeghen, *Prog.Part.Nucl.Phys.***47**, 401,2001 (2001); hep-ph/0106012.
9. H.F. Jones and M.D. Scadron, *Annals of Physics* **81**, 1 (1979).
10. P. Stoler, *Phys. Rev.* **D65**, 053013 (2002).
11. C.E. Carlson and F.L. Gross, *Phys. Rev.* **D36**, 2060 (1987).
12. L. Andivahis et al., *Phys. Rev.* **D50**, 5491 (1994).
13. R.G. Arnold et al., *Phys. Rev. Lett.* **57**, 174 (1986).
14. E.J. Brash et al. *Phys. Rev.* **C65**, 051001(R) (2002).
15. M.K. Jones et al. *Phys. Rev. Lett.* **84**,1398 (2000); O. Gayou et al. et al. *Phys. Rev.* **C64**,038202 (2001).

Q^2-EVOLUTION OF $\Delta N\gamma$ FORM FACTORS UP TO 4 $(GEV/C)^2$ FROM JLAB DATA

I.G.AZNAURYAN

Yerevan Physics Institute,
Alikhanian Brothers St. 2,
Yerevan, 375036 Armenia
E-mail: aznaury@jlab.org

We present the results on the ratios $E_{1+}^{(3/2)}/M_{1+}^{(3/2)}$ and $S_{1+}^{(3/2)}/M_{1+}^{(3/2)}$ for the $\gamma^* N \to \Delta(1232)$ transition at $Q^2 \leq 4 \ (GeV/c)^2$ extracted from the $p(e, e'p)\pi^0$ cross section using two approaches: dispersion relations and modified version of unitary isobar model. The obtained results are in good agreement with the results of other analyses obtained using truncated multipole expansion at $Q^2 = 0.4, \ 0.525, \ 0.65, \ 0.75, \ 0.9, \ 1.15, \ 1.45, \ 1.8 \ (GeV/c)^2$ and within dynamical and unitary isobar models at $Q^2 = 2.8, \ 4 \ (GeV/c)^2$. According to obtained results the ratio $E_{1+}^{(3/2)}/M_{1+}^{(3/2)}$ remains small in all investigated region of Q^2 with very unclear tendency to cross zero above 2 $(GeV/c)^2$. The absolute value of the ratio $S_{1+}^{(3/2)}/M_{1+}^{(3/2)}$ is clearly increasing with increasing Q^2, while it should be a constant value in the pQCD asymptotics. So, at $Q^2 \leq 4 \ (GeV/c)^2$ there is no evidence of approaching pQCD regime for these ratios. None of the soft approaches gives satisfactory description of the obtained results.

1. Introduction

It is known that for about 20 years the question: which is the scale of transition from soft to hard mechanism of QCD in exclusive processes, is the subject of controversy. Detail discussion of this problem can be found, for example, in papers [1,2,3]. The point of view, that this scale should be large, i.e. much larger than now available Q^2, is based mainly on the utilization of asymptotic wave function and is confirmed by the results obtained using local quark-hadron duality [4]. On the other hand there are arguments, based mainly on the utilization of the Chernyak-Zhitnitsky wave function [5], that hard mechanism of QCD can be observed at quite small Q^2. Experimental data on proton elastic form factors and form factors for the second and third resonance peaks extracted from inclusive data [3], indeed, manifest the features which are characteristic of pQCD starting with very small Q^2, about $2 - 3 \ GeV^2$. However, for $\gamma^* N \to \Delta(1232)$ the strong numerical suppression of the leading order amplitude is obtained using the wave functions of CZ type [6]. By this reason, in distinction to other form factors, the hard mechanism is expected for $\gamma^* N \to \Delta(1232)$ at much higher Q^2. Information on the Q^2- evolution of the ratios $E_{1+}^{(3/2)}/M_{1+}^{(3/2)}$, $S_{1+}^{(3/2)}/M_{1+}^{(3/2)}$ for the $\gamma^* N \to \Delta(1232)$ transition will allow to check this expectation, because the transition from soft to hard mechanism is characterized by a striking change in the behaviour of $E_{1+}^{3/2}/M_{1+}^{3/2}$, $S_{1+}^{3/2}/M_{1+}^{3/2}$ from

$$E_{1+}^{3/2}/M_{1+}^{3/2} \cong 0, \ S_{1+}^{3/2}/M_{1+}^{3/2} \cong 0 \tag{1}$$

at $Q^2 = 0$ to

$$E_{1+}^{3/2}/M_{1+}^{3/2} = 1, \ S_{1+}^{3/2}/M_{1+}^{3/2} = const \tag{2}$$

in the pQCD regime. By this reason, investigation of the Q^2-evolution of the ratios $E_{1+}^{(3/2)}/M_{1+}^{(3/2)}$, $S_{1+}^{(3/2)}/M_{1+}^{(3/2)}$ is very informative for understanding of the mechanisms and the scale of transition to pQCD regime.

In this report, we present the results on the ratios $E_{1+}^{(3/2)}/M_{1+}^{(3/2)}$, $S_{1+}^{(3/2)}/M_{1+}^{(3/2)}$ for the $\gamma^* N \to \Delta(1232)$ transition at $Q^2 = 0.4, 0.525, 0.65, 0.75, 0.9, 1.15, 1.45, 1.8, 2.8, 4 \ (GeV/c)^2$. These results are extracted from the JLab data [7,8] on $p(e, e'p)\pi^0$ cross section using two approaches: dispersion relations and modified version of the unitary isobar model of Ref. [9]. The detail description of the approaches is done in Ref. [10].

2. Results and discission

The obtained results for the ratios $E_{1+}^{(3/2)}/M_{1+}^{(3/2)}$, $S_{1+}^{(3/2)}/M_{1+}^{(3/2)}$ are presented in Figure 1.

In this Figure the results obtained by truncated multipole analysis of at $Q^2 = 0.4, 0.525, 0.65, 0.75, 0.9, 1.15, 1.45, 1.8 \ (GeV/c)^2$ [9] and using dynamical and unitary isobar models at $Q^2 = 2.8, 4 \ (GeV/c)^2$ [11] are also presented.

Let us note, that presented results correspond to the so called "dressed" $\gamma N \Delta$ vertex, i.e. are extracted from the whole magnitudes of multipoles $M_{1+}^{(3/2)}, E_{1+}^{(3/2)}, S_{1+}^{(3/2)}$ at the resonance position, where $\delta_{1+}^{(3/2)} = \pi/2$. Extraction of "bare" multipoles can be made only using models and is model-dependent.

Note also, that there are two sets of results at $Q^2 = 0.65, 0.75, 0.9 \ (GeV/c)^2$, which is connected with two kinds of measurements of $p(e, e'p)\pi^0$ cross section in [8] at different energies of initial electron.

From Figure 1 it is seen that the results obtained using different approaches agree with each other. The ratio $E_{1+}^{(3/2)}/M_{1+}^{(3/2)}$ remains small

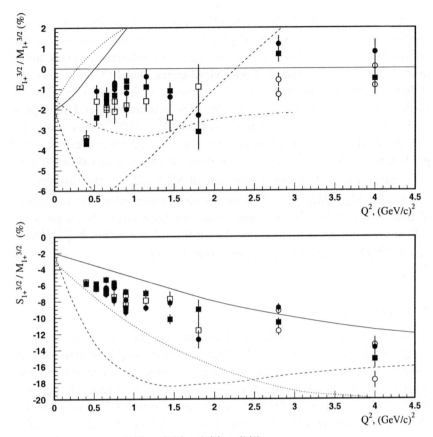

Figure 1. The ratios $E_{1+}^{(3/2)}/M_{1+}^{(3/2)}$, $S_{1+}^{(3/2)}/M_{1+}^{(3/2)}$ for the $\gamma^* N \to \Delta(1232)$ transition extracted from the $p(e,e'p)\pi^0$ cross section [7,8]. Results of our analysis are obtained using dispersion relations (full circles) and our modified version of unitary isobar model (full squares). Open squares correspond to truncated multipole analysis [8], open circles are obtained within dynamical and unitary isobar models in Ref. [11]. The predictions of the light-cone relativistic quark models are presented by solid [12] and dotted [13] curves, dashed-dotted curve corresponds to relativized quark model [14], and dashed curves are the results of Ref. [15].

in all Q^2 region up to 4 $(GeV/c)^2$, revealing some tendency to cross zero above $Q^2 > 2$ $(GeV/c)^2$. However, this tendency is not clear, and only measurements at higher Q^2 can clear up the situation here. The absolute value of the ratio $S_{1+}^{(3/2)}/M_{1+}^{(3/2)}$ is clearly increasing with increasing Q^2, and does not reveal the tendency of approaching the constant value. So, at $Q^2 \leq 4$ $(GeV/c)^2$ there is no indication of approaching pQCD regime for the ratios $E_{1+}^{(3/2)}/M_{1+}^{(3/2)}$ and $S_{1+}^{(3/2)}/M_{1+}^{(3/2)}$.

Therefore, it is reasonable to conclude, that the behaviour of the ratios

$E_{1+}^{(3/2)}/M_{1+}^{(3/2)}$ and $S_{1+}^{(3/2)}/M_{1+}^{(3/2)}$ at $Q^2 \leq 4$ $(GeV/c)^2$ is related to the soft mechanisms. By this reason in Figure 1 the predictions based on the soft approaches are presented. These are light-cone relativistic quark model predictions [12,13], predictions of relativized quark model [14], and the results of Ref. [15] obtained by interpolation between very low and very high Q^2. Both light-cone relativistic quark models give qualitatively good description of the ratio $S_{1+}^{(3/2)}/M_{1+}^{(3/2)}$, but in the case of $E_{1+}^{(3/2)}/M_{1+}^{(3/2)}$ they contradict the obtained results. In contrast with this, the relativized quark model predictions [14] for $E_{1+}^{(3/2)}/M_{1+}^{(3/2)}$ are close to the results extracted from experimental data, however, in the case of $S_{1+}^{(3/2)}/M_{1+}^{(3/2)}$ the prediction of this model: $S_{1+}^{(3/2)}/M_{1+}^{(3/2)} \simeq 0$, disagrees with these results. The predictions of Ref. [15] contradict the obtained results for both ratios. So, none of the soft approaches gives satisfactory description of the ratios $E_{1+}^{(3/2)}/M_{1+}^{(3/2)}$ and $S_{1+}^{(3/2)}/M_{1+}^{(3/2)}$ in the investigated region of Q^2.

3. Conclusion

In summary, we have analysed JLab data on $p(e,e'p)\pi^0$ cross section at $Q^2 \leq 4$ $(GeV/c)^2$ in the $\Delta(1232)$ resonance region within two approaches: dispersion relations and modified version of unitary isobar model. As a result, we have obtained information on Q^2-evolution of the ratios $E_{1+}^{(3/2)}/M_{1+}^{(3/2)}$ and $S_{1+}^{(3/2)}/M_{1+}^{(3/2)}$, which are of interest for understanding the scale and mechanisms of transition from soft to hard regime of QCD in exclusive processes. The obtained results show that there is no evidence yet of approaching pQCD regime for the $\gamma^* N \to \Delta(1232)$ transition, and for investigation of the scale, where hard mechanism for this transition begin to work, measurements at higher Q^2 are needed. We have compared the results on $E_{1+}^{(3/2)}/M_{1+}^{(3/2)}$ and $S_{1+}^{(3/2)}/M_{1+}^{(3/2)}$ with existing predictions obtained within soft approaches. It turned out, that none of the soft approaches gives simultaneously satisfactory description of these ratios. This means, that more detail investigations of soft mechanisms for $\gamma^* N \to \Delta(1232)$ are necessary.

4. Acknowledgements

I am thankful to V. D. Burkert, C. E. Carlson, A. V. Radyushkin and V. L. Chernyak for useful discussions. I express my gratitude for the hospitality at Jefferson Lab where this work was done.

References

1. N. Isgur and C. H. Lewellin Smith, *Nucl. Phys.* **B317**, 526 (1989).
2. V. M. Belyaev and A. V. Radyushkin, *Phys. Rev.* **D53**, 6509 (1996).
3. G. Sterman and P. Stoler, *Annu. Rev. Nucl. Part. Sci.* **47**, 193 (1997).
4. V. A. Nesterenko and A. V. Radyushkin, *Phys. Lett.* **B121**, 439 (1983).
5. V. L. Chernyak and A. R. Zhitnitsky, *Phys. Rep.* **112**, 173 (1984).
6. C. E. Carlson, M.Gari and N. G. Stefanis, *Phys. Rev. Lett.* **58**, 1308 (1987).
7. V. V. Frolov et al., *Phys. Rev. Lett.* **82**, 45 (1999).
8. K. Joo, L. C. Smith et al., *Phys. Rev. Lett.* **88**, 12201 (2002).
9. D. Drechsel, O. Hanstein, S. S. Kamalov and L. Tiator, *Nucl. Phys.* **A645**, 145 (1999).
10. I. G. Aznauryan, e-print, nucl-th/0206033, June 14, 2002.
11. S. S. Kamalov, S. N. Yang, D. Drechsel, O. Hanstein, and L. Tiator, *Phys. Rev.* **C64**, 033201 (2001).
12. I. G. Aznauryan, *Phys. Lett.* **B316**, 391 (1993).
13. F. Cardarelli, E. Pace, G. Salme and S. Simula, *Nucl. Phys.* **A623**, 361 (1997).
14. M. Warns and W. Pfeil, *Z. Phys..* **C45**, 627 (1990).
15. C. E. Carlson and N. C. Mukhopadhyay, *Phys. Rev. Lett.* **81**, 2646 (1998).

PROTON STRANGENESS VIA COMPTON SCATTERING

STEPHEN R. COTANCH

Department of Physics, North Carolina State University,
Raleigh, NC 27695-8202, USA
E-mail: cotanch@ncsu.edu

It is documented that $p(\gamma, e^+e^-)p$ measurements will yield new, important information about the off-shell time-like nucleon form factors, especially in the ϕ meson region ($q^2 = M_\phi^2$) governing the ϕN couplings $g_{\phi NN}^{V,T}$. The ϕN couplings are related to the proton's strangeness content and are determined independently from a combined analysis of the neutron electric form factor and recent high $|t|$ ϕ photoproduction data. Utilizing vector meson dominance, novel dual peaked resonances from ϕN coupling are predicted in the Compton scattering cross section.

1. Introduction

This work motivates time-like virtual Compton scattering [TVCS], $p(\gamma, e^+e^-)p$, measurements at high momentum transfer and reports cross section enhancements directly related to the strangeness content of the proton. Essentially assuming only vector meson dominance [VMD], distinctive cross section resonances are predicted for time-like virtual photon four-momentum spanning the vector meson masses ($q^2 \sim M_V^2$ for $V = \rho, \omega, \phi$). These order of magnitude peaks arise from vector mesons, associated with the outgoing virtual photon, coupling in the t channel to π, η and Pomeron $\mathcal{P}$ exchanges, and also to the proton through the time-like form factors (F_1^p, F_2^p or G_E^p, G_M^p) in the s and u channels. There is no nucleon form factor data for $0 \leq q^2 \leq 4M_p^2$ since all measurements have utilized the two-body annihilation processes $e^+e^- \leftrightarrow N\bar{N}$. However, $p(\gamma, e^+e^-)p$ permits kinematically accessing this region as it involves a three-body final state with an unrestricted virtual photon mass $q^2 \geq 4M_e^2 \sim 0$. For additional background consult ref. [1] and our previous study[2,3] of $p(\pi^-, e^+e^-)n$.

2. Theoretical details and results

The strategy is to first phenomenologically extract the ϕN couplings by fitting the sensitive neutron electric form factor, G_E^n, and ϕ photoproduction data and then predict $p(\gamma, \gamma_v)p$. The hadron form factors are computed

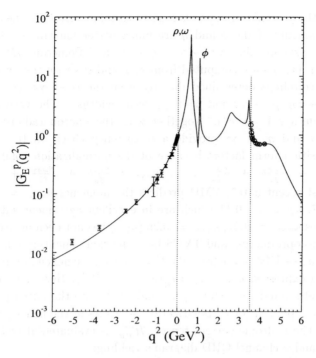

Figure 1.   Data and VMD (absolute value) for the proton electric form factor. Note the resonant peaks in the unmeasured time-like vector meson region.

utilizing a generalization[4,5] of the VMD model[6]. Sakurai's universality hypothesis and the $SU_f(3)$ symmetry relations are incorporated to give a good description of the baryon octet form factors, especially G_E^n which, due to its relatively small size, is very sensitive to the ϕN coupling. This yields the vector meson-nucleon couplings, $C_\rho(N) = 0.4$, $C_\omega(N) = 0.2$ and $C_\phi(N) = -0.1$, where $C_V(N) = g_{VNN}/f_V$ is the ratio of the vector meson-nucleon hadronic coupling, g_{VNN}, to the meson-leptonic decay constant, f_V. The $V \to e^+e^-$ decay widths determine the $V = \phi, \rho$ and ω decay constants, $f_\phi = -13.1$, $f_\rho = 5.0$, and $f_\omega = 17.1$, respectively, producing the ϕN vector coupling, $g_{\phi NN}^V = 1.3$. There is considerable uncertainty to both the vector and tensor, $g_{\phi NN}^T = \frac{\kappa_\phi^T}{M_\phi} g_{\phi NN}^V$, coupling constants. However, by analyzing recent ϕ photoproduction data (see below) much of this uncertainty can be reduced leading to the values $g_{\phi NN}^V = 1.3$, $g_{\phi NN}^T = 2.3$ and $\kappa_\phi^T = 1.8$. The vector ϕN coupling constant relative to ωN is $g_{\phi NN}^2/g_{\omega NN}^2 = 0.14$, slightly smaller but still consistent with refs. [7,11]. Figure 1 compares the VMD proton electric form factor to data. Note the dramatic, narrow resonance for $q^2 = M_\phi^2 \cong 1\, GeV^2$.

Although the proton's strangeness content remains uncertain, the u and d quark components of the ϕ and η are much better known. Again using VMD, the pseudoscalar, $\gamma\pi \to \gamma_v$, $\gamma\eta \to \gamma_v$ and Pomeron, $\gamma\mathcal{P} \to \gamma_v$ transition form factors are computed from ρ, ω and ϕ vector meson propagators with couplings determined directly from the $\phi \to \gamma\pi$, $\phi \to \gamma\eta$, $\omega \to \gamma\pi$, $\omega \to \gamma\eta$, $\rho \to \gamma\pi$ and $\rho \to \gamma\eta$ decay widths[8]. The vector meson leptonic decays $V \to e^+e^-$ together with the photon radiative decays $\pi \to \gamma\gamma$ and $\eta \to \gamma\gamma$ provide a consistency check on the VMD π and η transition form factors because of the normalization conditions $\kappa_{\pi\gamma\gamma} = \frac{\kappa_{\rho\pi\gamma}}{f_\rho} + \frac{\kappa_{\omega\pi\gamma}}{f_\omega} + \frac{\kappa_{\phi\pi\gamma}}{f_\phi}$, $\kappa_{\eta\gamma\gamma} = \frac{\kappa_{\rho\eta\gamma}}{f_\rho} + \frac{\kappa_{\omega\eta\gamma}}{f_\omega} + \frac{\kappa_{\phi\eta\gamma}}{f_\phi}$. With the most recent data[8], VMD predicts the moments $\kappa_{\pi\gamma\gamma} = 0.30$, $\kappa_{\eta\gamma\gamma} = 0.27$, $\kappa_{\mathcal{P}\gamma\gamma} = 0.11$ which are in excellent agreement with the observed values $\kappa_{\pi\gamma\gamma} = 0.27$, $\kappa_{\eta\gamma\gamma} = 0.26$ ($\kappa_{\mathcal{P}\gamma\gamma}$ has not been measured).

Since ϕ photoproduction and TVCS have the same quantum numbers, $\gamma(q,\lambda) + p(p,\sigma) \to V(q',\lambda') + p(p',\sigma')$, $V = \phi$ or γ_v, the helicity amplitude for both can be expressed as $T_{\lambda'\sigma'\lambda\sigma} = \epsilon_\mu(\lambda)\,\phi_\nu^*(\lambda')\,\mathcal{H}_{\sigma'\sigma}^{\mu\nu}$. Here q, p and λ, σ denote momentum and spin while $\epsilon_\mu(\lambda)$ and $\phi_\nu^*(\lambda')$ are the initial photon and final virtual photon (or ϕ) polarization 4-vectors in the helicity basis, respectively. The hadronic current tensor, $\mathcal{H}_{\sigma'\sigma}^{\mu\nu}$, is evaluated at tree level from the s, t and u channel QHD diagrams yielding:

t channel 0^+ Pomeron ($\mathcal{P}$) exchange:

$$\mathcal{H}_{\sigma'\sigma}^{\mu\nu} = \Gamma_{\mathcal{P}}\,\Pi_{\mathcal{P}}(t)\left(\frac{s - s_{th}}{s_0}\right)^{\alpha(t)}\bar{u}(p',\sigma')u(p,\sigma)\,[\,q \cdot q'\,g_{\mu\nu} - q'_\mu q_\nu\,]\,; \quad (1)$$

t channel 0^- meson ($x = \pi^0, \eta$) exchange:

$$\mathcal{H}_{\sigma'\sigma}^{\mu\nu} = \frac{\Gamma_x\,F_t(t;\lambda)}{M_\phi[(p'-p)^2 - M_x^2]}\,\bar{u}(p',\sigma')\,\gamma_5\,u(p,\sigma)\,\epsilon^{\mu\alpha\nu\beta}q_\alpha q'_\beta\,; \quad (2)$$

s channel proton (p) propagation:

$$\mathcal{H}_{\sigma'\sigma}^{\mu\nu} = \bar{u}(p',\sigma')\,[\Gamma_1'\,\gamma^\nu + i\Gamma_2'\sigma^{\nu\alpha}q'_\alpha\,]$$
$$\times \frac{(p+q)\cdot\gamma + M_p}{(p+q)^2 - M_p^2 + \Sigma_p(s)}\,[\,\Gamma_1\gamma^\mu + i\Gamma_2\sigma^{\mu\beta}q_\beta\,]\,u(p,\sigma)\,; \quad (3)$$

u channel channel proton (p) propagation:

$$\mathcal{H}_{\sigma'\sigma}^{\mu\nu} = \bar{u}(p',\sigma')\,[\Gamma_1\,\gamma^\mu + i\Gamma_2\sigma^{\mu\beta}q_\beta\,]$$
$$\times \frac{(p'-q)\cdot\gamma + M_p}{(p'-q)^2 - M_p^2 + \Sigma_p(u)}\,[\,\Gamma_1'\gamma^\nu + i\Gamma_2'\sigma^{\nu\alpha}q'_\alpha\,]\,u(p,\sigma)\,. \quad (4)$$

The scalar Pomeron coupling to the $\mathcal{P}NN$ and $\phi\mathcal{P}\gamma$ vertices is represented by $\Gamma_{\mathcal{P}}$, which is a product of Pomeron hadronic and electromagnetic coupling constants. The energy dependence, $[(s - s_{th})/s_0]^{\alpha(t)}$, is from Regge theory with Pomeron trajectory, $\alpha(t) = .999 + .27\ GeV^{-2}\ t$. The parameter s_{th} ($0 \leq s_{th} \leq s_0$) is introduced to describe the low energy dependence where $\sqrt{s_0}$ is the production threshold, $s_0 = (M_V + M_p)^2$ ($M_V = \sqrt{q'^2}$ or M_ϕ for $V = \gamma_v$ or ϕ). The ϕ photoproduction data clearly selects the maximum value, $s_{th} \rightarrow s_0$. The remaining Γ factors are products of hadronic and electromagnetic coupling constants (see ref. [1]) appropriate for either $V = \phi, \gamma_v$. Note in Fig. 2 the sensitivity and relative contributions from the $|t|$ channel Pomeron $\mathcal{P}$ (dense dotted curve), π (short dashed curve), η (sparce dotted curve) and s channel ϕN coupling (long dashed curve).

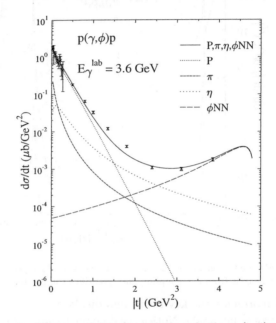

Figure 2.   Theory and data for ϕ photoproduction, $p(\gamma, \phi)p$.

The effective Pomeron propagator, $\Pi_{\mathcal{P}}(t)$, describes[9] scalar exchange, $\Pi_{\mathcal{P}}(t) = \frac{e^{\beta t}}{t - M_{\mathcal{P}}^2}$, and reproduces the known diffractive t dependence using the lightest scalar glueball mass, $M_{\mathcal{P}} = 1.7\ GeV$, and Pomeron slope $\beta = .27\ GeV^{-2}$. The pseudoscalar t channel form factor, $F_t(t; \lambda)$, governs hadronic structure and is necessary for the correct 4-momentum transfer dependence in meson photoproduction. Covariance and crossing symmetry are preserved using $F_t(t; \lambda) = \frac{\lambda^4 + t_{min}^2}{\lambda^4 + t^2}$, normalized to unity at $t_{min} = t$ ($\theta_{\gamma\phi}^{c.m.} = 0$). From $p(\gamma, \phi)p$ data, the optimum cutoff parameter is $\lambda = 0.7$ GeV. For the s and u channels we describe the off-shell proton by a self-energy correction, Σ_p, to the propagator. To maintain both gauge invariance and the correct on-shell proton mass, Σ_p must vanish at the proton pole and also be an odd function of $(s - M_p^2)$, $\Sigma_p(s) = \alpha_{off} \frac{(s-M_p^2)^3}{M_p^4}$. The dimensionless off-shell parameter, $\alpha_{off} = 1.29$, was also adjusted for optimal agreement with recent, and higher $|t|$, ϕ photoproduction data[10].

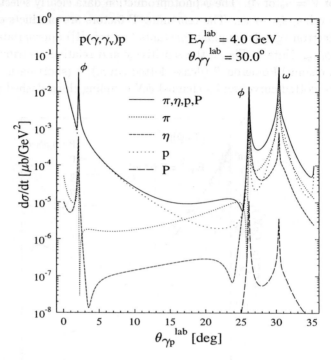

Figure 3. VMD prediction for TVCS, $p(\gamma, \gamma_v)p$, showing dual peak resonances. The smaller angle ϕ peak is from ϕN coupling and quantifies the proton's strangeness.

Figure 3 represents the key result and displays the TVCS cross section versus final proton lab angle. Notice the dual peak resonant signature due to the quadratic relation between q^2 and recoil proton lab angle. The smaller angle ϕ peak, corresponding to high $|t|$, is dominated by the u channel (proton propagator) $g_{\phi NN}$ coupling (sparce dotted curve labeled p). This is the prediction for measuring the proton strangeness content. The two other peaks near $25°$ and $30°$, have lower $|t|$ and respectively represent the more established ϕ and ω coupling to mesons in t channel exchange.

Measurements of the TVCS cross section ratio $R = \sigma(q^2 = M_\phi^2)/\sigma(q^2 = M_\omega^2)$ at high $|t|$ should yield a result proportional to $g_{\phi NN}^2/g_{\omega NN}^2$. For this model it has been confirmed numerically that $R = 0.14$ f (where f is a kinematic quantity of order unity), over an order of magnitude larger than the OZI prediction[11], $R = tan^2 \delta$ $f = 0.0042$ f, where $\delta = 3.7°$ is the deviation from the ideal quark flavor mixing angle in the ϕ.

Note also that TVCS probes the (half) off-shell proton form factors. Reference [12] addresses this issue further, including ambiguity in off-shell formulation. Here it is simply noted that both on and off-shell form factors contain important information about ϕN coupling in the time-like region and that the above ratio R will be less sensitive than the form factors to off-shell effects. A complete analysis also requires N^* resonances. However, by choosing large s and $|t|$ kinematics with $u \approx 0$, only the lightest baryon resonances will compete with the proton Born term $1/(u - M_p^2)$. Related, ref. [12] also details how the Bethe-Heitler process can be exploited to extract the TVCS amplitude by measuring the e^+e^- asymmetry.

3. Conclusion

Summarizing, it has been shown that VMD provides a good, comprehensive description of meson and nucleon electromagnetic observables and that both ϕ photoproduction and TVCS are sensitive to VMD parameters. Further, the TVCS process is ideal for investigating the unknown time-like nucleon form factor which will permit confirming the validity of VDM for the ρ and ω, which is anticipated, and also quantifying the ϕN coupling governing the proton's hidden strangeness.

Acknowledgments

This work was supported by grants DOE DE-FG02-97ER41048 and NSF INT-9807009. Contributions from R. A. Williams were significant.

References

1. S. R. Cotanch and R. A. Williams, nucl-th/0205061 (2002).
2. R. A. Williams and S. R. Cotanch, Phys. Rev. Lett. **77**, 1008 (1996).
3. S. R. Cotanch and R. A. Williams, Nucl. Phys. **A631**, 478 (1998).
4. R. A. Williams, S. Krewald, and K. Linen, Phys. Rev. C **51**, 566 (1995).
5. R. A. Williams and C. Puckett-Truman, Phys. Rev. C **53**, 1580 (1996).
6. M. Gari and W. Krümpelmann, Phys. Lett. **B274**, 159 (1992).
7. J. Ellis, E. Gabathuler, and M. Karliner, Phys. Lett. **B217**, 173 (1989).
8. Particle Data Group, Eur. Phys. J. C **15**, 1-877 (2000).
9. R. A. Williams, Phys. Rev. C **57**, 223 (1998).
10. CLASS Collaboration, E. Anciant *et al.*, Phys. Rev. Lett. **85**, 4682 (2000).
11. J. Ellis, M. Karliner, D. E. Kharzeev, and M. G. Sapozhnikov, Phys. Lett. **B353**, 319 (1995).
12. A. Yu. Korchin, O. Scholten, and F. de Jong, Phys. Lett. **B402**, 1 (1997).

MESON ELASTIC AND TRANSITION FORM FACTORS

P. MARIS

Dept. of Physics, North Carolina State University,
Box 8202, Raleigh, NC 27695-8202, USA
E-mail: pmaris@unity.ncsu.edu

The Dyson–Schwinger equations of QCD, truncated to ladder-rainbow level, are used to calculate meson form factors in impulse approximation. The infrared strength of the ladder-rainbow kernel is described by two parameters fitted to the chiral condensate and f_π; the ultraviolet behavior is fixed by the QCD running coupling. This obtained elastic form factors $F_\pi(Q^2)$ and $F_K(Q^2)$ agree well with the available data. We also calculate the $\rho \to \pi\gamma$ and $K^\star \to K\gamma$ transition form factors, which are useful for meson-exchange models.

1. Dyson–Schwinger equations

The set of Dyson–Schwinger equations [DSEs] form a useful tool to obtain a microscopic description of hadronic properties[1]. Here we use the DSEs to calculate elastic and transition form factors of the light mesons. The dressed quark propagator, as obtained from its DSE, together with the meson Bethe–Salpeter amplitude [BSA] and the dressed quark-photon vertex, form the necessary elements for calculations of form factors in impulse approximation, such as the pion elastic form factor[2].

The DSE for the renormalized quark propagator in Euclidean space is[1]

$$S(p)^{-1} = i\,Z_2\,\slashed{p} + Z_4\,m(\mu) + Z_1 \int \frac{d^4q}{(2\pi)^4}\, g^2 D_{\mu\nu}(p-q)\, \tfrac{\lambda^i}{2}\gamma_\mu\, S(q)\, \Gamma^i_\nu(q,p)\,, \quad (1)$$

where $D_{\mu\nu}(k)$ is the dressed-gluon propagator and $\Gamma^i_\nu(q;p)$ the dressed-quark-gluon vertex. The most general solution of Eq. (1) has the form $S(p)^{-1} = i\slashed{p}A(p^2) + B(p^2)$ and is renormalized at spacelike μ^2 according to $A(\mu^2) = 1$ and $B(\mu^2) = m(\mu)$ with $m(\mu)$ the current quark mass.

Mesons are described by solutions of the Bethe–Salpeter equation [BSE]

$$\Gamma_H(p_+, p_-; Q) = \int \frac{d^4q}{(2\pi)^4}\, K(p,q;Q)\, S(q_+)\,\Gamma_H(q_+, q_-; Q)\, S(q_-)\,, \quad (2)$$

at discrete values of $Q^2 = -m_H^2$, where m_H is the meson mass. In this equation, $p_+ = p + \eta Q$ and $p_- = p - (1 - \eta)Q$ are the outgoing and incoming quark momenta respectively, and similarly for q_+. The kernel K is the renormalized, amputated $q\bar{q}$ scattering kernel that is irreducible with respect to a pair of $q\bar{q}$ lines. Together with the canonical normalization condition for $q\bar{q}$ bound states, Eq. (2) completely determines the bound state BSA Γ_H. Different types of mesons, such as pseudoscalar or vector mesons, are characterized by different Dirac structures.

The quark-photon vertex, $\Gamma_\mu(p_+, p_-; Q)$, with Q the photon momentum and $p_\pm$ the quark momenta, is the solution of the inhomogeneous BSE

$$\Gamma_\mu(p_+, p_-; Q) = Z_2\, \gamma_\mu + \int \frac{d^4q}{(2\pi)^4}\, K(p, q; Q)\, S(q_+)\, \Gamma_\mu(q_+, q_-; Q)\, S(q_-)\,. \quad (3)$$

Solutions of the homogeneous version of Eq. (3) define vector meson bound states at timelike photon momenta $Q^2 = -m_V^2$. It follows that $\Gamma_\mu(p_+, p_-; Q)$ has poles at these locations[2,3].

To solve the BSE, we use a ladder truncation,

$$K(p, q; P) \to -\alpha^{\rm eff}\left((p-q)^2\right) D^0_{\mu\nu}(p-q)\tfrac{\lambda^i}{2}\gamma_\mu \otimes \tfrac{\lambda^i}{2}\gamma_\nu\,, \quad (4)$$

in conjunction with the rainbow truncation for the quark DSE, Eq. (1): $\Gamma^i_\nu(q, p) \to \gamma_\nu\lambda^i/2$ and $Z_1 g^2 D_{\mu\nu}(k) \to 4\pi\alpha^{\rm eff}(k^2)D^0_{\mu\nu}(k)$. Here, $D^0_{\mu\nu}(k)$ is the free gluon propagator in Landau gauge, and $\alpha^{\rm eff}(k^2)$ the effective quark-quark interaction, which reduces to the one-loop QCD running coupling $\alpha^{1-\rm loop}(k^2)$ in the perturbative region. For the infrared behavior of the interaction, we employ an Ansatz[4,5] that is sufficiently strong to produce a realistic value for the chiral condensate of about $(240\,{\rm GeV})^3$. The model parameters[5], along with the quark masses, are fitted to give a good description of the chiral condensate, $m_{\pi/K}$ and f_π. The obtained results for the light vector meson masses are within 5% of their experimental values, and the vector meson electroweak decay constants are within 9% of the data[5].

This truncation preserves both the vector Ward–Takahashi identity [WTI] for the $q\bar{q}\gamma$ vertex and the axial-vector WTI, independent of the details of the effective interaction. The latter ensures the existence of massless pseudoscalar mesons associated with dynamical chiral symmetry breaking[4,6]. In combination with impulse approximation, the former ensures electromagnetic current conservation[3].

2. Meson Form Factors

In impulse approximation, meson form factors are generically described by

$$I^{abc}(P, Q, K) = N_c \int \frac{d^4q}{(2\pi)^4}\, {\rm Tr}\big[S^a(q)\, \Gamma^{a\bar{b}}(q, q'; P)S^b(q')$$
$$\Gamma^{b\bar{c}}(q', q''; Q)\, S^c(q'')\, \Gamma^{c\bar{a}}(q'', q; K)\big]\,, \quad (5)$$

302

where $q - q' = P$, $q' - q'' = Q$, $q'' - q = K$, and momentum conservation dictates $P + Q + K = 0$. In Eq. (5), S^i is the dressed quark propagator with flavor index i, and $\Gamma^{ij}(k, k'; P)$ stands for a generic vertex function with incoming quark flavor j and momentum k', and outgoing quark flavor i and momentum k. Depending on the specific process under consideration, this vertex function could be a meson BSA or a quark-photon vertex. In the calculations discussed below, the propagators and the vertices are all obtained as solutions of their respective DSE in rainbow-ladder truncation, without adjusting any of the model parameters.

2.1. Pion and kaon elastic form factors

There are two diagrams that contribute to meson electromagnetic form factors: one with the photon coupled to the quark and one with the photon coupled to the antiquark respectively. With photon momentum Q, and incoming and outgoing meson momenta $P \mp Q/2$, we can define a form factor for each of these diagrams[3]

$$2 P_\nu F_{a\bar{b}\bar{b}}(Q^2) = I_\nu^{abb}(P - Q/2, Q, -(P + Q/2)).\qquad(6)$$

We work in the isospin symmetry limit, and thus $F_\pi(Q^2) = F_{u\bar{u}u}(Q^2)$. The K^+ and K^0 form factors are given by $F_{K^+} = \frac{2}{3}F_{u\bar{s}u} + \frac{1}{3}F_{u\bar{s}\bar{s}}$ and $F_{K^0} = -\frac{1}{3}F_{d\bar{s}d} + \frac{1}{3}F_{d\bar{s}\bar{s}}$ respectively.

Our result for $Q^2 F_\pi$ and F_{K^+} are shown in Fig. 1, together with the experimental data[7,8,9,10]. Up to about $Q^2 = 2\,\mathrm{GeV}^2$, our result for F_π can be described very well by a monopole with our calculated ρ-mass, $m_\rho = 742\,\mathrm{MeV}$ (note that our calculated ρ-mass is slightly below the experimental

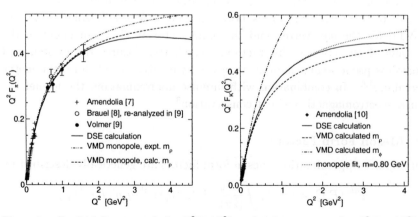

Figure 1. On the left, our result for $Q^2 F_\pi(Q^2)$, and right, our curve for $Q^2 F_{K^+}(Q^2)$.

value). Above this value, our curve starts to deviate more and more from this naive VMD monopole. Our result is in excellent agreement with the most recent JLab data[9]; it would be very interesting to compare with future JLab data in the 3 to 5 GeV^2 range, where we expect to see a significant deviation from the naive monopole behavior.

Also our results for F_K agree with the available experimental data[10], as do both the neutral and the charged kaon charge radius[3]. Again, our curve for F_K can be fitted quite well by a monopole up to about $Q^2 = 2\,\text{GeV}^2$, see Fig. 1; above 2 GeV^2, we see a clear deviation from a monopole behavior. Also for this form factor it would be interesting to compare with future JLab data at larger Q^2.

2.2. *Vector-pseudoscalar-photon transitions*

We can describe the radiative decay of the vector mesons using the same loop integral, Eq. (5), this time with one vector meson BSA, one pseudoscalar BSA, and one $q\bar{q}\gamma$-vertex[11]. The on-shell value gives us the coupling constant. For virtual photons, we can define a transition form factor $F_{VP\gamma}(Q^2)$, normalized to 1 at $Q^2 = 0$, which can be used in estimating meson-exchange contributions to hadronic processes[12,13].

In the isospin limit, both the $\rho^0\,\pi^0\,\gamma$ and $\rho^\pm\,\pi^\pm\,\gamma$ vertices are given by

$$\frac{1}{3} I^{uuu}_{\mu\nu}(P, Q, -(P+Q)) = \frac{g_{\rho\pi\gamma}}{m_\rho}\, \epsilon_{\mu\nu\alpha\beta} P_\alpha Q_\beta\, F_{\rho\pi\gamma}(Q^2)\,, \qquad (7)$$

where P is the ρ momentum. The $\omega\,\pi\,\gamma$ vertex is a factor of 3 larger, due to the difference in isospin factors. For the $K^\star \to K\gamma$ decay, we have to add two terms: one with the photon coupled to the $\bar{s}$-quark and one with the photon coupled to the u- or d-quark, corresponding to the charged or neutral $K^\star$ decay respectively[11].

As Eq. (7) shows, it is $g_{VP\gamma}/m_V$ that is the natural outcome of our calculations; therefore, it is this combination that we give in Table 1, together with the corresponding partial decay widths[11]. The agreement between theory and experiment for $g_{VP\gamma}/m_V$ is within about 10%, except for the discrepancy in the charged $K^\star \to K\gamma$ decay for which we have no explanation. Likewise the large difference between the neutral and charged ρ decay width is beyond the reach of the isospin symmetric impulse approximation.

Table 1. Vector meson radiative decays: g/m in GeV^{-1} and $\Gamma_{V \to P\gamma}$ in keV.

	g/m	$\Gamma_{\rho^\pm \pi^\pm \gamma}$	g/m	$\Gamma_{\omega\pi\gamma}$	g/m	$\Gamma_{K^{\star\pm} K^{\pm}\gamma}$	g/m	$\Gamma_{K^{\star 0} K^0 \gamma}$
calc.	0.69	53	2.07	479	0.99	90	1.19	130
expt.[14]	0.74	68	2.31	717	0.83	50.3	1.28	116

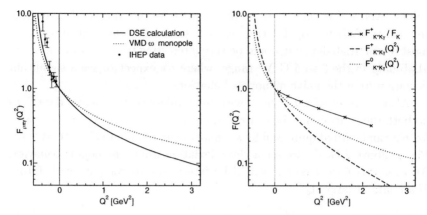

Figure 2. On the left, our result for $F_{\omega\pi\gamma}(Q^2)$ with experimental data[15], and right, our curve for the charged and neutral $F_{K^*K\gamma}(Q^2)$.

Note that part of the difference between the experimental and calculated decay width comes from the phase space factor because our calculated vector meson masses deviate up to 5% from the physical masses.

The corresponding transition form factors are shown in Fig. 2. In contrast to the elastic form factors, these transition form factors fall off significantly faster than a VMD-like monopole; our numerical results[11] for $F_{\rho\pi\gamma}(Q^2)$ suggest an asymptotic behavior of $1/Q^4$. Only in the timelike region, near the vector meson pole, do we see a clear VMD-like behavior.

For the $K^*K\gamma$ form factor the situation is more complicated due to the interference of the diagrams with the photon coupled to the s-quark and to the u- or d-quark. This causes the charged form factor to fall off much more rapidly than both the neutral $K^*K\gamma$ form factor and the elastic form factor $F_K(Q^2)$, as can be seen from Fig. 2. The latter implies that the contribution to charged kaon electroproduction from intermediate K^* exchange gets suppressed with increasing Q^2 compared to the contribution from virtual kaon exchange.

2.3. Remarks on meson electroproduction

We can use this approach to estimate the range of validity of meson-dominance models[11]. For off-shell pions, the meson-dominance assumption appears to be quite good for spacelike momentum transfer of the order of $t \sim 0.1$ GeV^2. On the other hand, for heavier mesons such as kaons or ρ-mesons, the naive meson-dominance assumption introduces significant errors, even at small spacelike values of $t \sim 0.2$ GeV^2. In addition, any meson-exchange model to describe meson electroproduction necessarily introduces off-shell ambiguities. Clearly, a microscopic description is needed.

Any microscopic description of meson-electroproduction requires a quark-gluon description of the nucleon. It is a difficult task to combine such a description with the meson form factors considered here. Recently, significant progress in describing processes involving four external particles has been made, using the rainbow-ladder truncation of the DSEs, in conjunction with an extention of the impulse approximation[16]. This approach incorporates the non-analytic effects of intermediate meson-exchange contributions while avoiding ambiguous off-shell definitions. It has succesfully been applied to π-π scattering, where we can identify the non-analytic contributions from σ- and ρ-exchange to the S-wave and P-wave scattering amplitudes respectively[17]. Furthermore, it was shown to reproduce the correct chiral limit[16]. We plan to extend this approach to describe processes such as $\gamma\pi\pi$, pion compton scattering, and meson electroproduction.

Acknowledgments

Most of this work was done in collaboration with Peter Tandy. This work was funded by the US Department of Energy under grants No. DE-FG02-96ER40947 and DE-FG02-97ER41048, and benefitted from the resources of the National Energy Research Scientific Computing Center.

References

1. C. D. Roberts and S. M. Schmidt, Prog. Part. Nucl. Phys. **45S1**, 1 (2000); R. Alkofer and L. von Smekal, Phys. Rept. **353**, 281 (2001);
2. P. Maris and P. C. Tandy, Phys. Rev. C **61**, 045202 (2000).
3. P. Maris and P. C. Tandy, Phys. Rev. C **62**, 055204 (2000).
4. P. Maris and C. D. Roberts, Phys. Rev. C **56**, 3369 (1997).
5. P. Maris and P. C. Tandy, Phys. Rev. C **60**, 055214 (1999).
6. P. Maris, C. D. Roberts and P. C. Tandy, Phys. Lett. B **420**, 267 (1998).
7. S. R. Amendolia *et al.*, Nucl. Phys. B **277**, 168 (1986).
8. P. Brauel *et al.*, Z. Phys. **C3**, 101 (1979).
9. J. Volmer *et al.*, Phys. Rev. Lett. **86**, 1713 (2001).
10. S. R. Amendolia *et al.*, Phys. Lett. B **178**, 435 (1986).
11. P. Maris and P. C. Tandy, Phys. Rev. C **65**, 045211 (2002).
12. P. C. Tandy, Prog. Part. Nucl. Phys. **39**, 117 (1997).
13. J. W. Van Orden, N. Devine and F. Gross, Phys. Rev. Lett. **75**, 4369 (1995).
14. Particle Data Group, C. Caso *et al.*, Eur. Phys. J. **C3**, 1 (1998).
15. R. I. Dzhelyadin *et al.*, Phys. Lett. B **102**, 296 (1981) [JETP Lett. **33**, 228 (1981)].
16. P. Bicudo *et al.*, Phys. Rev. D **65**, 076008 (2002).
17. S. Cotanch and P. Maris, in preparation.

THE PION FORM FACTOR

H. P. BLOK

Department of Physics, Vrije Universiteit, Amsterdam, The Netherlands
E-mail: henkb@nat.vu.nl

G. M. HUBER

Department of Physics, University of Regina, Regina SK, S4S 0A2 Canada
E-mail: huberg@uregina.ca

D. J. MACK

Physics Division, Jefferson Laboratory, Newport News, VA 23606, USA
E-mail: mack@jlab.org

The experimental situation with regard to measurements of the pion charge form factor is reviewed. Both existing data and planned experiments are discussed.

1. Introduction

The pion, and specifically its charge form factor, is of key interest in the study of the quark-gluon structure of hadrons. This is exemplified by the many calculations that treat the pion as one of their prime examples. One of the reasons is that the valence structure of the pion, being $\langle q\bar{q} \rangle$, is relatively simple. Hence it is expected that the value of the four-momentum transfer squared Q^2, down to which a pQCD approach to the pion structure can be applied, is lower than for the nucleon. Whereas, e.g., the proton form factors seem to be completely dominated by constituent quark properties [1] up to at least $Q^2 = 10$ - 20 $(\text{GeV/c})^2$, recent estimates [2] suggest that pQCD contributions start to dominate the pion form factor at $Q^2 \geq 5$ $(\text{GeV/c})^2$. Furthermore, the asymptotic normalization of the pion wave function, in contrast to that of the nucleon, is known from the pion decay. Within perturbative QCD one can then derive [3]

$$\lim_{Q^2 \to \infty} F_\pi = \frac{8\pi \alpha_s f_\pi^2}{Q^2}, \tag{1}$$

where f_π is the pion decay constant. The question is down to which finite value of Q^2 this relation is valid. Thus the interest is in the transition from the soft regime, governed by all kinds of quark-gluon correlations, at low Q^2, to the perturbative (including next-to-leading order and transverse corrections) regime at high Q^2.

The charge form factor of the pion at very low values of Q^2, which is governed by the charge radius of the pion, has been determined [4] up to $Q^2 = 0.28$ $(\text{GeV}/c)^2$ from scattering high-energy pions from atomic electrons. For the determination of the pion form factor at higher values of Q^2 one has to use high-energy electroproduction of pions on a nucleon, i.e., employ the $^1\text{H}(e, e'\pi^+)n$ reaction. For selected kinematical conditions this process can be described as quasi-elastic scattering of the electron from a virtual pion in the proton. The cross section for this process can be written as

$$\frac{d^3\sigma}{dE' d\Omega_{e'} d\Omega_\pi} = \Gamma_V \frac{d^2\sigma}{dt d\phi}, \tag{2}$$

where Γ_V is the virtual photon flux factor, ϕ is the azimuthal angle of the outgoing pion with respect to the electron scattering plane and t is the Mandelstam variable $t = (p_\pi - q)^2$. The two-fold differential cross section can be written as

$$2\pi \frac{d^2\sigma}{dt d\phi} = \epsilon \frac{d\sigma_{\text{L}}}{dt} + \frac{d\sigma_{\text{T}}}{dt} + \sqrt{2\epsilon(\epsilon+1)} \frac{d\sigma_{\text{LT}}}{dt} \cos\phi$$

$$+ \epsilon \frac{d\sigma_{\text{TT}}}{dt} \cos 2\phi, \tag{3}$$

where ϵ is the virtual-photon polarization parameter. The cross sections $\sigma_X \equiv \frac{d\sigma_X}{dt}$ depend on W, Q^2 and t. In the t-pole approximation the longitudinal cross section σ_{L} is proportional to the square of the pion form factor:

$$\sigma_L \propto \frac{-t\, Q^2}{(t - m_\pi^2)^2} F_\pi^2. \tag{4}$$

The ϕ acceptance of the experiment should be large enough for the interference terms σ_{LT} and σ_{TT} to be determined. Then, by taking data at two energies at every Q^2, σ_{L} can be separated from σ_{T} by means of a Rosenbluth separation.

2. Existing measurements

The pion form factor has been studied for Q^2 values from 0.4 to 9.8 $(\text{GeV}/c)^2$ at CEA/Cornell [5]. For Q^2 above 1.6 $(\text{GeV}/c)^2$ these are at present still the only existing data. In these experiments only in a few cases was an L/T separation performed, and even then the resulting uncertainties in σ_{L} were so large that the L/T separated data were not used.

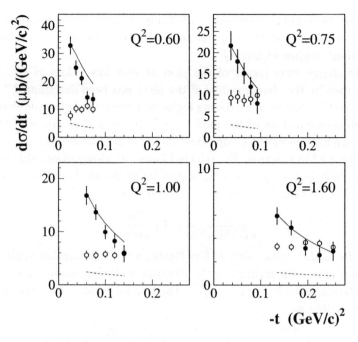

Figure 1. Separated cross sections σ_L and σ_T (full and open symbols, resp.) compared to the Regge model (full curve for L, dashed curve for T). The Q^2 values are in units of $(\mathrm{GeV}/c)^2$.

Instead, for the actual determination of the pion form factor, σ_L was calculated by subtracting from the measured (differential) cross section a σ_T that was assumed to be proportional to the total virtual photon cross section, and no uncertainty in σ_T was included in this subtraction. This means that the published values of F_π have large additional model uncertainties on top of the already relatively large statistical (and systematic) uncertainties.

The pion form factor was also studied at DESY [6] for $Q^2 = 0.7\ (\mathrm{GeV}/c)^2$. In this case a full separation of all structure functions was performed. We will come back to these data.

Recently the pion form factor was studied [7] at CEBAF for $Q^2 = 0.6$ - 1.6 $(\mathrm{GeV}/c)^2$. Using the High Momentum Spectrometer and the Short Orbit Spectrometer of Hall C and electron energies between 2.4 and 4.0 GeV, data for the reaction ${}^1\mathrm{H}(e, e'\pi^+)n$ were taken for central values of Q^2 of 0.6, 0.75, 1.0 and 1.6 $(\mathrm{GeV}/c)^2$, at a central value of the invariant mass W of 1.95 GeV. Because of the excellent properties of the electron beam and experimental setup, L/T separated cross sections could be determined with high accuracy.

The extracted cross sections are displayed in Figure 1. The error bars represent the combined statistical and systematic uncertainties. As a result of the Rosenbluth separation the total error bars on σ_L are enlarged considerably, resulting in typical error bars of about 10%.

In order to determine the value of F_π, the experimental data were compared to the results of a Regge model by Vanderhaeghen, Guidal and Laget (VGL) [8]. In this model the pion electroproduction process is described as the exchange of Regge trajectories for π and ρ like particles. The VGL model is compared to the data in Figure 1. Here the value of F_π, which is a parameter in the model, was adjusted at every Q^2 to reproduce the σ_L data at the lowest value of $-t$. The transverse cross section σ_T is underestimated, which can possibly be attributed to resonance contributions at $W = 1.95$ GeV that are not included in the Regge model.

The t-pole dominance for σ_L at small $-t$ was checked by studying the reactions ^{2}H$(e, e'\pi^+)nn$ and ^{2}H$(e, e'\pi^-)pp$, which gave within the uncertainties a ratio of unity for the longitudinal cross sections.

The comparison with the σ_L data shows that the t dependence in the VGL model is less steep than that of the experimental data. As suggested by the analysis [9] of older data, where a similar behaviour was observed, we attributed this discrepancy to the presence of a small negative background contribution to the longitudinal cross section, presumably again due to resonances. The values of F_π, extracted taking this into account, are shown in Figure 2.

For consistency we have determined F_π in the same way from the cross sections at $Q^2 = 0.7$ (GeV/c)^2, $W = 2.19$ GeV from DESY [6]. The background term in σ_L was found to be smaller than in the Jefferson Lab data, presumably because of the larger value of W. The resulting best value for F_π, also shown in Figure 2, is larger by 12% than the original result, which was obtained by using the Born term model by Gutbrod and Kramer [9]. Those authors used a phenomenological t-dependent function, whereas the Regge model by itself gives a good description of the t-dependence of the (unseparated) data from Ref. [5].

The data for F_π in the region of Q^2 up to 1.6 (GeV/c)^2 globally follow a monopole form obeying the pion charge radius [4]. It should be mentioned that the older Bebek data in this region suggested lower F_π values. However, as mentioned, they did not use L/T separated cross sections, but took a prescription for σ_T. Our measured data for σ_T indicate that the values used were too high, so that their values for F_π came out systematically low.

In Figure 2 the data are also compared to a sample of theoretical calculations. The model by Maris and Tandy [10] provides a good description of

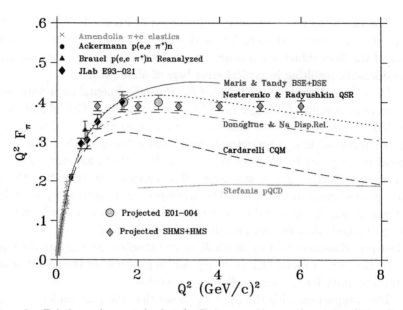

Figure 2. Existing and expected values for F_π in comparison to the results of several calculations. The model uncertainty is estimated to be about 5%. The (model-independent) data from Ref. [4] are also shown. A monopole behaviour of the form factor obeying the measured charge radius is almost identical to the Maris and Tandy curve.

the data. It is based on the Bethe-Salpeter equation with dressed quark and gluon propagators, and includes parameters that were determined without the use of F_π data. The data are also well described by the QCD sum rule plus hard scattering estimate of Ref. [11]. Other models [12,13] were fitted to the older F_π data and therefore underestimate the present data. Figure 2 also includes the results from a perturbative QCD calculation [14]. Apart from the basic dependence given by Eq. 1, but extended to next-to-leading order, it includes transverse momenta of the quarks, Sudakov factors, and a way to regularize the infrared divergence. As a result the value of $Q^2 F_\pi$ is about constant at 0.18 over the whole range of Q^2 shown. Other pQCD calculations yield similar results, but with a lower value of $Q^2 F_\pi$ [15]. Hence it is clear that in the region below $Q^2 \approx 2$ (GeV/c)^2, where accurate data exist, soft contributions are much larger than pQCD ones. For this reason it is highly interesting to get reliable data at higher values of Q^2.

3. Future experiments

The JLab experiment will be extended in the year 2003. Data will be taken at $Q^2 = 2.5$ (GeV/c)^2, the highest value compatible with the present set-up,

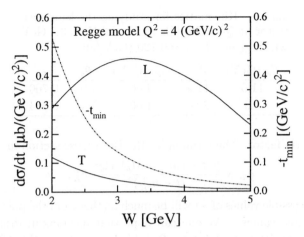

Figure 3. Dependence of the values of $-t_{min}$ and of σ_L and σ_T at t_{min}, calculated with the Regge model, on W.

which is determined by the combination of the maximum momentum of the SOS spectrometer (1.75 GeV/c) and the minium angle of the HMS spectrometer (10.5 degrees). The value of W will be 2.20 GeV. The increased value of W gives a smaller value of $-t_{min}$, closer to the pole, and is in a region where the Regge model is supposed to be more reliable. Data will also be taken at $Q^2 = 1.6$ (GeV/c)^2 with $W = 2.20$ GeV. By comparing those to the existing ones, taken at $W = 1.95$ GeV, the model dependence in the extraction of F_π will be gauged.

With the planned upgrade of CEBAF to 12 GeV the pion form factor can be studied [16] up to $Q^2 \approx 6$ (GeV/c)^2 with $W \geq 3.0$ GeV. The HMS spectrometer will now be used to detect the scattered electron, while the pion will be detected in the proposed SHMS spectrometer. Since the direction of $\vec{q}$ is rather forward at high Q^2 and W, the small angle capability of SHMS is essential. Taking advantage of the higher incoming energy, the value of W can be increased, with even more of the benefits described above. As a result, contrary to common belief, the ratio of σ_L over σ_T will still be favourable. This is illustrated in Figure 3. The value of σ_T decreases with W due to kinematical factors, but for σ_L this reduction is more than compensated for by the value of $-t_{min}$ getting smaller, i.e., closer to the pole.

Figure 2 shows what data will be obtained, with the expected experimental accuracy. The model uncertainty is estimated to be about 5%.

With the 12 GeV upgrade one can also start to think about studying

312

Table 1. Kinematics for studying the kaon form factor at $Q^2 = 2.0$ (GeV/c)^2 and $W = 3.3$ GeV, which gives $-t_{min} = 0.120$ (GeV/c)^2.

E_e (GeV)	$\theta_{e'}$	$E_{e'}$ (GeV)	θ_q	ϵ
11.0	11.4	4.60	7.98	0.700
8.0	22.8	1.60	5.43	0.364

the kaon form factor. The formula for the Born cross section in this case is

$$\sigma_L \propto \frac{[-t + (M_\Lambda - M_p)^2]\, Q^2}{(t - m_K^2)^2}\, F_K^2. \tag{5}$$

Clearly, accessible values of $-t$ will be much further from the pole. However, by using large values of W one can hope that meaningful results can be obtained up to $Q^2 \approx 2$ (GeV/c)^2. Possible kinematics at this Q^2 are given in table 1. In the analysis one has also to take into account two-step processes like forming first a K^* particle, which then decays into a K.

References

1. G. Miller, this workshop
2. W. Schweiger, *Nucl. Phys. Proc. Suppl.* **108**, 242 (2002)
3. G.R. Farrar, D.R. Jackson, *Phys. Rev. Lett.* **43**, 246 (1979)
4. S. R. Amendolia *et al.*, *Nucl. Phys.* **B277**, 168 (1986)
5. C. J. Bebek *et al.*, *Phys. Rev.* **D17**, 1693 (1978)
6. P. Brauel *et al.*, *Z. Phys.* **C3**, 101 (1979)
7. J. Volmer *et al.*, *Phys. Rev. Lett.* **86**, 1713 (2001); J. Volmer, PhD thesis, Vrije Universiteit, Amsterdam (2000), unpublished
8. M. Vanderhaeghen, M. Guidal and J.-M. Laget, *Phys. Rev.* **C57**, 1454 (1998); *Nucl. Phys.* **A627**, 645 (1997)
9. F. Gutbrod and G. Kramer, *Nucl. Phys.* **B49**, 461 (1972)
10. P. Maris and P. C. Tandy, *Phys. Rev.* **C62**, 055204 (2000)
11. V. A. Nesterenko and A. V. Radyushkin, *Phys. Lett.* **B115**, 410 (1982)
12. F. Cardarelli *et al.*, *Phys. Lett.* **B332**, 1 (1994); *Phys. Lett.* **B357**, 267 (1995)
13. J.F. Donoghue and E.S. Na, *Phys. Rev.* **D56**, 7073 (1997)
14. N.G. Stefanis, W. Schroers and H.-Ch. Kim, *Phys. Lett.* **B449**, 299 (1999); *Eur. Phys. J.* **C18**, 137 (2000)
15. R. Jakob and P. Kroll, *Phys. Lett.* **B315**, 463 (1993) and **B319**, 545 (1993); *J. Phys. G.* **22**, 45 (1996)
16. "The Science Driving the 12 GeV Upgrade of CEBAF", *Jefferson Lab Report*, Febr. 2001

PROTON FORM FACTORS FOR TIMELIKE MOMENTUM TRANSFERS

KAMAL K. SETH

Northwestern University, Evanston, IL 60208, USA
E-mail: kseth@nwu.edu

The significance and importance of the timelike form factors is reviewed. The existing experimental data for protons is discussed, and the plans for future measurements at squared momentum transfers up to 20 GeV^2 are presented. It is pointed out that the planned measurements of $e^+e^- \rightarrow p\bar{p}$ will allow simultaneous measurements of pion and kaon form factors, and also of heavier baryons like deltas and lambdas.

1. Introduction

One may be fascinated by neutrinos which oscillate, or kaons or B's which break one's cherished symmetries, or one may give in to de Beer's seductions, but the plain simple truth is that ONLY PROTONS ARE FOREVER. This being true, understanding the structure of the *proton* is the most important responsibility of nuclear and particle physicists. Nuclear physicists will likely put the effective carrier of nuclear force, the pion, at the same level of priority, and we will accept that.

Pion structure is simple. With only a quark and an antiquark, and zero spin, pions are easy for theorists. They are difficult for experimentalists because *pion targets do not exist.*

Proton structure is difficult. With three quarks and half integral spin, protons are difficult for theorists. They are easier for experimentalists because *proton targets exist.* Being an experimentalist let me concentrate on the proton first.

2. Form Factors

Electromagnetic form factors are the quintessential observables for understanding the structure of a composite object.[1] A real or virtual photon is used as a probe which couples to the distribution of charges and currents in the composite object, as it transfers a finite amount of four-momentum to it.

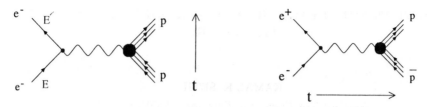

Figure 1. Diagrams illustrating spacelike momentum transfer in elastic scattering of electrons from protons (left), and timelike momentum transfer in e^+e^- annihilation into $p\bar{p}$ (right).

With the advent of QCD, the baseline model for form factors becomes part of the more general problem of *exclusive reaction in QCD*. In their pioneering work, Brodsky and Lepage,[2] introduced the idea of *factorization*, namely that the amplitude for an exclusive process can be factorized into two parts. One part contains the hadron distribution function, $\phi(x_i, Q^2)$ with all the non-perturbative aspects of the i constituents of the hadron, each carrying a fraction x_i of the hadron's total momentum Q^2. The other part is a hard scattering part T_H, which can be calculated in perturbative QCD, because the scattering consists of a series of short distance interactions between quarks with longitudinal momentum fractions x_i. This description leads to *QCD counting rules*,[3] which stipulate that for a large momentum transfer, Q, the form factors are proportional to $Q^{(2-2n)}\alpha_S^2(Q^2)$. This implies that the magnetic and electric form factors of a baryon (n=3), $G_M^p(Q^2)$ and $G_E^p(Q^2)$ are both proportional to $Q^{-4}\alpha_S^2(Q^2)$, and the meson (n=2) form factor is proportional to $Q^{-2}\alpha_S^2(Q^2)$.

Momentum transfers can be *spacelike* ($Q^2 > 0$) or *timelike* ($Q^2 < 0$), as Fig. 1 illustrates. The spacelike and timelike form factors are obviously related by crossing symmetry, i.e., they must be identical by virtue of the property of analytic functions at very large argument, i.e.,

$$\lim_{Q^2 \to -\infty} F(Q^2) = \lim_{Q^2 \to +\infty} F(Q^2)$$

We now specialize our discussion to protons.

3. Proton Form Factors
3.1. *Proton spacelike form factors*

These are measured by elastic scattering of electrons by protons (Fig. 1(a)) with four momentum transfer, $Q^2 = -q^2 = (4EE'/c^2)sin^2(\theta/2)$ at scattering angle θ. With $\tau = (Q/2m_pc^2)^2$, the differential cross section is given by

$$d\sigma/d\Omega = (d\sigma/d\Omega)_{Mott}[G_E^2(Q^2)+\tau G_M^2(Q^2)(1+2(1+\tau)tan^2(\theta/2))]/(1+$$

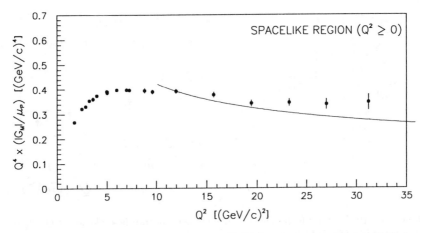

Figure 2. Proton form factors for spacelike momentum transfers, plotted at $Q^4 G^p_M/\mu_p$. The curve is a pQCD fit for α_s^2 variation.

τ)

The form factors are normalized as $G_E(Q = 0) = 1, G_M(Q = 0) = \mu_p = 2.79$. As is well known, because of the factors τ and $tan^2(\theta/2)$ the magnetic form factor $G_M(Q^2)$ dominates over the electric form factor G_E at large Q^2 and large scattering angles θ. This makes it difficult to measure G_E reliably at large Q^2.[4] However, $G_M(Q^2)$ has been measured quite relaibly upto Q^2 = 31.3 GeV^2.[5] Above 9 GeV^2, $G_E(Q^2) = G_M(Q^2)/\mu_p$ was assumed.[6] In Fig. 2, $Q^4 G_M(Q^2)/\mu_p$ is shown as a function of Q^2. Also shown is an arbitrarily normalized curve, $C\alpha_S^2(Q^2)$, with $C = 10.3$. We notice that above about 10 GeV^2 this curve describes the data fairly well, and one may very well conclude, as is indeed often done, that pQCD is applicable above ≈ 10 GeV^2.

3.2. Proton timelike form factors

These can be measured either by electron-positron annihilation creating a proton-antiproton pair, $e^+e^- \to p\bar{p}$, or by the inverse reaction $p\bar{p} \to e^+e^-$. Either way, Q^2 must be greater than $(2m_p)^2$. The differential cross sections are given by Ref. 7.

$$d\sigma_{e^+e^- \to p\bar{p}}/d\Omega = (d\sigma_{p\bar{p} \to e^+e^-}/d\Omega)\beta_p^2$$
$$= (\hbar c)^2 (\alpha m_p c^2/s)^2 \beta_p [G_E^2 sin^2\theta + \tau G_M^2(1 + cos^2\theta)]$$

The cross sections are so small that only cross sections integrated over the acceptance of the detector have ever been reported. In order to deter-

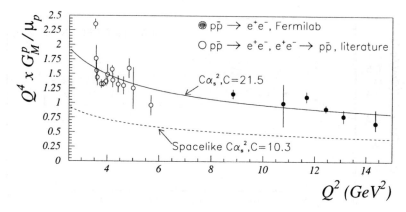

Figure 3. Proton form factors for timelike momentum transfers, plotted at $Q^4 G_M^p/\mu_p$. The curve is a pQCD fit for α_s^2 variation.

mine $G_M^p(Q^2)$ it is generally assumed that $G_E(Q^2) = G_M(Q^2)/\mu_p$. Experiments at Orsay (1979-1990)[8] measured $\sigma(e^+e^- \to p\bar{p})$ and reported $G_M(Q^2)$ in the range $Q^2 = 3.75$-5.69 GeV^2 with quoted errors of $\pm 10\%$ to 20%. Experiments at CERN (1977-1991) measured $\sigma(p\bar{p} \to e^+e^-)$ and reported $G_M(Q^2)$ in the range $Q^2 = 3.52 - 4.18$ GeV^2, with quoted errors of $\sim 5\% - 10\%$.[9] These measurements were generally fitted in the framework of the Vector Dominance Model (VDM), because, admittedly, Q^2 was not large enough to invoke pQCD.

Our E760 experiment at Fermilab was primarily devoted to the study of charmonium resonances by their formation in $p\bar{p}$ annihilation, and their decay into photons and/or electrons. Thus, whenever we were not scanning a vector resonance, J/ψ or ψ', measurement of e^+e^- yield gave us a direct measurement of the timelike form factor of the proton. In this way, E760 first measured timelike form factors at Q^2 (GeV^2) $= 8.840(\eta_c), 12.43(h_c)$ and $13.11(\eta_c')$.[10] The results (shown in Fig. 3, along with the latest averages) were gratifying in the sense that $Q^4 G_M^p(Q^2)$ appeared to show the same $C\alpha_S^2(Q)$ behavior as the spacelike form factors, but the results were also surprising in that the timelike magnetic form factor $G_M^p(Q^2)$ was nearly twice as large and the spacelike $G_M^p(Q^2)$ at the same Q^2.

Since the publication of the E760 results, we have made additional measurements of the timelike form factors. The results from our 1996-97 run of experiment E835 have been published.[11] There are newer preliminary results from our year 2000 run of experiment E835p. The averages of all measurements, including those from E835p are shown in Fig. 3, with the solid curve illustrating a $C\alpha_S^2(C = 21.5)$ fit. For comparison, the dashed curve which fit the spacelike form factors, with $C = 10.28$ is also shown.

3.3. Theoretical Predictions

The E760 results were the first to extend the data for the timelike form factors to above 5 GeV^2, and they made it possible to immediately test two preexisting calculations, both in the Vector Dominance Model. Compared to our experimental results, the predictions of Körner and Kureda[12] are factors 3 to 4 smaller, and those of Dubnicka[13] are factors 3 to 5 smaller. Dubnicka fits the spacelike form factors to tune her parameters, so that her result corresponds to the ratio timelike/spacelike = 1/3 to 1/5, whereas our experimental result is ≈ 2.

There are several calculations of spacelike form factors of nucleons based on QCD sum-rules. However, none are available in this framework for timelike form factors.

There are few QCD based predictions of the timelike form factors of the proton. Hyer[14] calculated the timelike formfactor $F_1(Q^2)$, or the Dirac form factor, using several 'candidate' proton distribution amplitudes and the pQCD hard scattering approximation with Sudakov supprexion a la Sterman.[15] With the assumption $G_E(Q^2) = G_M(Q^2)$, which was one of the assumptions we made in presenting our results, $G_M(Q^2) = F_1(Q^2)$, Hyer compared his predictions for $F_1(Q^2)$ with our results. Hyer obtained results for $Q^4 F_1(Q^2)$ which were essentially constant with Q^2 for each of the several assumed distribution functions for the proton. This is in disagreement with the trend of the data shown in Fig. 3.

The only calculation of both spacelike and timelike form factors within one single framework was made by Kroll et al.[16] Both $F_1(Q^2)$ and $F_2(Q^2)$ were calculated in the QCD factorization formalism with Sudakov suppression. The unique feature of this calculation was the assumption of a diquark-quark model for the proton wave function. This, of course, introduces spin-isospin scalar and vector diquarks with their own form factors and cut-offs, which are adjusted to simultaneously fit both the spacelike and timelike form factors. The model has been criticized for having succeeded by introducing extra parameters, but the fact remains that it is the only calculation so far which fits both spacelike and timelike data quite well. As an aside, we also note that the diquark-quark model allows violation of the Hadron Helicilty Conservation rule in $p\bar{p}$ annihilation. This rule is indeed strongly violated by the strong population of spin zero $\eta_c(^1S_0)$ and $\chi_{c0}(^3P_0)$ charmonium resonances in $p\bar{p}$ annihilation.[17]

There is no specific theoretical calculation

for the ratio $G_M^p(timelike)/G_M^p(spacelike) \approx 2$ observed in our data for protons, but there are several, often contradictory, suggestions of how this may arise, usually based on the simpler analysis of the *pion* form factors.

Gousset and Pire[18] have calculated timelike and spacelike form factors of pions in the pQCD factorization approach with evolution and Sudakov effects taken into account. Although their prediction for the absolute value of timelike F_π at $Q^2 \approx 9.6$ GeV^2 turns out to be approximately factor two smaller than the best experimental result, they do obtain $F_\pi(timelike)/F_\pi(spacelike) \approx 2$. They also find that the approach to the asymptotic equality of the timelike and spacelike form factors is very slow. Gousset and Pire have so far not delivered the promised corresponding calculation for protons. However, their pion calculation has been used by others to assume that the proton problem is also solved. Gousset and Pire themselves only claimed that in the simple diquark-quark model of the proton a result 'similar' to their pion result should follow.[18]

Bakulev, Radyushkin, and Stefanis[19] have recently reexamined various possible contributions to the continuation from spacelike to timelike momentum transfers. They conclude that there is no significant change in the running coupling constant or in the perturbative QCD contribution to the hadronic form factors. However, for the pion form factors they show that the Sudakov effects in the soft contributions do lead to a nearly factor two enhancement of the timelike form factors over spacelike form factors at large Q^2. They also fit the absolute values of both form factors.

One is tempted to believe that for pions the relationship between timelike and spacelike form factors is under theoretical control, but a note of caution is in order because this conclusion is based on experimental data which have $\approx \pm 50\%$ errors for spacelike $F_\pi(Q^2)$ for $Q^2 > 2$ GeV^2 and for timelike $F_\pi(Q^2)$ for $Q^2 > 5$ GeV^2. For protons we do not really even have such an explanation. So where do we go from here?

3.4. *The timelike future*

As experimentalists the only way we can contribute to this dialog is by providing results for timelike form factors at larger Q^2 and with higher precision. This is possible at CLEO-c/CESR-c, the new incarnation of the Cornell electron-positron facility, whose collaboration we have recently joined. For the reaction $e^+e^- \to p\bar{p}$, we expect the following integrated cross sections, and estimated number of identified events. (See Fig. 4)

$$Q^2 = 9.0 \text{ GeV}^2, \quad \sigma = 10,200 \text{ fb}, \quad N \approx 6000/\text{fb}^{-1},$$
$$Q^2 = 15 \text{ GeV}^2, \quad \sigma = 560 \text{ fb}, \quad N \approx 350/\text{fb}^{-1},$$

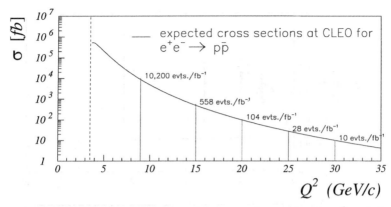

Figure 4. Calculated cross sections for $\sigma(e^+e^- \to p\bar{p})$ assuming the α_s^2 variation of $Q^4 G_M^p/\mu_p$.

$$Q^2 = 20 \text{ GeV}^2, \qquad \sigma = 100 \text{ fb}, \qquad N \approx 60/\text{fb}^{-1},$$

based on extrapolating our measured cross sections for $Q^{-4}\alpha_s^2$ variation of $G_M(Q^2)$, and a conservative estimate of 60% overall efficiency for good acceptance and identification.

CLEO-c plans to invest ≈ 1 fb^{-1} at J/ψ, and 0.25 fb^{-1} in background measurements in its vicinity, say at $\sqrt{s} = 3.0$ GeV, so that we should have easily an $N = 1500$ count measurement at $Q^2 = 9.0$ GeV^2. This should enable us, for the first time ever, to measure the angular distribution of the cross sections, and thus do a reasonable separation of G_E and G_M. At $Q^2 = 15$ GeV^2 ($\sqrt{s} = 3.87$ GeV), we expect to record ~ 100 events for a ~ 0.3 fb^{-1} investment of luminosity. To appreciate how great an improvement this would be over the present, we recall that barely 2 form factor events were recorded for $Q^2 = 14.36$ GeV^2 in our Fermilab experiments. It seems quite practical, given a similar investment of luminosity, to measure $G_M(Q^2 = 20$ GeV$^2)$ with a statistical precision of $\pm 12\%$. These measurements should go a long way in revealing how the ratio $G_M^p(timelike)/G_M^p(spacelike)$ is developing with Q^2, and establishing the validity of the present explanations based on pion form factors.

It is worth pointing out that with e^+e^- annihilation one simultaneously measures the timelike form factors of any particle-antiparticle pair that one can successfully identify. Thus, we have hope of measuring timelike form factors of Δ and Λ at the same time as we measure proton form factors. To top it all, there is the very special interest in measuring pion and kaon form factors.

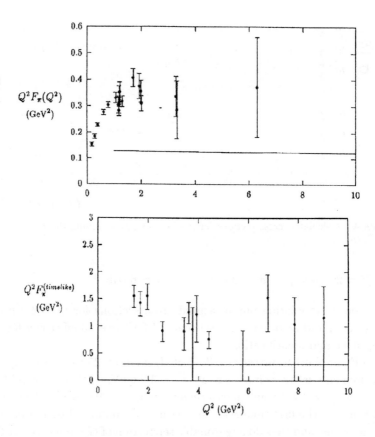

Figure 5. Top panel - Existing experimental data for the pion form factor for spacelike momentum transfers. Bottom panel - Existing experimental data for the pion form factor for timelike momentum transfers.

4. Pion Form Factors

Pion form factors have been at the center of a long standing controversy about where the perturbative regime of QCD begins.[20] It is rather ironic that the controversy is based on extremely scant and poor data for both spacelike and timelike form factors of the pion (see Fig. 5). Because pion targets are not available, spacelike form factors of pions at large Q^2 can be measured only in pion production reactions such as $p(e, e'\pi)n$, whose interpretation remains questionable.[21] The only timelike measurements with the reaction $e^+e^- \to \pi^+\pi^-$ for $Q^2 > 4$ GeV^2 were made at ADONE. They date back to 1975,[22] and have only three data points for $Q^2 > 4$ GeV^2 with huge errors. For $Q^2 = 6.76$, 7.84, and 9.00 GeV^2, the measured values of F_π are $0.23^{+0.06}_{-0.09}$, $0.13^{+0.06}_{-0.13}$, and $0.13^{+0.06}_{-0.13}$ respectively, based on 8, 2, and 4 events, respectively. In a striking contrast, at CLEO-c, for the 0.25 fb^{-1} measurement at $Q^2 = 9.0$ GeV^2, we expect ~ 3800 events/0.25 fb^{-1}! Even at $Q^2 = 20$ GeV^2 we expect ~ 350 events/0.25 fb^{-1}. This is a heady prospect indeed, and it should settle quite a few theoretical arguments.

Needless to reiterate, because CLEO-c identification between pions and kaons is excellent over the whole range of Q^2, we expect to measure the timelike formfactors of kaons with equal precision, at the same time.

5. Summary

To summarize this presentation, the measurements of timelike form factors of the proton in the $p\bar{p} \to e^+e^-$ reaction at Fermilab have revealed a large, nearly factor two, enhancement of the timelike form factors over the spacelike form factors for $Q^2 \geq 8$ GeV^2. Several proposed explanations of this observation have been presented, but none is generally accepted.

It is expected that precision measurement of timelike form factors of protons and pions (and hopefully of other mesons and baryons) for Q^2 upto ~ 20 GeV^2 will be made soon at CLEO-c/CESR-c.

Acknowledgments

This work was supported by the U.S. Department of Energy.

References

1. For form factor reviews, see e.g., G. Sterman and P. Stoler, *Ann. Rev. Nucl. Part. Sci.* **47**, 193-233 (1997); Kamal K. Seth, Proc. Internat. Workshop on Exclusive Reactions at High Momentum Transfer, Elba 1993, edited by C. Carlson, P. Stoler and M. Taiuti, World Scientific (Singapore) 1994, pp. 70-83.
2. S. J. Brodsky and G. P. Lepage, *Phys. Rev. Lett.* **43**, 545 (1979); *Phys. Lett.* **87B**, 359 (1979); G. P. Lepage and S. J. Brodsky, *Phys. Rev. D* **22**, 2157 (1980).
3. S. J. Brodsky and G. R. Farrar, *Phys. Rev. Lett.* **31**, 1153 (1973); *Phys. Rev. D* **11**, 1309 (1975); V. Matveev, R. Muradayan, and A. Tavkhelidze, *Nuovo Cimento Lett.* **7**, 719 (1973).
4. The JLAB Hall A Collaboration, M. K. Jones *et al.*, *Phys. Rev. Lett.* **84**, 1398 (2000).
5. A. F. Sill *et al.*, *Phys. Rev. D* **48**, 29 (1993).
6. R. C. Walker *et al.*, *Phys. Rev. D* **49**, 5671 (1994); L. Andivahis *et al.*, *Phys. Rev. D* **50**, 5491 (1994).
7. N. Cabibo and R. Gatto, *Phys. Rev.* **124**, 1577 (1961).
8. B. Delcourt *et al.*, *Phys. Lett.* **B86**, 395 (1979); D. Bisello *et al.*, *Nucl. Phys. B* **224**, 379 (1983); D. Bisello *et al.*, *Z. Phys.* **C48**, 23 (1990).
9. G. Bassompierre *et al.*, *Phys. Lett.* **68B**, 477 (1977); *Nuovo Cimento* **73A**,

322

477 (1983); G. Bardin *et al.*, *Phys. Lett. B* **255**, 149 (1991); *Phys. Lett. B* **257**, 514 (1991).

10. E760 Collaboration, T. A. Armstrong *et al.*, *Phys. Rev. Lett.* **70**, 1212 (1993).
11. E835 Collaboration, M. Ambrogiani *et al.*, *Phys. Rev. D* **60**, 032002 (1999).
12. J. G. Körner and M. Kureda, *Phys. Rev. D* **16**, 2165 (1977).
13. S. Dubnicka, *Nuovo Cimento* **103A**, 1417 (1990).
14. T. Hyer, *Phys. Rev. D* **47**, 3875 (1993).
15. J. Botts and G. Sterman, *Nucl. Phys. B* **325**, 62 (1989); H. Li and G. Sterman, *Nucl. Phys. B* **381**, 129 (1992).
16. P. Kroll *et al.*, *Phys. Lett. B* **316**, 546 (1993).
17. E760 Collaboration, T. A. Armstrong *et al.*, *Phys. Rev. D* **52**, 4839 (1995); E835 Collaboration, S. Bagnasco *et al.*, *Phys. Lett. B* **533**, 237 (2002).
18. T. Gousset and B. Pire, *Phys. Rev. D* **51**, 15 (1995).
19. A. P. Bakulev, A. V. Radyushkin, and N. G. Stefanis, *Phys. Rev. D* **62**, 113001 (2000).
20. N. Isgur and C. H. Llewellyn Smith, *Phys. Rev. Lett.* **52**, 1080 (1984); *Phys. Lett. B* **217**, 535 (1989); *Nucl. Phys. B* **317**, 526 (1989).
21. C. E. Carlson and J. Milana, *Phys. Rev. Lett.* **65**, 1717 (1990); S. J. Brodsky *et al.*, *Phys. Rev. D* **57**, 245 (1998).
22. D. Bollini *et al.*, *Lett. Nuovo Cimento* **14**, 418 (1975).

DUALITY IN SEMI-EXCLUSIVE PROCESSES[*]

PAUL HOYER[†]

Nordita
Blegdamsvej 17
DK - 2100 Copenhagen, Denmark
E-mail: hoyer@nordita.dk

Bloom-Gilman duality relates parton distributions to nucleon form factors and thus constrains the dynamics of exclusive processes. The quark electric charge dependence implies that exclusive scattering is incoherent on the quarks even at high momentum transfers. Data on semi-exclusive meson production exceeds the duality prediction by more than an order of magnitude and violates quark helicity conservation. This suggests that the subprocess is dominated by soft 'endpoint' contributions which obey dimensional scaling. The large transverse size of the subprocess may explain the absence of color transparency in fixed angle processes.

1. Bloom-Gilman Duality

The remarkable relation between DIS $eN \to eX$ and exclusive resonance production $eN \to eN^*$ known as Bloom-Gilman duality[1] has been confirmed and extended by data from JLab[2]. Empirically,

$$\int_{\delta x} dx F_2^{scaling}(x) \propto \frac{d\sigma}{dQ^2}(eN \to eN^*) \propto |F_{pN^*}(Q^2)|^2 \qquad (1)$$

where $F_{pN^*}(Q^2)$ is the exclusive $p \to N^*$ electromagnetic form factor. The Bjorken variable $x = Q^2/(W^2 + Q^2 - M_N^2)$ is given by the photon virtuality Q^2 and the hadron mass $W = M_{N^*}$. On the lhs of (1) the leading twist structure function $F_2^{scaling}(x)$ is integrated over an interval δx covering the N^* mass region. This semi-local duality relation is approximately satisfied for each nucleon resonance region including the Born term:

---

[*]Transparencies of this talk including figures may be found at http://www.nordita.dk/~hoyer .
[†]On leave from the Department of Physical Sciences, University of Helsinki, Finland. Adjoint Senior Scientist, Helsinki Institute of Physics.

$N^* = P_{11}(938)$, $P_{33}(1232)$, $S_{11}(1535)$ and $F_{15}(1680)$. The magnitude and x-dependence of the scaling structure function is thus related to the magnitude and Q^2-dependence of the N^* electromagnetic form factors. The duality relation is approximately satisfied even at low Q^2.

Bloom-Gilman duality means that the DIS scaling function is coded into the N^* form factors. This is surprising because hard inclusive and exclusive processes are usually thought to be determined by separate parts of the nucleon wave function. The F_2 structure function is (at lowest order in α_s) an incoherent sum, weighted by e_q^2, of inclusive quark distributions built from non-compact multiparton Fock states. The exclusive form factor on the other hand is believed to be governed by the wave function of the valence Fock state $|qqq\rangle$ of transverse size $1/Q$ [3]. The virtual photon then couples *coherently* to the valence quarks, implying a dependence on the quark electric charge of the form $(\sum_q e_q)^2$. Such a different dependence on e_q of the two sides in Eq. (1) contradicts the observed fact that duality is satisfied in a semilocal sense for both proton and neutron targets[2].

Data thus indicates that the Fock states in the nucleon electromagnetic form factors have size $\gg 1/Q$ so that the contribution of each quark to $|F_{pN^*}(Q^2)|^2$ is $\propto e_q^2$. With the standard scaling laws $F_{pN^*}(Q^2) \propto 1/Q^4$ and $F_2^{scaling}(x) \propto (1-x)^3$ both sides of Eq. (1) have the same Q^2-dependence. Duality and incoherent exclusive scattering will then hold at arbitrarily high Q^2. We shall find further evidence below that exclusive processes at large momentum transfer involve non-compact Fock states.

2. Semi-Exclusive Processes

Further information on the relation between inclusive and exclusive scattering can be obtained from generalizations of Bloom-Gilman duality. It was already observed that the spin[4] and nuclear target A dependence[5] of the resonances agrees with that of the DIS scaling region as required by duality.

Semi-exclusive processes such as $\gamma p \to \pi^+ Y$ (Fig. 1) provide a qualitatively new testing ground of duality. In the kinematic limit where the total energy $s = (q+p)^2$ is much larger than the mass of the inclusive system Y ($s \gg M_Y^2 \gg \Lambda_{QCD}^2$) the produced π^+ meson is separated from the hadrons in Y by a rapidity gap. When also the momentum transfer $t = (q-q')^2$ is large the π^+ is expected to be produced via a hard subprocess such as $\gamma u \to \pi^+ d$. We have then a generalisation of ordinary DIS, with the $eq \to eq$ subprocess replaced by $\gamma q \to \pi^+ q'$ and with the physical

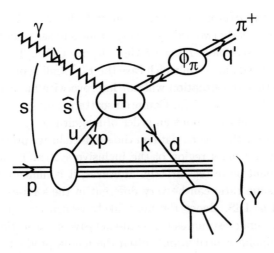

Figure 1. Semi-exclusive scattering. In the limit $\Lambda^2_{QCD} \ll |t|, M^2_Y \ll s$ the cross section factorizes into a hard subprocess cross section $\hat{\sigma}(H)$ times a target parton distribution.

cross section given by[6]

$$\frac{d\sigma}{dx\,dt}(\gamma p \to \pi^+ Y) = \sum_{q,q'} q(x) \frac{d\sigma}{dt}(\gamma q \to \pi^+ q') \qquad (2)$$

where $q(x)$ is the inclusive distribution of the struck quark in the target. From the point of view of the target physics, there is a one-to-one correspondence between the semi-exclusive process and ordinary DIS, with $Q^2 \leftrightarrow -t$ and $W^2 \leftrightarrow M^2_Y$. The momentum fraction of the struck quark in the semi-exclusive process is thus $x = -t/(M^2_Y - t - M^2_N)$.

The close analogy with DIS makes it natural to study Bloom-Gilman duality for semi-exclusive processes[7,8]. This links large momentum transfer exclusive cross sections to standard DIS structure functions and, via Eq. (1), to exclusive form factors, e.g.,

$$\int_{\delta x} [u(x) + \bar{d}(x)] \frac{d\sigma}{dt}(\gamma u \to \pi^+ d) \simeq \frac{d\sigma}{dt}(\gamma p \to \pi^+ N^*) \qquad (3)$$

With the standard behavior $u(x) \sim (1-x)^3$ we find, for $N^* = n$,

$$\frac{d\sigma}{dt}(\gamma p \to \pi^+ n) \propto \frac{1}{s^2 t^5} \qquad (4)$$

which is consistent with data[9].

Combining Eqs. (1) and (3) one obtains[8]

$$\frac{d\sigma}{dt}(\gamma p \to \pi^+ n) = 16\pi^2 \frac{\alpha \alpha_s^2 f_\pi^2}{|t| s^2} \frac{G^2_{Mp}(-t)}{(1 + r_{du}/4)} \qquad (5)$$

where r_{du} is the d/u-quark distribution ratio for $x \to 1$. This estimate turns out to be nearly two orders of magnitude smaller than the measured[9] $\gamma p \to \pi^+ n$ cross section at $E_\gamma = 7.5$ GeV, $|t| \simeq 2$ GeV^2. The discrepancy is even worse at fixed angle – in which case the $1/s^7$ scaling of both theory and data implies that the situation will not improve with momentum transfer. A similar result was found[8] for Compton scattering $\gamma p \to \gamma p$, where duality underestimates data by about an order of magnitude.

Unfortunately there is no data on the semi-exclusive process $\gamma p \to \pi^+ Y$ in the continuum mass region of the inclusive system Y, where the prediction (2) could be tested directly. However, it is unlikely that the resonance/continuum ratio can be very different in the semi-exclusive process as compared to DIS. Hence the most likely reason for the failure of Eq. (5) is that (2) underestimates the semi-exclusive cross section. This could happen if endpoint contributions soften the meson production subprocess. In the previous section we saw that standard DIS duality also points in this direction.

3. Spin Dependence of Semi-Exclusive ρ Production

The ZEUS Collaboration recently published[10] striking data on polarization effects in semi-exclusive vector meson production $\gamma + p \to V + Y$, for $V = \rho^0, \phi, J/\psi$. The ρ^0 cross section scales as $d\sigma/dt \propto (-t)^n$, with $n = -3.21 \pm 0.04 \pm 0.15$ in close agreement with $n = -3$ expected from dimensional counting for the subprocess $\gamma g \to q\bar{q} + g$. The corresponding data on ϕ production gives a power $n = -2.7 \pm 0.1 \pm 0.2$. The ratio $\sigma(\phi)/\sigma(\rho) \simeq 2/9$ for $-t \gtrsim 3$ GeV^2 is in accordance with flavor SU$_3$. These features suggest that the subprocess is hard and described by PQCD.

Helicity conservation at the level of the hard subprocess requires that the quark and antiquark are created with opposite helicities. Hence the ρ meson they form is predicted to be longitudinally polarized, $\lambda_\rho = 0$. At the level of the external particles, on the other hand, helicity conservation implies that the ρ meson has the same helicity as the (real) projectile photon, i.e., $\lambda_\rho = \pm 1$. We thus have a situation where helicity conservation at the external particle level is in conflict with helicity conservation at the quark level.

The ZEUS data[10] shows that s-channel helicity is nearly conserved in the entire measured range ($-t \leq 6$ GeV^2) for both ρ and ϕ mesons. Hence helicity is violated at the subprocess level. In PQCD this brings a suppression factor proportional to the quark mass squared, $m_q^2/(-t)$. The cross section is then expected to scale with a power $n = -4$, whereas the data is closer to the dimensional counting rule $n = 3$. Taken at face value, this suggests that the subprocess is soft and endpoint dominated (one quark carrying most of the meson momentum) yet obeys the dimensional counting rule.

4. Endpoint Behavior of Parton Subprocesses

We have seen that Bloom-Gilman duality (in DIS as well as in semi-exclusive production) and quark helicity violation in vector meson production suggest a dominance of endpoint effects in exclusive hadron processes. PQCD estimates of the Lepage-Brodsky[3] hard exclusive scattering dynamics likewise show[11] that configurations where one quark carries most of the hadron momentum contribute importantly due to (nearly) on-shell internal propagators.

A possible reason for the failure to observe color transparency in large angle elastic ep[12] and pp[13] scattering on nuclear targets is that the relevant nucleon Fock states are not compact, again due to endpoint contributions. We recently studied[14] the size of the $\gamma u \to \pi^+ d$ subprocess at large momentum transfer t in PQCD, based on the derivative of the cross section wrt. the virtuality Q^2 of the photon at $Q^2 = 0$. It turns out that while the amplitude itself is regular at the endpoints, $A \propto (e_u - e_d) \int dz\, \phi_\pi(z)/(1-z)$, the Q^2 derivative brings another factor of $1 - z$ in the denominator. The integral over the quark momentum fraction z is then singular at $z = 1$, even though the pion distribution amplitude $\phi_\pi(z) \propto 1 - z$. The singularity of the Q^2-derivative implies an *infinite* size for the photoproduction subprocess!

The situation for the quark helicity flip subprocess $\gamma g \to q\bar{q} + g$ (*cf.* section 3) is similar. The $\lambda_\rho = 1$ amplitude has the form[14]

$$A \propto \frac{m_q}{\sqrt{-t}} \int dz \frac{\phi_\rho(z)}{z^2(1-z)^2} \left[1 + O\left(\frac{m_q^2}{t}\right) \right] \tag{6}$$

and is thus endpoint sensitive. This may explain the dominance of the quark helicity flip amplitude in the ZEUS data[10]. Dimensional scaling can be understood by noting that the amplitude is not light-cone (LC) dominated at the endpoints: the soft quark moves with non-relativistic speed for $z \lesssim \Lambda_{QCD}/\sqrt{-t}$. In this region the amplitude is not proportional to the LC distribution amplitude ϕ_ρ and there is no reason to expect that the numerator of the integrand in Eq. (6) vanishes. The linearly divergent integral then gives a factor $\sqrt{-t}/\Lambda_{QCD}$ which precisely compensates the t-dependence induced by the quark helicity flip. Details will be presented elsewhere[14].

5. Conclusions

There are several indications that large momentum transfer exclusive processes are not given by compact PQCD subprocesses and hadron distribution amplitudes. Rather, it appears that exclusive production is dominated by configurations where one quark carries most of the hadron momentum. The cross section scales dimensionally but does not obey quark helicity conservation nor color transparency.

Acknowledgments

The results presented here were obtained in collaborations with Patrik Edén, Alexander Khodjamirian, Jonathan Lenaghan, Kimmo Tuominen and Carsten Vogt. I am also grateful for discussions with Stan Brodsky and Jim Crittenden. Research supported in part by the European Commission under contract HPRN-CT-2000-00130.

References

1. E. D. Bloom and F. J. Gilman, *Phys. Rev. Lett.* **25**, 1140 (1970) and *Phys. Rev.* **D4**, 2901 (1971).
2. I. Niculescu *et al.*, *Phys. Rev. Lett.* **85**, 1182 and 1186 (2000).
3. G. P. Lepage and S. J. Brodsky, *Phys. Rev.* **D22**, 2157 (1980).
4. A. Fantoni, Talk at the XL Int. Winter Meeting on Nuclear Physics, Bormio, Italy (January 2002), HERMES report 02-012 available at http://www-hermes.desy.de/notes/pub/doc-public-subject.html#G1 .
5. J. Arrington, J. Crowder, R. Ent, C. Keppel and I. Niculescu, Jefferson Laboratory preprint PHY02-13 (2002).
6. S. J. Brodsky, M. Diehl, P. Hoyer and S. Peigne, *Phys. Lett.* **B449**, 306 (1999) [hep-ph/9812277].
7. A. Afanasev, C. E. Carlson and C. Wahlquist, *Phys. Rev.* **D62**, 074011 (2000) [hep-ph/0002271].
8. P. Edén, P. Hoyer and A. Khodjamirian, *JHEP* **0110:040** (2001) [hep-ph/0110297].
9. R. L. Anderson *et al.*, *Phys. Rev. Lett.* **30**, 627 (1973); *Phys. Rev.* **D14**, 679 (1976).
10. ZEUS Collaboration: S. Chekanov *et al.*, hep-ex/0205081.
11. N. Isgur and C. H. Llewellyn Smith, *Phys. Rev. Lett.* **52**, 1080 (1984) and *Nucl. Phys.* **B317**, 526 (1989));
    A.V. Radyushkin *Nucl. Phys.* **A527**, 153c (1991) and *Nucl. Phys.* **A532**, 141c (1991);
    J. Bolz, R. Jakob, P. Kroll, M. Bergmann and N.G. Stefanis, *Z. Phys.* **C66**, 267 (1995) [hep-ph/9405340].
12. T. G. O'Neill *et al.*, *Phys. Lett.* **B351**, 87 (1995) [hep-ph/9408260].
13. A. Leksanov *et al.*, *Phys. Rev. Lett.* **87**, 212301 (2001) [hep-ex/0104039].
14. P. Hoyer, J. T. Lenaghan, K. Tuominen and C. Vogt, in preparation.

VIOLATIONS OF PARTON-HADRON DUALITY IN DIS SCATTERING

S. LIUTI

University of Virginia
Charlottesville, VA 22904, USA
E-mail: sl4y@virginia.edu

We discuss the limits of validity of parton-hadron duality in *ep* scattering, in the large *x* region. The uncovered pattern of violations of duality can be explained by introducing a distribution of color neutral clusters in the intial stage of evolution, whose mass spectrum is predicted within the preconfiment property of QCD.

1. Introduction

The evolution of partonic structure as the relevant scale of the interaction is increased from a hadronic scale $\lesssim 1$ GeV^2 to asymptotic values, has been for long a subject of intensive studies. Deep Inelastic Scattering (DIS) experiments, from the very first observations of partons and of perturbative QCD (pQCD) evolution, have been providing through the years a continuous out-pouring of information on the detailed structure of hadrons in a wide kinematical regime.

At large enough values of Q^2, namely well above 1 GeV^2, the scattering processes is described as a one-on-one collision between the probe and a parton, including in this description pQCD radiative corrections. A rich dynamics ensues from the asymptotic nature of the pQCD series itself whose uncertainty is controlled by power corrections [1,2]. Power corrections are indeed a necessary contribution in order to reproduce the behavior of the data at intermediate and low values of Q^2. At low values of Q^2, however, partonic components become less likely to be resolved, and a transition to a regime eventually dominated by *non-partonic* degrees of freedom, outside the range of applicability of pQCD, and of the first few terms of the twist expansion itself, is expected to occur.

The concept of duality is rooted within this widely accepted picture, and can be summarized in the following few points: *i)* Duality is a consequence of the existence of two scales: Q^2, regulating the hard probe-parton interaction, and Λ^2, the hadronic scale; *ii)* In our description of inclusive hard processes we implicitly assume that the hard, short distance, probe-parton scattering, calculable within pQCD, will determine the final outcome, modulo a Q^2 independent part describing the hadronic contribution. This is is formally embodied by the "practical" Operator Product Expansion (OPE); *iii)* Due to the asymptotic nature of both the perturbative and local operators expectation values series, the uncertainty of this prediction is controlled by higher order terms in α_s and power corrections; *iv)* If the theoretical curve agrees with experiment within the limits of these corrections, duality sets in.

The same idea of duality is observed perhaps more clearly for inclusive and infrared stable quantities such as the ratio: $R(s) = \sigma(e^+e^- \to$ hadrons$)/\sigma(e^+e^- \to \mu^+\mu^-)$. As first observed in [3], at sufficiently high energy, the behavior of R is reproduced solely by the partonic cross section, slightly modified by pQCD corrections, the formation of hadrons playing no significant role. More recently, similar arguments have been put in the framework of QCD and extended to the calculation of τ and heavy flavors semileptonic decays (for a review see [4]).

DIS differs from e^+e^- annihilation because of the presence of an initial hadron (formally, DIS is not a priori infrared stable). We define however duality in a similar though "extended" way, following points *i)* - *iv)* above, as the ability of describing the data at finite Q^2, in terms of the probe-parton process only, with no interaction with the proton remnant. An important unsolved question to which this paper is dedicated is the problem of controlling the extent to which parton-hadron duality is at work.

In particular, both in DIS and in e^+e^- annihilation, a resonance structure appears and persists, up to relatively high energies. In DIS this happens when Bjorken x is large, $Q^2 \approx$ few GeV^2, and the final state invariant mass $W^2 \lesssim 4$ GeV^2. Quark-hadron duality however sets in, albeit in an averaged way. What is observed in the data is, in fact, a similarity between the partonic cross sections at large Q^2 or s – a smooth function embodying parton dynamics – and an "underlying structure" in the resonance region, the latter being obtained by averaging the resonances over *e.g.* x (or equivalently over a finite energy range). The similarity between the partonic curve and the averaged-resonances one is known as *global* duality, as opposed to a stricter point-by-point correspondence with the exact experimental cross section in the resonance region, which is defined as *local* duality. Local duality is manifestly violated in the resonance region, whereas in order to detect violations of global duality a careful pQCD analysis is needed.

The question arises of whether the behavior of these violations can be explained and controlled by addressing the asymptotic nature of the pQCD series, following a path initiated with the study of the QCD vacuum condensates, and summarized in [5]. There, the presence of resonances as oscillations in Minkowski space is shown to be induced by the replacement of the free quark Green functions with the Green functions in *e.g.* a background instanton field, the resonance strengths being damped at large energies by appropriate power coefficients. Open questions in this approach and elucidated in [5] include however both the order in α_S and the inclusion of non-perturbative terms, defining the curve which the resonances average to. Following this logic, we performed in [6] an accurate analysis of recent data [7], with the purpose of extracting the contribution of power corrections to global duality in the resonance region. By exploiting the high precision of the data it was found that $1/Q^2$ terms are sensibly smaller than the ones found at large W^2 [8] – an effect not attributed to large x resummation – and that they do not seem to follow a "$1/Q^2, 1/Q^4$..." type expansion. In other words, the data in the resonance region average to a curve that has a "non-standard" power correction pattern. In the language of [5], *global* duality is violated as well.

Here we discuss a dynamical model, based on large N_c, as a potential candidate for explaining the complex Q^2 dependence of the theory. We introduce in our description of the proton structure function a distribution of color neutral clusters in the intial stage of evolution, whose mass spectrum is predicted within the preconfinent property of QCD. We are able to predict the behavior of the average curves, and the transition to the DIS regime proper. The model does not include a description of local duality violations at this stage, although it allows for a connection with the models discussed in [4,5] to be pursued later on.

In the following sections we summarize the results found in [6], we introduce our model, and we present some initial results.

2. Quantitative Analysis of the Scale Dependence of Parton-Hadron Duality

In [6] the Q^2 dependence in the resonance region was extracted by first considering the average of the resonances over $\xi = 2x/(1 + \sqrt{1 + 4M^2x^2/Q^2})$, which properly takes into account Target Mass Corrections (TMC). The averaging procedure, described in detail in [7], yields a smooth curve in ξ which fits the resonance data with a $\chi^2/d.o.f.$ between 0.8 and 1.1. It represents an alternative analysis to the ones using moments. The fit was performed in bins of W^2, centered at: $W_R^2 = 1.6, 2.3, 2.8, 3.4$ GeV^2, respectively. Our subsequent study was aimed at establishing whether it was possible to define a breakpoint where pQCD no longer applies and a transition occurs, similar to what observed in the low x regime [9]. The analysis involved a number of steps similar to recent extractions of power corrections from inclusive data [8,10,11], where the form

$$F_2^{exp}(x, Q^2) = F_2^{pQCD+TMC}(x, Q^2) + \frac{H(x, Q^2)}{Q^2} + O(1/Q^4), \qquad (1)$$

was adopted, and where $F_2^{pQCD+TMC}(x, Q^2)$ is the twist-2 contribution, including the kinematical TMC; the other terms in the formula are the dynamical power corrections, formally arising from higher order terms in the twist expansion. Both $F_2^{pQCD+TMC}$ and H were extracted from the data at large x.

The shape of the initial NS PDFs was found to be well constrained at variance with the singlet and gluon distributions at low Q^2, whose shape is strongly correlated with the value of α_S. NNLO corrections were not included. These would introduce further theoretical uncertainties. In fact, the question of whether they can "mimick" the contributions of higher twists, including the uncertainties due to the well known scale/scheme dependence of calculations, within the current precision of data is still a subject of intense investigations [10]. Large x resummation was performed directly in x space by replacing the Q^2 scale with a z-dependent one, $\widetilde{W}^2 = Q^2(1-z)/z$ [12]. It was found that over the range $0.45 \leq x \leq 0.85$, large x resummation, and TMC improve the agreement with the data. We parametrized the remaining discrepancies through the Higher Twist (HT) coefficient $H(x, Q^2) = F_2^{pQCD+TMC}(x, Q^2)C_{HT}(x)$. In Fig.1 we show C_{HT}, extracted from: DIS data with $W^2 \geq 4$ GeV^2, from the resonance region ($W^2 < 4$ GeV^2), and over the entire range of W^2. The figure also shows the range of extractions previous to the current one [8]. While the large W^2 data track a curve that is consistent with the $1/W^2$ behavior expected from most models, the low W^2 data yield a much smaller value for C_{HT} and they show a bend-over of the slope vs. x. This surprising effect is not a consequence of the interplay of higher order corrections and the HT terms, but just of the extension of our detailed pQCD analysis to the large x, low W^2 kinematical region. In other words, we unraveled a Q^2 dependence that seems to deviate from the common wisdom developed since the pioneering analysis of [13], or we observe a violation of *global* duality.

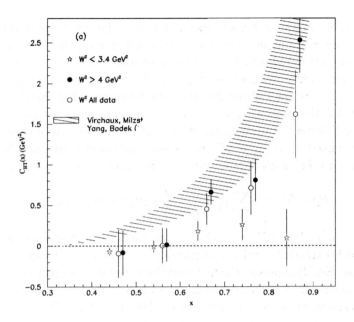

Figure 1. Extracted values of higher twist coefficient C_{HT}, as explained in text.

3. A Large N_c Based Dynamical Model for Parton-Hadron Duality

We propose a simple dynamical model for the structure function in the low W^2 ($W^2 \leq 4$ GeV^2) and low Q^2 ($Q^2 < 10$ GeV^2) regime, where non-partonic configurations are expected to be dominant. The DIS cross section for ep scattering is proportional to the structure functions $F^p_{1(2)}(x, Q^2)$ which in turn, measure combinations of the parton longitudinal momentum distributions, $q_i(x, Q^2)$, $i = u, d, s, ...$ at the scale Q^2. In the standard approach to DIS the Q^2 dependence of F_2 is described by the pQCD evolution equations [14], whose numerical solution requires parametrizing the input distributions at an initial scale Q^2_o where pQCD is believed to be still applicable. Q^2_o serves as a boundary between the perturbative and non-perturbative domains, although its value is somewhat arbitrary (in current parametrizations $Q^2_o \approx 0.4 - 10$ GeV^2). We refer to this situation as the "fixed initial scale" description, and we write explicitly the dependence of $q_i(x, Q^2, Q^2_o)$ on Q^2_o. A simple kinematical argument shows that Q^2_o is related to the invariant mass squared of the proton remnant after a parton is

334

emitted, by: $M_X^2 \approx Q_o^2/x$. At large x, $M_X^2 \approx Q_o^2$ (at low x this simple kinematical observation, and the factorization properties of the diffractive part of F_2, support the idea that the parton is emitted from a large mass object, indentified with a soft pomeron [15]). We, therefore, explore the possibility that partons are not emitted directly from the nucleon, but that, before the pQCD radiative processes are initiated, a semi-hard phase occurs where the dominant degrees of freedom are color neutral clusters with a mass distribution peaked at: $\mu_{peak}^2 \approx Q_o^2$. As a result, the nucleon structure function is related to the quark distribution by a smearing of the initial Q_o^2, namely

$$F_2(x, Q^2) = x \sum_i e_i^2 \int_{\mu_o^2 > \Lambda^2}^{W^2} \frac{d\mu^2}{\mu^2} P(\mu^2) q_i(x, Q^2, \mu^2), \qquad (2)$$

where $P(\mu^2)$ $(P_{peak}(\mu^2) \approx P(Q_o^2))$ is the clusters' mass distribution, and the sum is extended to valence quarks only since we are describing the large x region. Eq.(2) expresses the fact that the initial stage of pQCD evolution is characterized by color neutral clusters of variable mass, from which the hard scattering parton will emerge, in a subsequent stage of the interaction.

Our model is derived within the framework of the large N_c approximation, in analogy to the cluster hadronization schemes implemented in the HERWIG Monte Carlo simulation [16]. In this scheme, hadronization proceeds as prescribed by the pQCD property of preconfinement of color: [17] at the end of the parton's pQCD evolution, color singlets are formed with a Q^2– independent mass (and spatial) distribution. In practical implementations [16], all gluons left at the hadronization scale, are "forcibly", or non-perturbatively, split into $q\bar{q}$ pairs. It is this modification of the evolution equations that allows for the local proton-hadron conversion through preconfinement of colour: each color line "color–connects" e.g. a quark to an anti-quark, forming a color singlet. The color-singlet clusters are then fragmented into hadrons. In DIS the transition hadrons → quarks → hadrons, is complicated both by initial state radiation and by the presence of the beam cluster formed from the remnant of the initial hadron, Fig.2. This produces an additional rescattering term in the function P in Eq.(2) [18].

As a preliminary study, we considered both the low and the very large W^2 limits of Eq.(2). At low W^2, $F_2(W^2, Q^2) \approx P(Q^2)$, namely it is described by the behavior of the cluster distribution function. At very large W^2, P_{peak} determines the value of the initial Q_o^2. In Fig.3 we present our result for F_2 at $W^2 = 1.6$ GeV^2, where the cluster distribution was obtained from HERWIG [16] (triangles), whereas the red curve is an analytic calculation of the Sudakov-type behavior valid at large Q^2.

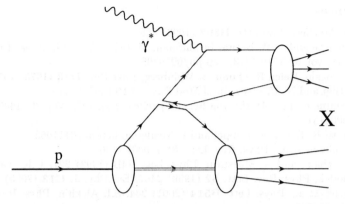

Figure 2. The different stages of conversion hadron $\to$ parton $\to$ hadron, through a cluster stage described schematically, at large N_c.

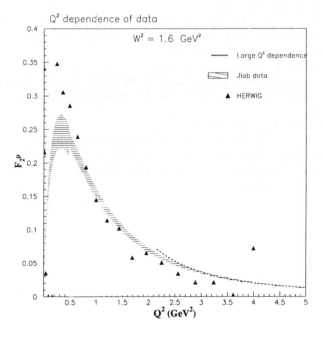

Figure 3. Q^2 dependence of Eq.(2) at fixed $W^2 = 1.6$ GeV^2

Acknowledgments

This work is supported by a research grant from the U.S. Department of Energy under grant no. DE-FG02-01ER41200.

References

1. A. H. Mueller, Phys.Lett. **B308** (1993) 355.
2. M. Beneke and V.M. Braun, In Shifman, M. (ed.): At the frontier of particle physics, vol. 3* 1719-1773; hep-ph/0010208.
3. E.C. Poggio, Helen R. Quinn, S. Weinberg, Phys.Rev. **D13** (1976) 1958.
4. I.Y. Bigi and N. Uraltsev, Int.J.Mod.Phys. **A16** (2001) 5201.
5. M. Shifman, in *"At the frontier of particle physics"*, Vol. 3, 1447, hep-ph/0009131.
6. S. Liuti, R. Ent, C.E. Keppel and I. Niculescu, hep-ph/0111063.
7. I. Niculescu *et al.*, Phys. Rev. Lett. **85** (2000) 1186.
8. M. Virchaux and A. Milzstajn, Phys. Lett. 74B (1992) 221; U.K. Yang and A. Bodek, Phys. Rev. Lett. **82** (1999) 2467; Eur.Phys.J. **C13** (2000) 241; S. Schaefer *et al.*, Phys. Lett. **B514** (2001) 284; S.I. Alekhin, Phys. Rev. **D63** (2001) 094022.
9. ZEUS Collaboration, J. Breitweg *et al.* Phys. Lett.**B487** (2000) 53.
10. A.L. Kataev, G. Parente, A.V. Sidorov, hep-ph/0106221; *ibid* Nucl.Phys. **B573** (2000) 405; A.L. Kataev, A.V. Kotikov, G. Parente, A.V. Sidorov, Phys. Lett. **B417** (1998) 374.
11. I. Niculescu, *et al.*, Phys. Rev. D **60** (1999) 094001; S. Liuti, Nucl. Phys. Proc. Suppl. **74** (1999) 380; X. Ji, and P. Unrau, Phys. Rev. **D52** (1995) 72.
12. R.G. Roberts, Eur.Phys.J. **C10** (1999) 697.
13. A. De Rujula, H. Georgi and H.D. Politzer, Ann. Phys,. **103** (1977) 315.
14. L.V. Gribov and L.N. Lipatov, Yad.Fiz. **20** (1975) 181; G. Altarelli and G. Parisi, Nucl.Phys. **B126** (1977) 298; Yu. Dokshitzer, Sov.Phys.JETP **46** (1977) 641.
15. F.M. Liu *et al.*, hep-ph/0109104.
16. G. Corcella *et al.*, HERWIG 6.4 release notes, hep-ph/0201201.
17. D. Amati and G. Veneziano, Phys. Lett. **B83** (1979) 87; A. Bassetto, M. Ciafaloni and G. Marchesini, Phys.Rept. **100** (1983) 201.
18. S. Liuti, *in preparation*.

QUARK-HADRON DUALITY STUDIES AT JEFFERSON LAB: AN OVERVIEW OF NEW AND EXISTING RESULTS

C. E. KEPPEL

Hampton University / Jefferson Laboratory
E-mail: keppel@jlab.org

Recent results from Jefferson Lab (JLab) have significantly increased our understanding of the fundamental nature of quark-hadron duality. JLab proton F_2 structure function data had been shown to surprisingly exhibit duality as far down as $Q^2 = 0.5$ $(GeV/c)^2$. More recently, duality has been observed also for the structure functions F_1, F_L, and g_1. Furthermore, duality seems to hold even to lower values of Q^2 for nuclei. All of these recent experimental results indicate clearly that parton-hadron duality, though not understood completely, is an intrinsic property of nucleon structure.

Understanding the structure and interaction of hadrons in terms of the quark and gluon degrees of freedom of Quantum ChromoDynamics (QCD) is perhaps the greatest unsolved problem in nuclear and particle physics. While at present we cannot describe the physics of hadrons directly from QCD, we know that in principle it should just be a matter of convenience in choosing to describe a process in terms of quark-gluon or hadronic degrees of freedom. This fact is referred to as *quark-hadron duality*, and means that one can use either set of complete basis states to describe physical phenomena. At high energies, where the interactions between quarks and gluons become weak and quarks can be considered asymptotically free, an efficient description of phenomena would be afforded in terms of quarks; at low energies, where the effects of confinement make strongly-coupled QCD highly non-perturbative, it would be more efficient to work in terms of collective degrees of freedom, the physical mesons and baryons.

Duality between quark and hadron descriptions reflects the relationship between confinement and asymptotic freedom, and is intimately related to the nature of the transition from non-perturbative to perturbative QCD. Although the duality between quark and hadron descriptions is formally exact in principle, how this reveals itself specifically in different physical processes and under different kinematical conditions is the key to understanding the consequences of QCD for hadronic structure. The phenomenon of duality is in fact quite general in nature and can be studied in a variety of processes, such as $e^+e^- \rightarrow$ hadrons, or semi-leptonic decays of heavy quarks [1]. For the latter, the results in terms of quark and hadronic variables are identical in the limit of infinitely heavy quarks [2]. One of the more intriguing examples, initially observed some 30 years ago by Bloom and Gilman at SLAC[3], is in inclusive inelastic electron–nucleon scattering.

Inclusive electron scattering provides a clean way to probe the quark distributions of a nucleon or nucleus. In the Bjorken limit of deep inelastic scattering (DIS), the interaction can be described in terms of scattering from quasifree partons. Data in the DIS region are well described in terms of quasifree partons, and observed scaling violations are understood in terms of QCD corrections. At lower momenta or energy transfer, however, the data are no longer obviously consistent with a simple quasifree parton picture. Resonance structures appear in the cross section, confusing the scaling observed at higher energies. Also, the required non-perturbative QCD corrections increase. This latter point has been challenged, however, in recent works allowing for orders in addition to leading order pQCD [4,5].

While a smooth scaling curve does not exist in the resonance region, it was observed that the F_2 structure function in the resonance and the DIS regions can nonetheless still be related [3]. Bloom and Gilman observed that when the structure function was taken as a function of $x' = [1 + W^2/Q^2]^{-1}$, and averaged over the resonance region, the result was identical to the DIS structure function averaged over the same region in x'. This duality between the resonance region and scaling limit structure functions was observed not just for the entire resonance region, but also locally around each prominent resonance. Indeed, the Q^2-dependence of the resonance structure functions seemed to be determined by the scaling limit curve, as the resonances "slid" along it in Q^2. So, while the structure function in the resonance region does not directly exhibit scaling, the *duality averaged* structure function curve does.

Data taken in Hall C at Jefferson Lab [6] were used to precisely test this duality in the resonance–averaged structure function. These data were analyzed in terms of the Nachtmann variable ξ, ($\xi = 2x/[1 + \sqrt{1 + 4m^2x^2/Q^2}]$), which has been shown to be the correct variable for studies of scaling violations [7]. Because $\xi \rightarrow x$ in the limit of $Q^2 \rightarrow \infty$, xi–scaling is identical to x–scaling in the Bjorken limit. At finite Q^2, the scaling violations that come from target mass corrections are removed by the utilization of ξ. The new data show that scaling–like behavior extends beyond the DIS and well into the resonance region, surprisingly as far down as $Q^2 = 0.5$ $(GeV/c)^2$. The above observations are typified by Figure 1. Here, the curve is the NMC DIS scaling curve parameterization [8]. This curve, representing the larger (x, Q^2) DIS data, also describes on average the resonance data at smaller (x, Q^2).

Recently, JLab Experiment E94-110 measured inclusive nucleon resonance electroproduction cross sections spanning the four-momentum transfer range $0.75 < Q^2 < 5.0$ $(GeV/c)^2$ to perform Rosenbluth separations to extract the ratio $R = \sigma_L/\sigma_T$, the ratio of the longitudinal to transverse cross sections. Final uncertainties are projected to be ultimately ± 0.05, representing a substantial improvement over the current errors on R which are an order of magnitude larger. These new data allow for the extraction of the structure functions F_1 and F_L and, for the first time ever, the study of duality in all unpolarized structure functions.

Preliminary results, as shown in Figures 2 and 3, indicate that duality is holding in these structure functions. The data points are from Rosenbluth-type separation measurements. The solid resonance curves are from a global cross section fit to R and F_2 in the resonance region – another product of this new analysis. The DIS scaling limit curve is here represented by the dashed line obtained from global fits to SLAC [9,10,11] F_2 and R data. From these, it is straightforward to obtain F_1, F_L. The deviation of the DIS curve at very low Q^2 is of no concern, as no data exists in this regime and the fit is therefore unconstrained.

As with the F_2 structure function, the new F_1, F_L data agree on average with the DIS scaling curve, apparently manifesting quark-hadron duality. This data is currently being finalized and submitted for publication. Also, work is in progress to extract the F_1 and F_L structure function moments for precision tests of duality in light of the OPE. I also note that JLab Hall B g_1 spin structure function data seem to manifest duality, but perhaps starting at a higher Q^2 [12].

New results are now being prepared for publication that demonstrate duality in nuclear structure functions as well. These data show that duality holds for nuclei in fact even better than for the nucleon. Fermi smearing flattens out the resonance region naturally in the nuclear environment, providing a smooth curve that is consistent with the DIS predictions. Perhaps the most striking result from this work is a comparison of the iron to deuterium F_2 structure function data at large x. This addresses the well-known DIS EMC effect, but in the resonance region at lower Q^2. As shown in Figure 4, the resonance region data (small, closed circles) are consistent with the DIS data from SLAC (all other points). This is quite dramatic, and places constraints on explanations for the EMC effect. What the exact modification mechanism is of the parton distributions in the nuclear medium is a subject of much debate. However, whatever it is must now also hold true in the confined resonance region.

In summation, duality appears to be a non-trivial and fundamental property of nucleon structure. It appears as a property of all tested structure functions, and in a variety of targets. Understanding duality will provide the basis for understanding the transition from perturbative to nonperturbative QCD.

References

1. M.B. Voloshin and M.A. Shifman, Sov. J. Nucl. Phys. **47**, 511 (1988); N. Isgur, Phys. Rev. D **40**, 101 (1989); Phys. Lett. **B448**, 111 (1999)
2. N. Isgur and M.B. Wise, Phys. Rev. D **43**, 819 (1991)
3. E.D. Bloom and F.J. Gilman, Phys. Rev. Lett. **16**, 1140 (1970)
4. U-K Yang and A. Bodek, Phys. Rev. Lett. **82**, 2467 (1999)
5. H.L. Lai, et al., Eur. Phys. J. **C12**, 375 (2000)
6. I. Niculescu, *et al.*, Phys. Rev. Lett. 85, 1186 (2000); I. Niculescu, *et al.*, Phys. Rev. Lett. 85, 1182 (2000)
7. H. Georgi and H. D. Politzer, Phys. Rev. D 14, 1829 (1976)
8. M. Arneodo *et al*, Nucl. Phys. **B483** (1997) 3
9. S. Dasu, *et al.*, Phys. Rev. D **49**, 5641 (1994)
10. L.W. Whitlow, *et al.*, Phys. Rev. Lett. **B282**, 475 (1992)
11. K. Abe, et al., Phys. Lett. **B452**, 194 (1999)
12. T. Forest, et al., Baryons 2002 Conference Proceedings; private communication

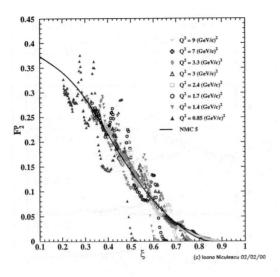

Figure 1. Jefferson Lab resonance region data, at a variety of Q^2 values as indicated, compared to the NMC DIS parameterization at $Q^2 = 5$. The statistical uncertainty of the data is smaller than the symbols.

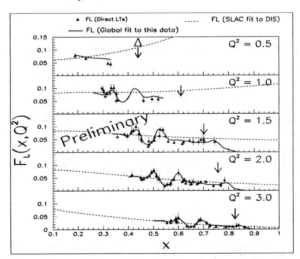

Figure 2. Comparison of the new E94-110 resonance region F_L data to the SLAC DIS curve.

342

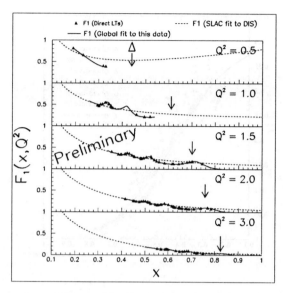

Figure 3. Comparison of the new E94-110 resonance region F_1 data to the SLAC DIS curve.

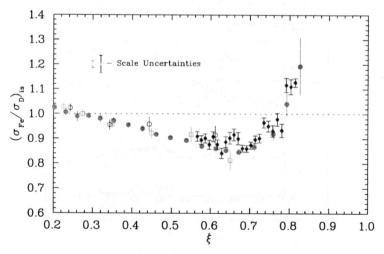

Figure 4. Comparison of the new E94-110 resonance region data to the SLAC DIS curve.

LIGHT FRONT TREATMENT OF THE DEUTERON

GERALD A. MILLER

Department of Physics, University of Washington
Seattle, WA 98195-1560
E-mail: miller@phys.washington.edu

Our recent calculations of deuteron properties using light front dynamics are discussed.

1. Introduction and Outline

This talk is based on the Ph. D. thesis work of Jason R. Cooke [1,2,3]. The aim is to discuss the philosophy and some of the results.

I will begin by defining light front physics, and why one should consider doing it. Then I will discuss how the calculations of computing deuteron binding energies and form factors are done.

2. Light Front

Light-front dynamics is a relativistic many-body dynamics in which fields are quantized at a "time" $=\tau = x^0 + x^3 \equiv x^+$. The τ-development operator is then given by $P^0 - P^3 \equiv P^-$. These equations show the notation that a four-vector A^μ is expressed as $A^\pm \equiv A^0 \pm A^3$. One quantizes at $x^+ = 0$ which is a light-front, hence the name "light front dynamics". The canonical spatial variable must be orthogonal to the time variable, and this is given by $x^- = x^0 - x^3$. The canonical momentum is then $P^+ = P^0 + P^3$. The other coordinates are $\mathbf{x}_\perp$ and $\mathbf{P}_\perp$.

The most important consequence of this is that the relation between energy and momentum of a free particle is given by: $p_\mu p^\mu = m^2 = p^+ p^- - p_\perp^2 \rightarrow p^- = \frac{p_\perp^2 + m^2}{p^+}$, a relativistic kinetic energy which does not contain a square root operator. This allows the separation of center of mass and relative coordinates, so that the computed wave functions are frame independent.

The use of the light front is particularly relevant for calculating form factors, which here are probability amplitudes for an deuteron to absorb a four momentum q and remain a deuteron. The initial and final deuterons have different total momenta. This means that the final deuteron is a boosted deuteron, with different wave function than the initial deuteron. In general, performing the boost is difficult for large values of $Q^2 = -q^2$. However the light front technique allows one to set up the calculation so that the boosts are independent of interactions. Indeed, the wave functions are functions of relative variables and are independent of frame.

2.1. *Light Front Deuteron Calculations Pluses and Minuses*

Let us briefly list the advantages and disadvantages of the calculations discussed here. The there are several advantages. One can start with a Lagrangian, $\mathcal{L}$, use it determine a nucleon-nucleon potential, and use the very same potential to determine the wave function Ψ. One can also examine the meson-nucleon Fock space components. Furthermore, the often noted simplicity of the vacuum seems to really occur for the meson-baryon $\mathcal{L}$. Of importance here is that the generators of boosts in the $x, y, +$ directions are kinematic.

The principle disadvantage is that rotational invariance not manifest, and can not be used to simplify the calculation. However, a correctly performed calculation should lead to results consistent with rotational invariance: for example, states with different magnetic quantum numbers should be degenerate. The failure to obey such requirements provides useful checks on the dynamical approximations. Thus the failure to have manifest rotational invariance can be turned into an advantage.

2.2. *Two-Body Variables and Boost*

The three-momentum of a nucleon i is characterized by $\mathbf{p}_i = (p_i^+, \mathbf{p}_{i\perp})$ with $p_i^- = (\mathbf{p}_{i\perp}^2 + m^2)/p_i^+$. The total momentum of a two-nucleon system is the sum of the nucleon momentum $\mathbf{P} = \mathbf{p_1} + \mathbf{p_2}$, $\mathbf{P}^+ = \mathbf{p_1^+} + \mathbf{p_2^+}$. We define relative coordinates:

$$\alpha = \frac{p_1^+}{P^+}, \quad \mathbf{p}_\perp = (1 - \alpha)\mathbf{p}_{1\perp} - \alpha\mathbf{p}_{2\perp}. \tag{1}$$

The wave function $\Psi(\mathbf{p}_\perp, \alpha)$ is expressed in terms of these relative variables, and therefore is independent of frame. In our impulse approximation treatment of elastic electron scattering a nucleon absorbs a photon of four momentum q^μ with $q^+ = 0, -q^2 = q_\perp^2 = Q^2$. The plus momentum of the struck nucleon does not change and the final relative perp-momentum is

$$\mathbf{p'}_\perp = (1 - \alpha)(\mathbf{p}_{1\perp} + \mathbf{q}_\perp) - \alpha\mathbf{p_2} = \mathbf{p}_\perp + (1 - \alpha)\mathbf{q}_\perp \tag{2}$$

Ignoring spin, a schematic formula for the form factor $F(Q^2)$ is

$$F(Q^2) = \int d^2\mathbf{p}_\perp \frac{d\alpha}{\alpha} \Psi(\mathbf{p}_\perp, \alpha) \Psi(\mathbf{p}_\perp + (1-\alpha)\mathbf{q}_\perp, \alpha) \qquad (3)$$

$$= \int d^2\mathbf{b} \frac{d\alpha}{\alpha} |\Psi(b, \alpha)|^2 e^{i(1-\alpha)\mathbf{q}_\perp \cdot \mathbf{b}}. \qquad (4)$$

Note that $F(Q^2)$ is not a Fourier transform of density.

2.3. Light Front Quantization LITE

Our motto here is that we need a $\mathcal{L}$ no matter how bad! The point is we rely on more than symmetries. We do the canonical procedure of converting a $\mathcal{L}$ to an energy-momentum tensor $T^{\mu\nu}$. Then the relevant total momentum operators are

$$P^\mu = \frac{1}{2} \int d^2 x_\perp dx^- T^{+\mu}. \qquad (5)$$

Then it is necessary to express $T^{+\mu}$ in terms of independent variables. For example the nucleon is a four-component spinor to represent a spin 1/2 particle. There are therefore only two independent degrees of freedom, and one must express the other two dependent one in terms of these. A similar procedure must be employed to handle vector mesons. The details have been in the literature for a while[4,5,6,7]. We'll only mention the essentials here.

In our case we use a chiral $\mathcal{L}$ which includes $N\pi, \sigma, \omega, \rho, \eta, \delta$ as degrees of freedom. We note that many technical difficulties are associated with massless particles, but these are not present here.

2.4. Light Front Quantization- not so LITE

Here is the flavor of the procedure of the previous section. We start with the chiral Lagrangian of Gurzey:

$$\mathcal{L} = \bar\psi' \left(\gamma^\mu (i\partial_\mu - g_v V_\mu) - MU - g_s\phi \right) \psi' + \text{Meson terms}. \qquad (6)$$

This is invariant under the transformation $\psi' \rightarrow e^{-i\gamma_5 \boldsymbol{T} \cdot \boldsymbol{a}} \psi'$, $U \rightarrow e^{-i\gamma_5 \boldsymbol{T} \cdot \boldsymbol{a}} U e^{i\gamma_5 \boldsymbol{T} \cdot \boldsymbol{a}}$, with $U \equiv e^{i\gamma_5 \boldsymbol{T} \cdot \boldsymbol{\pi}}$. Then it is necessary to separate the independent and dependent Fermion fields. The projection operators $\Lambda_\pm$ $\Lambda_\pm \equiv \gamma^0 \gamma^\pm / 2$, $\psi'_\pm \equiv \Lambda_\pm \psi'_\pm$ are used, so that the nucleon equation of motion is expressed as:

$$(i\partial^- - g_v V^-)\psi'_+ = \alpha_\perp \cdot (\mathbf{p}_\perp - g_v \mathbf{V}_\perp) + \beta(MU + g_s\phi))\psi'_-$$
$$(i\partial^+ - g_v V^+)\psi'_- = \alpha_\perp \cdot (\mathbf{p}_\perp - g_v \mathbf{V}_\perp) + \beta(MU + g_s\phi))\psi'_+. \qquad (7)$$

The calculation of the dependent term ψ_- is simplified through the use of the Soper-Yan transformation $\psi' \equiv e^{-ig_v\Lambda(x)}\psi$, $\partial^+\Lambda = V^+$, so that

$$\partial^+\bar{V}^\mu \equiv \partial^+V^\mu - \partial^\mu V^+ \tag{8}$$

$$(i\partial^- - g_v\bar{V}^-)\psi_+ = [\alpha_\perp \cdot (\mathbf{p}_\perp - g_v\bar{\mathbf{V}}_\perp) + \beta(MU + g_s\phi)] \frac{1}{i\partial^+}$$
$$\times [\alpha_\perp \cdot (\mathbf{p}_\perp - g_v\bar{\mathbf{V}}_\perp) + \beta(MU + g_s\phi)] \psi_+ \tag{9}$$

The previous equation displays the Hamiltonian. The term ψ_+ can be evaluated as a perturbation expansion. There are two kinds of denominators. There are denominators determined by the $i\partial^-$ which are ordinary denominators, and there are those determined by $\frac{1}{i\partial^+}$, which are the instantaneous propagators. These lead to interactions of the form $\pi \frac{\gamma^+}{p^+} \cdot \pi$, which are important for maintaining chiral symmetry in pion-nucleon interactions.

3. Light Front One Boson Exchange Potential

If one is considering the scattering of two free nucleons it is easy to show that light front dynamics leads to the ordinary one boson exchange amplitudes. This, for the exchange of a scalar meson of four-momentum q^μ and mass μ is given by $V_{OBE} \propto \mathcal{M} \propto 1/(q^2 - \mu^2)$ The exchange of a meson between two bound nucleons requires the inclusion of the correct energy denominator. In that case,

$$V_{OBE} \propto \frac{1}{q^+(P^- - \sum_n k_n^-)}, \tag{10}$$

where P^- is the deuteron mass (in the rest frame) and the sum over n is over the two nucleons and one meson in the intermediate state. The calculation of nucleon-nucleon phase shifts using this method is explained elsewhere[8]. One can obtain a good description of the phase shifts.

4. Deuteron Calculation–Jason Cooke

The first place one encounters problems with rotational invariance is that the OBEP of Eq. (10) leads to a dependence of the binding energy on the magnetic quantum number M_z. This problem is ameliorated by including the effects of the two meson exchange terms. Figs. 6 and 7 of Ref. [2] show how unwanted binding energy differences disappear as one includes more and more terms in the nucleon-nucleon potential.

The two-pion-exchange terms are interesting because it is important to maintain chiral symmetry. The effective πN interaction (to second order) is given by

$$T_{\pi N} = \bar{\psi} \left[-i\frac{M}{f}\gamma_5 \tau \cdot \pi - \frac{M}{2f^2}\pi^2 + i\gamma_5\frac{M}{f}\tau \cdot \pi \left(\frac{\gamma^+}{2p^+}\right) i\gamma_5\frac{M}{f}\tau \cdot \pi \right] \psi, \quad (11)$$

which maintains the soft-pion theorems of chiral symmetry in $\pi-$nucleon scattering. In particular, the unwanted effects of nucleon pair terms are suppressed by the third operator which arises from instantaneous fermion exchange. The operator $T_{\pi N}$ also appears in the midst of the TPEP, and is included in our calculations. The results shown in[1,2,3] are that the binding energy respects rotational invariance.

We note that the OBEP violation of rotational invariance is observed in calculations of deuteron form factors. Let us define $\langle \lambda' | J^+ | \lambda \rangle \equiv I_{\lambda' \lambda}$. The use of kinematic parity (parity transformation followed by a rotation by π about the $x-$axis) and time reversal invariance shows that there are four independent values of $\lambda' \lambda$. But there are only three form factors. The $I_{\lambda' \lambda}$ are not independent, as they must satisfy the so-called angular condition

$$(1 + 2\eta)I_{11} + I_{1,-1} - \sqrt{8\eta}I_{10} - I_{00} = 0, \qquad \eta = Q^2/4M_d^2.$$

which arises from the rotational invariance of the charge density. In practice we define

$$(1 + 2\eta)I_{11} + I_{1,-1} - \sqrt{8\eta}I_{10} - I_{00} \equiv \Delta \neq 0. \quad (12)$$

We compute the value of Δ, and if it is very small we say the calculation of the form factors respects rotational invariance. It turns out that Δ is very small. See Figs. 11,12 of Ref. [2]. A standard procedure has been to choose one of the four $I_{\lambda' \lambda}$ to be dependent, and compute other three to obtain the observables A, B, T_{20}. There are four different prescriptions, but there is little dependence of the results on the prescription; see Figs. 15,16 of Ref. [2].

The biggest problem is that the computed form factors depend strongly on the different parametrization of the nucleon form factors. See Fig. 17 of Ref. [2].

5. Summary and Discussion

I have discussed the calculations of Refs. [1,2,3]. Our procedure is to start with a Lagrangian, compute a Hamiltonian P^- and then determine the deuteron wave function using that very same Hamiltonian. We find that using a one meson exchange OME potential violates rotational invariance

348

as defined by the difference between the binding energy of states with $M_d = 0, 1$. Including the effects of the two meson exchange TME potential greatly reduces the values of the binding energy differences. We also explicitly show that the rotational non-invariance of the computed form factors is small, as measured by the generally small values of the parameter Δ of Eq. (12). It turns out that Δ is smaller for OME than for TME. This indicates that our impulse approximation is not sufficient and that meson exchange currents needed. In any case the uncertainties in our knowledge of the isoscalar form factor F_1 are much bigger than the effects caused by having $\Delta \neq 0$. Thus future measurements of the nucleon form factors are necessary to understand deuteron physics.

References

1. J. R. Cooke, arXiv:nucl-th/0112029.
2. J. R. Cooke and G. A. Miller, arXiv:nucl-th/0112037.
3. J. R. Cooke and G. A. Miller, Phys. Rev. C **65**, 067001 (2002) [arXiv:nucl-th/0112076].
4. D. E. Soper, Phys. Rev. D **4**, 1620 (1971).
5. T. M. Yan, Phys. Rev. D **7**, 1760 (1973).
6. T. M. Yan, Phys. Rev. D **7** (1973) 1780.
7. G. A. Miller, Phys. Rev. C **56**, 2789 (1997) [arXiv:nucl-th/9706028].
8. G. A. Miller and R. Machleidt, Phys. Lett. B **455**, 19 (1999); Phys. Rev. C **60**, 035202 (1999).

DEUTERON π^0 PHOTOPRODUCTION AT HIGH MOMENTUM TRANSFER

CHUENG-RYONG JI

Department of Physics,
North Carolina State University,
Raleigh, NC 27695, USA
E-mail: ji@ncsu.edu

We discuss the deuteron reduced nuclear amplitude using a simple field theoretic model and extend the formalism to the pion photoproduction process $(\gamma D \to \pi^0 D)$ at high momentum transfer. It provides a new improved factorization of reduced nuclear amplitude and predicts the $\gamma D \to \pi^0 D$ amplitude with the input of the $\gamma p \to \pi^0 p$ data. A comparison with the recent JLAB data for $\gamma D \to \pi^0 D$ and the available $\gamma p \to \pi^0 p$ shows good agreement between the perturbative QCD prediction and experiment at high momentum transfer. We also discuss a possible zero-mode contribution in the light-front dynamics to the helicity zero-to-zero matrix element of the plus current which is expected to dominate at high momentum transfer according to the perturbative QCD. In a simple model calculation, we find that the zero-mode contribution is largely suppressed at high momentum transfer.

1. Introduction

The leading twist predictions from the perturbative quantum chromodynamics(PQCD) are consistent with the constituent counting rule(CCR) and yield the dominance of helicity $0 \to 0$ amplitude. The electric structure function $A(Q^2)$ of the deuteron, which is a combination of the charge, quadrupole and magnetic form factors, has been measured at the Jefferson Laboratory (JLAB) for $2 \le Q^2 \le 6$ GeV^2 and the data[1] show the consistency with the CCR, *i.e.* $Q^{10}A(Q^2) \sim constant$. The reduced nuclear amplitude (RNA) formalism in general improves the agreement between the CCR predictions and the experimental data. Analogous to the deuteron reduced form factor defined by[2]

$$f_d(Q^2) = \frac{F_D(Q^2)}{F_p(Q^2/4)F_n(Q^2/4)} \sim \frac{constant}{Q^2},\tag{1}$$

the RNA for deuteron photodisintegration is given by[3]

$$m_{\gamma d \to np} = \frac{M_{\gamma D \to np}}{F_n(\hat{t}_n)F_p(\hat{t}_p)} \sim p_T^{-1} f(\theta_{c.m.}).\tag{2}$$

Comparison with the data shows this prediction is rather successful at $\theta_{c.m.} = 90°$ for $E_\gamma > 1\text{GeV}$ [4,5,6,7]. At the smaller angles such as $\theta_{c.m.} = 52°$ and 36°, however, the experimental data [4,5,6,7] deviate from the CCR and RNA predictions. In principle, this is not unexpected since the presence of the large nuclear mass and the large number of partons involved can be expected to delay the onset of leading-twist scaling. Nevertheless, it has been argued that the data are also well described by conventional calculations including meson-exchange currents. Moreover, the most recent measurements[8] of the proton form factors exhibit that the ratio of the Pauli form factor F_2 to the Dirac form factor F_1 behaves as $F_2/F_1 \sim Q^{-1}$ rather than Q^{-2} for $2 \leq Q^2 \leq 6$ GeV^2, indicating a delay in the evolution to the region applicable to the leading twist prection from PQCD. Another JLAB data presented as an example inconsistent with both CCR and RNA predictions are the Hall C data on the π^0 photoproduction from a deuteron target. Both data at $\theta_{c.m.} = 136°$ and 90° were interpreted [9] as being inconsistent with the RNA approach. This is in sharp contrast to the deuteron electric structure function $A(Q^2)$ which is consistent with both the CCR and RNA predictions as discussed above.

In this talk, we discuss an improvement of RNA factorization for the π^0 photoproduction from a deuteron target(Section 2). With this improvement, a comparison with the recent JLAB data[9] for $\gamma D \to \pi^0 D$ and the available $\gamma p \to \pi^0 p$ data[10,11,12,13,14] shows good qualitative agreement between the perturbative QCD prediction and experiment over a large range of momentum transfers and center-of-mass angles. We also discuss a possible zero-mode contribution in the light-front dynamics to the PQCD favorite helicity $0 \to 0$ amplitude(Section 3). In a simple model calculation, the zero-mode contribution is suppressed at high momentum transfer. Conclusions follow in Section 4.

2. Deuteron RNA and Improved Factorization for π^0 Photoproduction

The impulse approximation for the deuteron form factor given by

$$F_D(Q^2) = F_d^{body}(Q^2)F_N(Q^2), \tag{3}$$

where F_N is the on-shell nucleon form factor, cannot be valid at the scale $Q^2 \geq 2m_D\epsilon_D \approx (100\text{MeV})^2$. The struck nucleon is necessarily off-shell for high momentum transfer. The deuteron form factor $F_D(Q^2)$ is the probability amplitude for the deuteron to stay intact after absorbing momentum transfer Q. If the deuteron is taken as a lightly bound cluster of two nucleons, then the form factor contains the probability amplitudes for each nucleon to remain intact after absorbing momentum transfer $\sim q^\mu/2$. Thus, it is natural to factorize F_D in the form

$$F_D(Q^2) = f_d(Q^2)F_N^2(Q^2/4), \tag{4}$$

which defines the reduced form factor $f_d(Q^2)$. Using a simple field theoretic model, we have indeed shown[15] that the relativistic deuteron wavefunction has a cluster decomposition; $i.e.$ it factors into two seperate nucleon wavefunctions convoluted with a body wavefunction in the zero-binding limit. Assuming a quark-interchange mechanism, we then derived the deuteron reduced form factor at high momentum transfer given by Eq.(4), while recovering the impulse approximation given by Eq.(3) at small momentum transfer. The QCD predicts $Q^2 f_d(Q^2) \sim constant$ modulo logarithmic corrections due to the running coupling constant and anomalous dimensions of the nuclear wavefunction[2,16], which is in excellent agreement with the experimental data from JLAB[1].

Recently, we extended[17] the RNA analysis to the $\gamma D \to \pi^0 D$ process. The cluster decomposition of the deuteron wave function at small binding only allows the nuclear coherent process to proceed if each nucleon absorbs an equal fraction of the overall momentum transfer. Furthermore, each nucleon must scatter while remaining close to its mass shell. Thus the nuclear photo-production amplitude $M_{\gamma D \to \pi^0 D}(u,t)$ factorizes as a product of three factors:

(1) the nucleon photo-production amplitude $M_{\gamma N_1 \to \pi^0 N_1}(u/4, t/4)$ at half of the overall momentum transfer,

(2) a nucleon form factor $F_{N_2}(t/4)$ at half the overall momentum transfer, and

(3) the reduced deuteron form factor $f_d(t)$, which according to perturbative QCD, has the same monopole fall-off as a meson form factor.

Note that the on-shell condition requires the center of mass angle of pion photo-production on the nucleon N_1 to be identical to the center of mass angle of pion photo-production on the deuteron; the directions of incoming and outgoing particles in the nucleon subprocess must be the same as those of the deuteron process. The exchanged gluon carries half of the momentum transfer to the spectator nucleon. Thus as in the case of the deuteron form factor the nuclear amplitude contains an extra quark propagator at an approximate virtuality $t/3$ in addition to the on-shell nucleon amplitudes. This structure predicts the reduced amplitude scaling

$$M_{\gamma D \to \pi^0 D}(u,t) = C' f_d(t) M_{\gamma N_1 \to \pi^0 N_1}(u/4, t/4) F_{N_2}(t/4), \tag{5}$$

352

where C' is expected to be around unity. A comparison with elastic electron scattering then yields the following proportionality of amplitude ratios:

$$\frac{M_{\gamma D \to \pi^0 D}}{M_{eD \to eD}} = C' \frac{M_{\gamma p \to \pi^0 p}}{M_{ep \to ep}}. \qquad (6)$$

From the recent JLAB $\gamma D \to \pi^0 D$ data [9], we computed the corresponding invariant amplitudes[17] both for $\theta_{c.m.} = 90°$ and $136°$. We then used our factorization formula Eq. (5) to predict $|M_{\gamma D \to \pi^0 D}|$ with input from the available $\gamma p \to \pi^0 p$ data [10,11,12,13,14]. The results are presented in Figs. 1 and 2. In Fig. 1, the normalization of our prediction is fixed (at $C' = 0.8$)

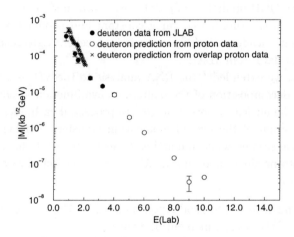

Figure 1. $|M_{\gamma D \to \pi^0 D}|$ versus photon lab energy E_{lab} at $\theta_{c.m.} = 90°$. The filled circles are obtained directly from the recent JLAB $\gamma D \to \pi^0 D$ data [9], and the crosses and open circles are our predictions from the $\gamma p \to \pi^0 p$ data presented in Ref. [10] and Refs. [11,12]. Note that an open circle is overlaid on top of a filled circle for the overlapping data point at $E_{\text{lab}} = 4$ GeV.

by the overlapping data point at $E_{\text{lab}} = 4$ GeV, which is the highest photon lab energy used in the JLAB $\gamma D \to \pi^0 D$ experiment [9]. It is interesting to find that the general trend of our prediction (the open circles) is very similar to that of the direct result from the JLAB data [9], shown as filled circles. Moreover, our prediction in the E_{lab} overlap region, denoted by crosses, mimics the shape of the direct result. Similar behaviors can be seen in Fig.2. Both amplitudes shown in Figs.1 and 2 are also consistent with the scaling behavior[17] predicted by PQCD, $P_T^{11}|M_{\gamma D \to \pi^0 D}| \sim constant$ at fixed $\theta_{c.m.}$. See more details in Ref.17.

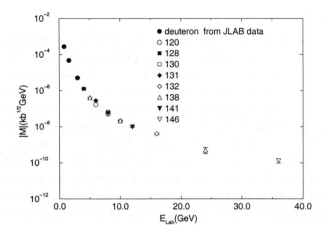

Figure 2. $|M_{\gamma D \to \pi^0 D}|$ versus photon lab energy E_{lab} at $\theta_{c.m.} = 136°$. The filled circles are obtained directly from the recent JLAB $\gamma D \to \pi^0 D$ data [9], and the other symbols are our predictions based on $\gamma p \to \pi^0 p$ data at angles near $\theta_{c.m.} = 136°$, presented in Refs. [10,11,12].

3. Caveats from Zero-Modes

While the factorization theorem in PQCD is based on the dominance of helicity zero-to-zero amplitude at high momentum transfer, we find[18] that the helicity zero-to-zero matrix element of the plus current is most contageous from the zero-mode contribution in the light-front dynamics. The zero-mode[19] occurs when the integrand is divergent although the integration range goes to zero. If the zero-mode occurs in the form factor analysis, the off-diagonal matrix elements in the Fock-state expansion cannot be neglected at $q^+ = 0$ frame. Since the PQCD factorization theorem relies only on the diagonal matrix elements in the Fock expansion at $q^+ = 0$ frame, the zero-mode contribution complicates the PQCD analysis. Indeed, the plus component of the helicity-zero polarization vector is not zero, *i.e.* $\epsilon^+(0) \neq 0$, so that the minus component of the internal momentum (*e.g.*, $k^- = \frac{k_\perp^2 + m^2}{k^+}$) can survive in the integrand for the helicity zero-to-zero amplitude just enough to make the integrand be divergent even though the integration range goes to zero. In our recent model calculation[18], we showed that the zero-mode contribution exists only in the helicity zero-to-zero amplitude. However, the zero-mode contribution gets diminished at the high momentum transfer. This indicates that the PQCD factorization at high momentum transfer may be protected from the zero-mode contamination.

4. Conclusions

With the new improved RNA, it may be possible to understand the JLAB(Hall C) $\gamma D \to \pi^0 D$ data based on PQCD. The JLAB $\gamma D \to \pi^0 D$ data seem to be consistent with the available $\gamma p \to \pi^0 p$ data without requiring any anomalous contribution such as odderon. Also, the CCR scaling is not inconsistent to these data for $E_{\text{lab}} \geq 10$ GeV. We also obtain a new prediction given by Eq.(6). However, we find a possible zero-mode contribution to the helicity $0 \to 0$ amplitude favored by PQCD in spin-one system. In our model calculation, the zero-mode contribution was largely suppressed at high Q^2 region. Nevertheless, it would be interesting to investigate any correlation from the zero-mode to the delay in PQCD applicability shown in the recent observation of $QF_2/F_1 \sim constant$ for the proton form factors.

Acknowledgments

This work is in collaboration with Stan Brodsky, John Hiller and Jerry Miller. Ben Bakker and Ho-Meoyng Choi are also collaborated in the zero-mode issue of spin-one form factors. This work was supported in part by a grant from the U.S.Department of Energy (DE-FG02-96ER 40947) and the National Science Foundation (INT-9906384).

References

1. L.C.Alexa et al., *Phys. Rev. Lett.* **82**, 1374 (1999).
2. S.J.Brodsky, C.-R.Ji and G.P.Lepage *Phys. Rev. Lett* **51**, 83 (1983).
3. S. J. Brodsky and J. R. Hiller, *Phys. Rev.***C28**, 475 (1983); **C30**, 412E (1984).
4. J.Napolitano et al., *Phys. Rev. Lett.* **61**, 2530 (1988); S.J. Freeman et al., *Phys. Rev.***C48**, 1864 (1993).
5. J.E.Belz et al., *Phys. Rev. Lett.* **74**, 646 (1995).
6. C.W.Bochna et al., *Phys. Rev. Lett.* **81**, 4576 (1998).
7. E.C.Schulte et al., *Phys. Rev. Lett.* **87**, 102302 (2001).
8. A. Nathan's presentation in this workshop.
9. D.G.Meekins et al., *Phys. Rev.* **C60**, 052201 (1999).
10. A.Imanishi et al., *Phys. Rev. Lett.* **54**, 2497 (1985).
11. M.A.Shupe et al., *Phys. Rev. Lett.* **40**, 271 (1978).
12. G.C.Bolon et al., *Phys. Rev. Lett.* **18**, 926 (1967).
13. R.L.Anderson et al., *Phys. Rev. Lett.* **30**, 627 (1973).
14. R.L.Anderson et al., *Phys. Rev. Lett.* **26**, 30 (1971).
15. S.J.Brodsky and C.-R.Ji, *Phys. Rev.* **D33**, 2653 (1986).
16. C.-R.Ji and S.J.Brodsky, *Phys. Rev.* **D34**, 1460 (1986).
17. S.J.Brodsky, J.R.Hiller, C.-R.Ji and G.A.Miller, *Phys. Rev.***C64**, 055204 (2001).
18. B.L.G.Bakker,H.-M.Choi and C.-R.Ji,*Phys. Rev.* **D65**, in press.
19. H.-M.Choi and C.-R.Ji,*Phys. Rev.***D58**, 071901 (1998).

PROBING THE DEUTERON AT HIGH −T

R. GILMAN

Rutgers University, Piscataway, NJ 08855 USA and
Thomas Jefferson National Accelerator Facility, Newport News, VA 23606 USA
E-mail: rgilman@physics.rutgers.edu

FOR THE JEFFERSON LAB HALL A COLLABORATION

The deuteron is probed at both higher s and higher momentum transfer to the nucleon with photodisintegration than with elastic scattering. Kinematics makes the construction of a meson-baryon theory more difficult, and the potential success of a quark-based theory more likely. I review the latest experimental results, and discuss how well various quark models agree with the data.

1. Low-Energy Photodisintegration

At low energies, deuteron photodisintegration is understood with meson-baryon models, in which the incoming photon couples to a nucleon, and resonance excitation plus final state interactions are included. Good quantitative medium-energy predictions have been available since the work of Laget,[1] although discrepancies in the cross-section data were not resolved until later.[2] More refined, modern calculations[3] have good agreement with differential cross sections, with the Σ linearly polarized photon asymmetry,[4,5] and with many other, less extensively determined, polarization observables.[6] The difficulty in understanding large induced polarizations, for energies just above the Δ resonance, remains, and illustrates the difficulty in constructing high energy theories of photodisintegration.

2. Quark-Gluon vs. Meson-Baryon Degrees of Freedom

Meson-baryon theories and quark theories are in principle equivalent, as each set of degrees of freedom provides a complete basis. But one does not know in advance whether a satisfactory theoretical description of the reaction can actually be constructed in either framework.

Construction of a microscopic meson-baryon theory becomes an intractable problem at energies of a few GeV. Photodisintegration from the Born term is too small; resonances and final state interactions are needed.

There are 286 channels,[6] pairs of the 24 well established baryon resonances, needed to describe the reaction dynamics. Such a calculation has obvious problems with mathematical precision, plus unknown baryon-baryon interactions and contributions of resonances that are not yet well established. An asymptotic meson-baryon theory, with adjustable parameters that in effect average over resonances, can be constructed,[7] but it will be unsatisfying unless it is widely applicable or the origin of its parameters is understood.

Perturbative QCD (pQCD) is the appropriate quark-gluon theory for high momentum transfers, but its applicability to exclusive reactions in accessible kinematics is questionable.[8,9] Although exclusive reactions follow the constituent counting rules,[10,11] absolute calculations are too small, indicating dominance of soft, rather than perturbative physics. Hadronic helicity conservation (HHC), predicted by pQCD[12,13] is also taken as evidence against the applicability of pQCD. HHC is violated in all of the few cases in which it has been tested. Spin effects are prominent at all energies.

The importance of orbital angular momentum[14] in reactions has become more recognized. High energy vector interactions conserve quark helicity, but orbital angular momentum effects can violate HHC. These ideas are related to the dominance of quark interchange diagrams seen experimentally[15] and to independent scattering mechanisms. It may be that pQCD applies over a wider range than generally believed, once hard scattering is separated from soft wave function physics, which can be modelled.

3. High-Energy Photodisintegration Cross Sections

Over 15 years, our collaborations have measured high-energy photodisintegration in experiments SLAC NE8[16,17] and NE17,[18] and JLab E89-012,[19] E89-019,[20] E96-003,[21] and E99-008.[22] Figure 1 shows our high-energy cross section data together with low energy Mainz data.[23] The cross section approximately follows the constituent counting rule behavior, $d\sigma/dt \propto s^{-11}$, once $p_T \approx 1.3$ GeV.

Several calculations are shown. The long-short dashed curve results from a product of nucleon form factors times phase space factors; normalized by a single overall factor, it largely reproduces the trends of the data. This was suggested starting from pQCD,[24] and by considering the absorption of a hard photon on two quarks being exchanged between the nucleons.[25] A calculation[26] of the quark interchange idea relates photodisintegration to the nucleon-nucleon scattering amplitude, taken from data, along with a hard photo-quark coupling and deuteron wave function. Uncertainties, interpolations, and extrapolations of the data lead to the shaded

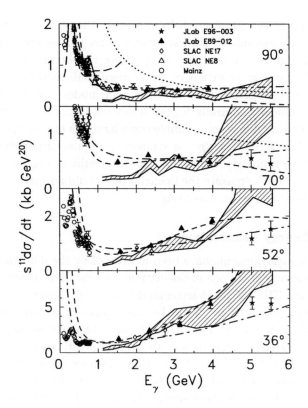

Figure 1. High-energy, forward-angle deuteron photodisintegration cross sections.

region shown, which agrees well with the data. The Reduced Nuclear Amplitudes (RNA) model,[27] the dotted curve, attempts to extend the pQCD scaling by including nucleon form factors. It uses one overall normalization; it is too large at forward angles, but there is an additional unknown angle dependent normalization. The quark-gluon string (QGS) model,[28] the short dashed curve, uses non-linear Regge trajectory techniques to evaluate photodisintegration as 3 quark exchange. Meson-baryon calculations include coupled channels and front form calculations,[29] the dashed curve at low energies for 90°, and a calculation[30], the solid curve at low energies for 90° that overlaps the data very well, and is notable for its inclusion of 17 baryon resonances, though in a static approximation.

The generally good agreement of most models with the data and with each other makes interpretation difficult. Clearly, calculations are preferred if they have no adjustable parameters, but many of the calculations shown have been adjusted to the data. More extensive angular distributions and polarization observables would help to identify the reaction mechanism.

We recently measured[22] the asymmetry of the cross-section angular distribution about 90°; a clear forward peaking was observed. Preliminary results of a much more extensive set of angular distributions from JLab Hall B E93-017 are consistent.[31] Meson-baryon models[7] and the quark hard rescattering model[26] naturally suggest a forward-backward asymmetry, due to proton-neutron differences. Simple considerations that use nucleon form factors, such as RNA,[27] suggest a symmetric angular distribution. QGS[28] predicts the asymmetry of our data, and interestingly suggests local minima at 0° and 180°, opposite the behavior one would normally expect from form factors or diffractive scattering.

4. High-Energy Polarizations

Polarization measurements have been motivated by the difference between HHC and meson-baryon model calculations, but recall the discussion above on pQCD and HHC. Meson-baryon models generally predict, e.g., large induced proton polarizations, from the interference of resonant amplitudes with the Born term.[30,32] Figure 2 compares two meson-baryon[3,30] and the hard rescattering[33] calculations to induced polarization, p_y, data from "Stanford",[34] "Tokyo",[35,36] and our "JLab" measurements,[20]. The magnitude of p_y drops rapidly above 0.5 GeV, and is consistent with vanishing above 1 GeV. The six highest energy data average to -0.02 ± 0.02. The sensitivity of p_y to interfering amplitudes is obvious. The calculated peak in the Δ-resonance region is offset from the data by $\approx$100 MeV, and much smaller. In the meson-baryon framework, the combined effects of all resonances plus final state interactions appear to generate small polarizations.

As both the scaling of the cross section and the vanishing of p_y start near 1 GeV, and are consistent with pQCD expectations, further tests are required. Figure 2 shows the polarization transfer observables in both lab and c.m. frames, along with a recent meson-baryon calculation.[3] The boost effects are not significant here. There are no other data for these observables. If HHC is valid, both $C_{x'}$ and $C_{z'}$ vanish. Since $C_{x'}$ is consistently non-zero, hadron helicity is *not* conserved. Recent Σ asymmetry measurements from Yerevan[37] lead to a similar conclusion, for $E_\gamma = 1 - 1.6$ GeV.

5. Conclusions

There are now many high-energy deuteron photodisintegration measurements; the behavior of the cross section is fairly well determined. The energy dependence of the cross section does not distinguish well between

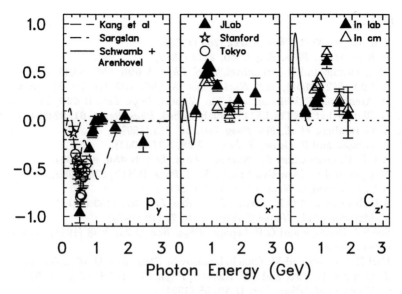

Figure 2. Recoil proton polarization observables for $\vec{\gamma}d \rightarrow \vec{p}n$ at $\theta_{cm} = 90°$.

the models; several of the models are unsatisfactory in that they adjust parameters, such as the overall normalization of their predictions. The polarizations are small at higher energies, but hadron helicity conservation is violated. The polarizations have generally not been estimated; the hard rescattering model[33] does qualitatively reproduce their behavior.

The data are difficult to understand in a conventional microscopic meson-baryon model. Nonperturbative quark models show promise, but are insufficiently tested at present. Future measurements of spin observables and other reactions can help to better establish these models. The proposed Hall A MAD spectrometer and Jefferson Lab 12 GeV upgrade could expand measurements over a much larger kinematic range.

Acknowledgments

I thank particularly R. Holt, who deserves primary credit for our studies over the years, and K. Wijesooriya and E. Schulte for their lead roles in the analyses of the recent experiments. I also thank A. Afanasev, F. Gross, T.-S.H. Lee, A. Radyushkin, and M.M. Sargsian, for their efforts to help my understanding of theoretical issues. Southeastern Universities Research Association manages Thomas Jefferson National Accelerator Facility under DOE contract DE-AC05-84ER40150. The U.S. National Science Foundation grant PHY 00-98642 supports my research at Rutgers University.

360

References

1. J.M. Laget, *Nucl. Phys.* A **312**, 265 (1978).
2. P. Rossi *et al.*, *Phys. Rev.* C **40**, 2412 (1989).
3. M. Schwamb and H. Arenhövel, *Nucl. Phys.* A **696**, 556 (2001), *Nucl. Phys.* A **690**, 682 (2001), and *Nucl. Phys.* A **690**, 647 (2001); M. Schwamb, H. Arenhövel, P. Wilhelm, and Th. Wilbois, *Phys. Lett.* B **420**, 255 (1998).
4. G. Blanpied *et al.* (The LEGS Collaboraton), *Phys. Rev.* C **61**, 024604 (1999).
5. S. Wartenberg *et al.*, *Few Body Syst.* **26**, 213 (1999).
6. R. Gilman and F. Gross, *J. Phys.* G **28**, R37 (2002).
7. A.E.L. Dieperink and S.I. Nagorny, *Phys. Lett.* B **456**, 9 (1999).
8. N. Isgur and C. Llewellyn Smith, *Nucl. Phys.* B **317**, 526 (1989).
9. A.V. Radyushkin, *Nucl. Phys.* A **523**, 141c (1991).
10. S.J. Brodsky and G.R. Farrar, *Phys. Rev. Lett.* **31**, 1153 (1973).
11. V. Matveev *et al.*, *Nuovo Cimento Lett.* **7**, 719 (1973).
12. See S.J. Brodsky and G.P. Lepage, *Phys. Rev.* D **24**, 2848 (1981), and references therein.
13. Carl E. Carlson and M. Chachkhunashvili, *Phys. Rev.* D **45**, 2555 (1992).
14. T. Gousset, B. Pire, and J.P. Ralston, *Phys. Rev.* D **53**, 1202 (1996).
15. C. White *et al.*, *Phys. Rev.* D **49**, 58 (1994).
16. J. Napolitano *et al.*, *Phys. Rev. Lett.* **61**, 2530 (1988).
17. S.J. Freedman *et al.*, *Phys. Rev.* C **48**, 1864 (1993).
18. J.E. Belz *et al.*, *Phys. Rev. Lett.* **74**, 646 (1995).
19. C. Bochna *et al.*, *Phys. Rev. Lett.* **81**, 4576 (1998).
20. K. Wijesooriya *et al.*, *Phys. Rev. Lett.* **86**, 2975 (2001).
21. E. Schulte *et al.*, *Phys. Rev. Lett.* **87**, 102302 (2001).
22. E. Schulte *et al.*, submitted to Phys. Rev. C.
23. R. Crawford *et al.*, *Nucl. Phys.* A **603**, 303 (1996).
24. Carl E. Carlson, *Nucl. Phys.* A **508**, 481c (1990).
25. A.V. Radyushkin, private communication.
26. L.L. Frankfurt, G.A. Miller, M.M. Sargsian, and M.I. Strikman, *Phys. Rev. Lett.* **84**, 3045 (2000) and *Nucl. Phys.* A **663**, 349 (2000).
27. S.J. Brodsky and J.R. Hiller, *Phys. Rev.* C **28**, 475 (1983).
28. V.Yu Grishina *et al.*, *Eur. Phys. J* A **10**, 355 (2001).
29. T.-S.H. Lee, *Few Body Syst. Supplement* **6**, 526 (1992) and Argonne National Laboratory preprints PHY-6886-TH-91 (1991) and PHY-6843-TH-91 (1991).
30. Y. Kang *et al.*, *Abstracts of the Particle and Nuclear Intersections Conference*, (MIT, Cambridge, MA 1990); Y. Kang, Ph.D. dissertation, Bonn (1993).
31. P. Rossi, private communication.
32. M. Schwamb, H. Arenhövel, and P. Wilhelm, *Few Body Syst.* **19**, 121 (1995).
33. M.M. Sargsian, private communication and talk at Institute for Nuclear Theory program INT-01-1, Correlations in Nucleons and Nuclei, May 2001.
34. F.F. Liu *et al.*, *Phys. Rev.* **165**, 1478 (1968).
35. T. Kamae *et al.*, *Phys. Rev. Lett.* **38**, 468 (1977).
36. T. Kamae *et al.*, *Nucl. Phys.* B **139**, 394 (1978); H. Ikeda *et al.*, *Phys. Rev. Lett.* **42**, 1321 (1979); H. Ikeda *et al.*, *Nucl. Phys.* B **172**, 509 (1980).
37. F. Adamian *et al.*, *Eur. Phys. J* A **8**, 423 (2000).

NOVEL HARD SEMIEXCLUSIVE PROCESSES AND COLOR SINGLET CLUSTERS IN HADRONS

LEONID FRANKFURT

Physics Department, Tel Aviv University, Tel Aviv, Israel

M.V. POLYAKOV

Institut für Theoretische Physik II, Ruhr–Universität Bochum, D–44780 Bochum, Germany

M. STRIKMAN

Department of Physics, Pennsylvania State University, University Park, PA 16802, USA

D.ZHALOV

Department of Physics, Pennsylvania State University, University Park, PA 16802, USA

M.ZHALOV

Petersburg Nuclear Physics Institute, Gatchina, Russia

Hard scattering to a three cluster final state is suggested as a method to probe configurations in hadrons containing small size color singlet cluster and a residual quark-gluon system of a finite mass. Examples of such processes include $e + N \to e + p + M_X(\Lambda + M'_X), p + p \to p + p + M_X(p + \Lambda + M'_X)$ where $M_X(M'_X)$ could be a pion(kaon) or other state of finite mass which does not increase with momentum transfer (Q^2). We argue that different models of the nucleon may lead to very different qualitative predictions for the spectrum of states M_X. We find that in the pion model of nonperturbative $q\bar{q}$ sea in a nucleon the cross section of these reactions is comparable to the cross section of the corresponding two-body reaction. Studies of these reactions are feasible using both fixed target detectors (EVA at BNL, HERMES at DESY) and collider detectors with a good acceptance in the forward direction.

1. Introduction

The QCD analyzes of the eighties have demonstrated that cross sections of hard two-body exclusive processes are expressed through the minimal Fock space components of hadrons involved in the reaction (for recent reviews and references see [1,2]). The need to establish at what Q^2 minimal Fock space components start to dominate in these processes has stimulated searches for the color transparency phenomena.

It is natural to move one step further and ask a question whether collapsing of three valence quarks to a small size color singlet configuration in a nucleon or of valence quark and antiquark in a meson would result in disappearance of other constituents? Such scenario would be natural in quantum electrodynamics for the case of positronium - the photon field disappears in the case when electron and positron are close together (that is the ratio of the amplitude for positronium to be in the Fock state $e^+e^-\gamma$ and in the Fock state e^+e^- decreases with decrease of of the distance between e^+ and e^-). However in QCD where interactions at large distances are strong it is possible that non minimal Fock components of the hadron contain configurations with small color singlet clusters without an additional smallness as compared to the minimal Fock component. In fact there is no need to restrict the question to the case of clusters build of the valence quarks in a hadron - one can as well consider any small color singlet cluster - for example a $q\bar{q}$ cluster in the nucleon, a three quark cluster with strangeness or charm in the nucleon, etc.

In this talk we will consider the knock out of the clusters induced both by electrons and by hadrons[a].

2. Exclusive production of forward baryons off nucleons

Over the last few years $q\bar{q}$ clusters in the nucleons and mesons were implicitly considered in the context of the study of exclusive DIS processes: $\gamma_L^* + N \to$ "$meson$"+$baryon$" for which the factorization theorem is valid [4,5] which states that in the limit of large Q^2 the amplitude of the process at fixed x is factorized into the convolution of a hard interaction block calculable in perturbative QCD, the short-distance $q\bar{q}$ wave function of the meson, and the generalized/skewed parton distribution (GPD) in the nucleon. The HERA data have confirmed a number of the key predictions of [4] including shrinkage of the t-distribution the light meson production with increase of Q^2 to the limiting value which is close to the slope of the J/ψ production cross section. The Q^2 dependence of the slope is consistent with predictions of [6] (which extended the analysis of [4] to account for the geometrical higher twist effects which arise due to the finite transverse size of the virtual photon wave function) indicating that at $Q^2 \geq 4 GeV^2$ small transverse size configurations in ρ-meson are selected and that the suppression of the color dipole - nucleon interaction occurs already for transverse distances ≤ 0.4 fm.

---

[a]As far as we know these processes were first discussed in [3] where they were referred to as star dust processes.

The proof of the factorization for the meson exclusive production [5], is essentially based on the observation that the cancellation of the soft gluon interactions is intimately related to the fact that the meson arises from a quark-antiquark pair generated by the hard scattering. Thus the pair starts as a small-size configuration and only substantially later grows to a normal hadronic size, to a meson. Similarly, the factorization theorem should be valid for the production of leading baryons

$$\gamma^*(q) + p \to B(q + \Delta) + M(p - \Delta), \tag{1}$$

and even leading antibaryons

$$\gamma^*(q) + p \to \bar{B}(q + \Delta) + B_2(p - \Delta), \tag{2}$$

where B_2 is a system with the baryon charge of two. For example in the case of the process 1 the dominant diagram is given by Fig.1:

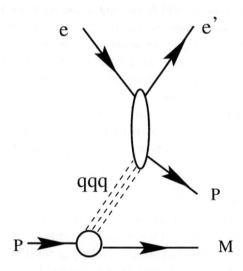

Figure 1. *Production of a fast baryon and recoiling mesonic system.*

In QCD to describe the hard exclusive processes one needs to use generalized (skewed) parton distributions. Since the objects one has to introduce for description of $N \to N$ transitions and non-diagonal transitions like $N \to \Lambda, \Delta$ are pretty different we suggest to refer to the first type of distributions as as generalized parton distributions (GPD), while in the case of non-diagonal transitions use the term skewed PD.

To describe in QCD process (1), one needs to introduce a new non-perturbative mathematical object [7] in addition to the GPDs and SPDs. It can be called a super skewed parton distribution (amplitude). It is defined as a non-diagonal matrix element of the tri-local quark operator between a meson M and a proton:

$$\int \prod_{i=1}^{3} dz_i^- \exp[i \sum_{i=1}^{3} x_i \, (p \cdot z_i)] \cdot$$

$$\langle M(p - \Delta) | \varepsilon_{abc} \, \psi_{j_1}^a(z_1) \, \psi_{j_2}^b(z_2) \, \psi_{j_3}^c(z_3) | N(p) \rangle \Big|_{z_i^+ = z_i^\perp = 0}$$

$$= \delta(1 - \zeta - x_1 - x_2 - x_3) \, F_{j_1 j_2 j_3}(x_1, x_2, x_3, \zeta, t) , \qquad (3)$$

where
a, b, c are color indices, j_i are spin-flavor indices, and $F_{j_1 j_2 j_3}(x_1, x_2, x_3, \zeta, t)$ are the new superSPDs. They can be decomposed into invariant spin-flavor structures which depend on the quantum numbers of the meson M. They depend on the variables x_i (which are contracted with the hard kernel in the amplitude), on the skewedness parameter $\zeta = 1 - \Delta^+/p^+$ and the momentum transfer squared $t = \Delta^2$. In some sense, with this definition of ζ the limit $\zeta \to 0$ corresponds to the usual distribution amplitude, $i.e.$ skewedness $\to 0$ means (for appropriate quantum numbers of the current) superSDP $\to$ "nucleon distribution amplitude".

Though quantitative calculations of processes (1, 2) will take time, some qualitative predictions could be made right away. First we observe that in the Bjorken limit the light cone fraction of the slow meson satisfies condition:

$$\alpha_h = \frac{p_{M-}}{p_{N-}} = \frac{E_h - p_{3M}}{E_N - p_{N3}} = \frac{E_M - p_{3M}}{m_N} = 1 - x \qquad (4)$$

and its transverse momentum p_t relative to the $\vec{q}$ direction been fixed. To ensure an early onset of scaling it is natural to consider the process as a function of Q^2 for fixed α_h, p_t. This way we can make a natural link to the picture of removing a cluster from the nucleon leaving the residual system undisturbed. If the color transparency suppresses the final state interaction between the fast moving nucleon and the residual meson state early enough it would be natural to expect an early onset of the factorization of the cross section to a function which depends on α_h, p_t and the cross section of the electron-nucleon elastic scattering:

$$\frac{d\sigma(e + N \to e + N + M)}{d\alpha_M d^2 p_t / \alpha_M} = f_M(\alpha_M, p_t)(1 - \alpha_M)\sigma(eN \to eN), \qquad (5)$$

where $(1 - \alpha_M)$ is the flux factor and $\sigma(eN \to eN)$ in the cross section of the elastic eN scattering in the appropriate kinematics.

In the case of the pion production the soft pion limit corresponding to $1 - x \sim m_\pi/m_N, p_t \leq m_\pi$ is of special interest because one could use the factorization theorem and the chiral perturbation theory similar to consideration of the process $eN \to eN\pi$ at large Q^2 and small W [8]. However reaching this kinematics would require extremely high Q^2. At the same time reactions with leading nucleon for $x \leq 0.3$ could be studied at sufficiently large W already at Jlab and HERMES.

Reaction (1) provides also a promising avenue to look for exotic meson states including gluonium. Indeed, if one would consider, for example, the MIT bag model, the removal of three quarks from the system could leave the residual system looking like a bag made predominantly of glue. It is natural to expect that such a system would have a large overlapping integral with gluonium states.

An interesting example of a related process where cluster structure can manifest itself is the deep inelastic exclusive diffraction at large enough t:

$$e + p \to e + leading \ \rho + M + B \tag{6}$$

where $-t = -(p_{\gamma^*} - p_\rho)^2 \geq 2GeV^2$, $Q^2 \geq few \ GeV^2$ and $p_t(M) \approx p_t(\rho)$ and $p_t(M) \gg p_t(B)$. The t- dependence of the process $e + p \to e + \rho + p$ is predominantly determined by the two-gluon nucleon form factor and it can be fitted as $\approx 1/(1 - t/m^2)^4$ with $m^2 \sim 1GeV^2$ at intermediate energies and $m^2 \sim 0.6GeV^2$ at HERA energies. In the case of scattering off a meson a natural guess would be that the t-dependence is much slower - perhaps $\approx 1/(1 - t/m^2)^2$ with similar m^2. So for large enough t this process may have cross section comparable to the cross section of the $e + p \to e + \rho + p$ process. Study of this process could provide a test of the interpretation of the smaller gluon radius of the nucleon indicated by our recent analysis [9].

3. Hadron induced hard semiexclusive processes

A natural extension of the processes discussed for electron scattering is hadron scattering process:

$$A + B \to C_{int} + C_{sp} + D, \tag{7}$$

where sufficiently large momentum is transfered to C_{int} and D (scattering at finite c.m. scattering angles in the c.m. frame of C_{int} and D), while C_{sp} similar to the case of the process 1 is produced in the fragmentation region of either A or B.

Taking for certainty D in the target fragmentation of B we can expect that in the color transparency approximation the process will proceed via scattering of a hadron A in the minimal Fock space configuration off a color singlet cluster in the hadron B with minimal number of constituents allowed for the process $A + ``cluster'' \rightarrow C_{int} + D$. An obvious practical advantage of these processes as compared to the processes 7 is that one can use different beams - pions, kaons, hyperons to probe the clustering structure of different hadrons while the processes 1 in practice are restricted to the case of proton targets.

The two-body large angle hadron scattering processes are known to satisfy to a good approximation dimensional counting rules, for a review see [10]. At sufficiently large momentum transfer the small size configurations should give the dominant contribution and hence the rescattering effects should be small. Hence we expect the scaling relations for these processes of the similar kind for a fixed value of $\alpha_{C_{sp}}, p_t$ C_{sp}:

$$\frac{d\sigma(A + B \rightarrow C_{int} + C_{sp} + D)}{d\alpha_{sp}d^2p_t \, {}_{sp}/\alpha_{sp}} = \phi(\alpha_{sp}, p_t \, {}_{sp})R(\theta_{c.m.})\left(s_o/s'\right)^n \qquad (8)$$

where $s' = (p_{C_{int}} + p_D)^2$, $\theta_{c.m.}$ is the c.m. angle in the $C_{int} - D$ system, and n is expressed through the number of constituents involved in the subprocess in the same way as in the two-body large angle scattering:

$$n = n_q(A) + n_q(cluster) + n_q(C_{int}) + n_q(D) - 2. \qquad (9)$$

There is a number of the processes where the hard subprocess resembles the scattering off two hadrons for which the cross section is known, like the process $p + p \rightarrow p + p + M_{spect}$, or $p + p \rightarrow p + \pi + N_{spect}$ presented in Fig.2.

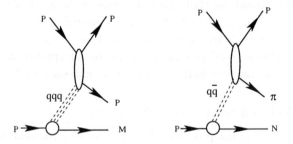

Figure 2. (a) Production of two high p_t baryons and recoiling mesonic system, (b) production of a high p_t pion and a nucleon and recoiling baryonic system.

In this case we can write an interpolation formulae similar to the ones for the electron scattering: For example,

$$\frac{d\sigma(p + p \rightarrow p + p + \pi^0)}{d\alpha_{\pi^0}d_t^p/\alpha_{\pi^0}} = F(\alpha_\pi, p_t)(1 - \alpha_p i)d\sigma^{pp \rightarrow pp}(s', \theta_{c.m.}) \qquad (10)$$

with $s' \approx= (1 - \alpha_\pi)s$. Since the hard cross section decreases strongly with increase of s' at fixed $\theta_{c.m.}$ one expects a strong enhancement of production of mesons with relatively high values of α_π in these processes. Also, assuming that the distributions of the small color clusters contributing to the electron reaction and to the hadron reactions are about the same we can get scaling relations between the cross section of proton and electron induced processes. For example,

$$\frac{\frac{d\sigma(p+p\to p+p+\pi^0)}{d\alpha_{\pi 0}d^2p_t/\alpha_{\pi 0}}}{\frac{d\sigma(e+N\to e+N+\pi^0)}{d\alpha_{\pi 0}d^2p_t/\alpha_{\pi 0}}} \approx \frac{\sigma(p+p\to p+p)}{\sigma(eN\to eN)}, \qquad (11)$$

We have performed first estimates of the rate of the production of pions using the discussed mechanism and using a simple model for the $\Gamma_{NN\pi}$ vertex. Enhancement of the scattering off three quark clusters which carry $(1-x_\pi)$ fraction of the total light cone momentum of the nucleon as compared to the scattering off the nucleon as a whole by a factor $(1-x_\pi)^{-10}$ leads to enhanced role of the pion cloud and results in the cross sections of the same magnitude as the elastic pp scattering. We are currently performing more detailed studies to determine the contribution of this mechanism to the cross section measured by EVA [12].

We also found that at the intermediate energies E$\leq$ 10 GeV studied at EVA [12] an important background to the discussed mechanism is provided by production of excited nucleon states. Indeed, the cross section of the processes like $pp \to N^* + p$ at $\theta_{c.m.} \sim 90^0$ is comparable to the elastic pp scattering. Moreover if we would sum over a sufficient range of masses of N^*'s we should expect to find a cross section which is enhanced as compared to the elastic scattering since it does not contain a smallness for three quarks to transform to a particular state (nucleon). If we use the quark counting rules as a guide, the ratio of quasielastic and elastic processes should increase with s as $\propto s^2$. At sufficiently high energies kinematics of the meson production in these processes is qualitatively different from the process 7 since the average pion transverse momenta in this case increase with s approximately as $\propto \sqrt{s}$ (for the case of the $N\pi$ final state). However we have checked that for $E_{inc} \leq 10 GeV$ it is difficult to separate fragmentation in the initial and final states even in the simplest reaction $pp \to pp\pi^0$ except for production of pions at very small p_t.

In the previous discussion we assumed that hard two body subprocess is dominated by the scattering of constituents in small size configurations, so that the interaction with residual system could be neglected. However the current data on transparency in high-energy large angle $(p,2p)$ reactions suggests a rather complicated interplay of the contributions of large and small size configurations. In this case the initial and final state interactions with the would be spectator would be possible. For example in the process $pp \to pp\pi^0$ with π^0 a spectator both the proton in the initial state and both protons in the final

368

state can rescatter off the pion. Presence of multiple rescatterings will lead to rather complicated patterns of angular correlations similar to those found for the large angle $p^2H \to ppn$ process in [11]. These rescatterings will weakly affect distribution over α_M though they would change overall absolute value of the cross section and lead to broadening of the p_t distribution of the spectator.

Another possible source of angular asymmetries is postselection of the initial state which could be different at intermediate energies when the size of three quark cluster is not too small. The requirement of the elastic scattering off the three quark cluster may select the alignment of the cluster relative to the reaction axis hence modifying the angular distribution of the spectator system. In particular this effect can emerge as a kind of a hadron level Sudakov radiation, see discussion in [12].

To summarize, a systematic study of the lepton and hadron induced hard semiexclusive reactions with leading baryons is necessary. It would provide a qualitatively new information about correlations of partons in hadrons and as well as about the dynamics of the large angle elastic scattering.

4. Acknowledgments

This work has been supported in part by the USDOE, the Humbold foundation, GIF, NSF, CRDF.

References

1. A. V. Radyushkin, arXiv:hep-ph/0101227.
2. S. J. Brodsky, SLAC-PUB-8649 *At the frontier of particle physics, vol.2 1343-1444, ed. Shifman,M.*.
3. L.Frankfurt and M.Strikman, In proceedings of Conference Baryons 1995, Santa Fe 1995, pp 211-220; M. Strikman and M. Zhalov, Nucl. Phys. A **670**, 135 (2000).
4. S. J. Brodsky, L. Frankfurt, J. F. Gunion, A. H. Mueller and M. Strikman, Phys. Rev. D **50**, 3134 (1994).
5. J. Collins, L. Frankfurt, and M. Strikman, Phys. Rev. **D56**, 2982 (1997).
6. L. Frankfurt, W. Koepf and M. Strikman, Phys. Rev. D **54**, 3194 (1996); ibid **57**, 512 (1998).
7. L. L. Frankfurt, P. V. Pobylitsa, V. Polyakov and M. Strikman, Phys. Rev. D **60**, 014010 (1999).
8. P. V. Pobylitsa, . V. Polyakov and M. Strikman, Phys. Rev. Lett. **87**, 022001 (2001).
9. L. Frankfurt and M. Strikman, arXiv:hep-ph/0205223; Phys. Rev. D **66** (2002) in press.
10. D. W. Sivers, S. J. Brodsky and R. Blankenbecler, Phys. Rept. **23**, 1 (1976).
11. L. L. Frankfurt, E. Piasetzky, M. M. Sargsian and M. I. Strikman, Phys. Rev. C **56**, 2752 (1997).
12. S.Heppelman, talk at this workshop; D. Zhalov, PhD Thesis, PSU 2001.

HARD RESCATTERING MECHANISM IN HIGH ENERGY PHOTODISINTEGRATION OF THE LIGHT NUCLEI

M.M. SARGSIAN

Department of Physics, Florida International University Miami, FL 33199

We discuss the high energy photodisintegrataion of light nuclei in which the energy of the absorbed photon is equally shared between two nucleons in the target. For these reactions we investigate the model in which photon absorption by a quark in one nucleon followed by its high momentum transfer interaction with a quark of the other nucleon leads to the production of two nucleons with high relative momentum. We sum the relevant quark rescattering diagrams, and demonstrate that the scattering amplitude can be expressed as a convolution of the large angle NN scattering amplitude, the hard photon-quark interaction vertex and the low-momentum nuclear wave function. Within this model we calculate the cross sections and polarization observables of high energy $\gamma + d \to pn$ and $\gamma + {}^3He \to pp + n$ reactions.

1. Introduction

The reactions of high energy photodisintegration of light nuclei in which the energy of the incoming photon is equally shared by two outgoing nucleons is unique in a sense that it deposits to the nuclear system a large amount of energy which should be shared at least by two nucleons in the nuclei (see e.g.[1,2]). When this energy exceeds the typical meson mass (*sevearl hundred MeV*) one expects that the interaction picture based on a meson exchange currents should break down. In this limit the hard contribution is expected to dominate. As a result one expects that the quark gluon picture of interaction will become a relevant framework for description of the reaction.

The experiments on high energy two-body photodisintegration of the deuteron[3,4] demonstrated that starting at $E_\gamma \geq 1$ GeV the conventional mesonic picture of nuclear interactions is indeed breaking down. One of the first predictions for $\gamma d \to pn$ reactions within QCD was based on the quark counting rule, which predicted $d\sigma/dt \sim s^{-11}$. This prediction was based on the hypothesis that the Fock state with the minimal number of partonic constituents will dominate in two-body large angle hard collisions[5]. Although successful in describing energy dependences of number of hard processes, this hypothesis does not allow to make calculation of the absolute values of the cross sections. Especially for reactions involving baryons, the calculations within perturbative QCD underestimate the measured cross sections by orders of magnitude see e.g.[6]. This may be an indication that in the accessible range of energies bulk of the interaction is in the domain of the nonperturbative QCD[6,7], for which the theoretical methods of calculations are very limited.

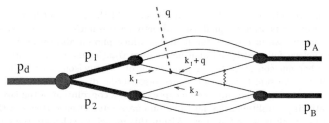

Figure 1.   Quark Rescattering diagram.

2. Hard Rescattering Model

Recently we suggested a model in which the absorption of the photon by a quark of one nucleon, followed by a high-momentum transfer (hard) rescattering with a quark from the second nucleon, produces the final two nucleon state of large relative momenta. The typical diagram representing such a scenario is presented in Figure 1. Based on the analysis of these type of diagrams we find that:

- the dominant contribution comes from the soft vertices of $d \to NN$ transition, while quark rescattering proceeds trough hard gluon exchange
- the $d \to NN$ transition can be evaluated based on the conventional deuteron wave function calculated using the realistic nucleon-nucleon potentials.
- the structure of hard interaction for the rescattering part of the reaction is similar to that of hard NN scattering.
- as a result the sum of the multitude of diagrams with incalculable nonperturbative parts of the interaction can be expressed through the experimentally measured amplitude of hard NN scattering.

3. Kinematics and the Cross Section

We consider the kinematics of sufficiently large energies $E_\gamma \geq 2\ GeV$ and momentum transfer $-t, -u \geq 2\ GeV^2$. These restrictions allow to make

a two important approximations: first is that the mass of the intermediate hadronic state produced in the γN interaction is in deep inelastic
continuum, thus the partonic picture is relevant. The second one is that
the quark rescattering is hard and can be factorized from the soft nuclear
wave function. We evaluate Feynman diagrams such as Fig. 1, in which
quarks are exchanged between nucleons via the exchange of a gluon. All
other quark-interactions are included in the partonic wave function of the
nucleon, ψ_N. We use a simplified notation in which only the momenta of
the interacting quarks require labeling. The scattering amplitude T for
photo-disintegration of the deuteron (of four-momentum p_d and mass M_d)
into two nucleons of momentum p_A and p_B is given by:

$$T = -\sum_{e_q} \int \left(\frac{\psi_N^\dagger(x_2', p_{B\perp}, k_{2\perp})}{x_2'} \bar{u}(p_B - p_2 + k_2) \left[-ig T_c^F \gamma^\nu \right] \right.$$

$$\frac{u(k_1 + q)\bar{u}(k_1 + q)}{(k_1 + q)^2 - m_q^2 + i\epsilon} \left[-ie_q \epsilon^\perp \cdot \gamma^\perp \right] u(k_1) \left. \frac{\psi_N(x_1, p_{1\perp}, k_{1\perp})}{x_1} \right)$$

$$\left\{ \frac{\psi_N^\dagger(x_1', p_{A\perp}, k_{1\perp})}{x_1'} \bar{u}(p_A - p_1 + k_1) \left[-ig T_c^F \gamma_\mu \right] u(k_2) \frac{\psi_N(x_2, p_{2\perp}, k_{2\perp})}{x_2} \right\}$$

$$G^{\mu\nu} \frac{\Psi_d(\alpha, p_\perp)}{1-\alpha} \frac{dx_1}{1-x_1} \frac{d^2 k_{1\perp}}{2(2\pi)^3} \frac{dx_2}{1-x_2} \frac{d^2 k_{2\perp}}{2(2\pi)^3} \frac{d\alpha}{\alpha} \frac{d^2 p_\perp}{2(2\pi)^3}, \tag{1}$$

where p_1 and p_2 are the momenta of the nucleons in the deuteron, with
$\alpha \equiv \frac{p_{1+}}{p_{d+}}$, $p_2 = p_d - p_1$ and $p_{1\perp} = -p_{2\perp} \equiv p_\perp$. Each nucleon consists
of one active quark of momenta k_1 and k_2: $x_i \equiv \frac{k_{i+}}{p_{i+}} = \frac{k_{i+}}{\alpha p_{d+}}$ $(i = 1, 2)$.
$G^{\mu\nu}$ describes the gluon exchange between interchanged quarks. We use
the reference frame where $p_d = (p_{d0}, p_{dz}, p_\perp) \equiv (\frac{\sqrt{s'}}{2} + \frac{M_d^2}{2\sqrt{s'}}, \frac{\sqrt{s'}}{2} - \frac{M_d^2}{2\sqrt{s'}}, 0)$,
with $s = (q + p_d)^2$, $s' \equiv s - M_D^2$, and the photon four-momentum is $q = (\frac{\sqrt{s'}}{2}, -\frac{\sqrt{s'}}{2}, 0)$.

We first observe that the denominator of the knocked-out quark propagator, when recoil quark-gluon system with mass m_R is on mass shell, has
a pole at $\alpha_c \equiv \frac{x_1 m_R^2 + k_{1\perp}^2}{(1-x_1)x_1 \bar{s}}$:

$$(k_1 + q)^2 - m_q^2 + i\epsilon \approx x_1 s'(\alpha - \alpha_c + i\epsilon), \tag{2}$$

where $\bar{s} \equiv s'(1 + \frac{M_d^2}{s'})$. Next we calculate the photon-quark hard scattering
vertex and integrate over the α using only the pole contribution in Eq.(2).
Note that the dominant contribution arises from the soft component of the
deuteron when $\alpha_c = \frac{1}{2}$, which requires $k_{1\perp}^2 \approx \frac{(1-x_1)x_1\bar{s}}{2}$.

Summing over the struck quark contributions from photon scattering
off neutron and proton one can express the scattering amplitudes through

the pn hard scattering amplitude within the quark interchange mechanism (QIM)-$A_{pn}^{QIM}(s, l^2)$ as follows:

$$T \approx \frac{ie(\epsilon^+ + \epsilon^-)(e_u + e_d)}{2\sqrt{s'}} \int f(\frac{l^2}{s}) A_{pn}^{QIM}(s, l^2) \Psi_d(\frac{1}{2}, p_\perp) \frac{d^2 p_\perp}{(2\pi)^2}. \quad (3)$$

where $l = p_p - p_1$, $\epsilon^\pm = \frac{1}{\sqrt{2}}(\epsilon_x \pm i\epsilon_y)$ and e_u and e_d are the electric charges of u and d quarks. The factor $f(l^2/s)$ accounts for the difference between the hard propagators in our process and those occurring in wide angle pn scattering. Within the Feynman mechanism[8], the interacting quark carries the whole momentum of the nucleon ($x_1 \to 1$), thus $f(l^2/s) = 1$. Within the minimal Fock state approximation, $f(l^2/s)$ is the scaling function of the θ_{cm} only with $f(\theta_{cm} = 90^0) \approx 1$[9].

We compute the differential cross section averaging $|T|^2$ over the spins of initial photon and deuteron and summing over the spins of the final nucleons. Then we use the observation that the quark interchange topologies are the dominant for fixed angle, $\theta_{cm} = 90^0$ high momentum transfer (non strange) baryon-baryon scattering. Thus in the region of $\theta_{cm} \approx 90^0$ we replace A_{pn}^{QIN} by the experimental data - A_{pn}^{Exp} and obtain[9a]:

$$\frac{d\sigma^{\gamma d \to pn}}{dt} = \frac{8\alpha}{9} \pi^4 \cdot \frac{1}{s'} C(\frac{\tilde{t}}{s}) \frac{d\sigma^{pn \to pn}(s, \tilde{t})}{dt}$$
$$\times \left| \int \Psi_d^{NR}(p_z = 0, p_\perp) \sqrt{m_N} \frac{d^2 p_\perp}{(2\pi)^2} \right|^2, \quad (4)$$

where $\tilde{t} = (p_B - p_d/2)^2$. Here Ψ_d^{NR} is the nonrelativistic deuteron wave function which can be calculated using realistic NN interaction potentials. The function $C(\frac{\tilde{t}}{s}) \approx f^2(\tilde{t}/s) \approx 1$ at $\theta_{cm} \sim 90^0$ and is slowly varying function of θ_{cm}.

Within the hard rescattering model we can calculate also the high energy photodisintegration of ^3He in the specially chosen kinematics, in which the absorbed photon's energy is equally shared between the two outgoing protons while the final neutron is very slow $p_n \leq 100 MeV/c$. In this case the cross section of the reaction is expressed through the nonrelativistic wave function of the ^3He and the amplitude of hard pp scattering, with small corrections coming from soft pn rescattering.

---

[a]The Eq.(6) in Ref.[9] contains a misprint: 4 in front of Eq.(6) should be 8. Note that the numerical calculations in [9] are done with the correct factor.

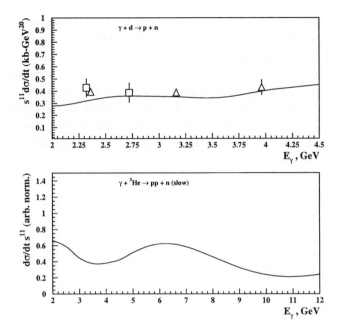

Figure 2. The $d\sigma/dts^{11}$ as a function of E_γ at $\theta_{cm} = 90^0$. Data are from [3] (triangles) and [4] (squares).

4. Comparison with the Data

4.1. *Energy dependence*

One can estimate the energy dependences of the differential cross sections of the $\gamma d \to pn$ and $\gamma^3 He \to pp + n$ reactions at $\theta_{cm} = 90^0$ using the experimental data on hard pn and pp scattering. Figure 2 represents the calculations and the comparison with existing data on deuteron target. Our calculation of $\gamma d \to pn$ has $\approx 25\%$ accuracy because of the rather large experimental errors in high momentum transfer pn scattering cross section. The prediction for $\gamma^3 He \to pp + n$ cross section (which should be considered preliminary) demonstrates the oscillation which is related to the observed oscillations in high momentum transfer pp cross section[10].

4.2. *Angular Dependence*

As we mentioned above, the hard rescattering mechanism can predict the value of $C(t/s) \approx 1$ only at $\theta_{cm} = 90^2$. One expects that this function should be a smooth function of θ_{cm}. As a result the angular dependence of photodisintegration reaction should in general reflect the angular dependence of hard NN scattering cross section.

In particular it is interesting to observe that the pn scattering cross section exhibits a strong angular asymmetry with the cross sections being dominate at forward angles. As Figure 3 demonstrates, within hard rescattering model this feature is reflected also in the angular dependence of $\gamma d \rightarrow pn$ cross section. This result is in a qualitative agreement with the preliminary JLab data.

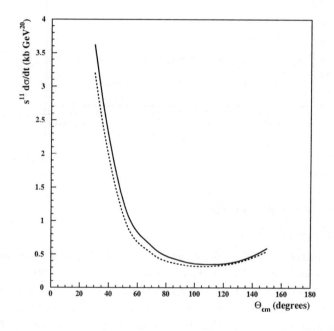

Figure 3. Angular dependence, for $\gamma d \rightarrow pn$ reaction at $E_\gamma = 2 GeV$ (dashed curve) and 3 GeV (solid curve).

5. Polarization Observables

Hard rescattering model allows also to calculate the different polarization observables of the photodisintegration reaction. Writing down explicitly the helicity indexes of interacting particles, for the scattering amplitude of Eq.(3) one obtains:

$$T^{h,m,\eta_2,\eta_1} \approx \qquad\qquad\qquad\qquad\qquad (5)$$

$$\frac{i(e_u + e_d)}{2\sqrt{s'}} \int f(\frac{l'^2}{s}) < \eta_1, \eta_2 |A_{pn}(s, l^2)| h, \lambda_2 > \Psi_d^{m,h,\lambda_2}(\frac{1}{2}, p_\perp) \frac{d^2 p_\perp}{(2\pi)^2}.$$

where h, m, η_2 and η_1 represent the helicities of incoming photon, target deuteron and two outgoing nucleons respectively. In derivation of Eq.(6) it is assumed additionally that the quark which interacts with the photon carries the total helicity of the parent nucleon. The helicity amplitudes, A_{pn} in Eq.(6) can be expressed through the five independent helicity amplitudes:

$$\phi_1(s,t) = < + + |A| + + > \qquad \phi_2(s,t) = < + + |A| - - >$$
$$\phi_3(s,t) = < + - |A| + - > \qquad \phi_4(s,t) = < + - |A| - + >$$
$$\phi_5(s,t) = < + + |A| + - > \qquad (6)$$

for which the following hierarchy relation can be stated in a model independent way[11]:

$$\phi_2 < \phi_5 < \phi_1, \phi_3, \phi_4 \qquad (7)$$

Using the definitions of Eq.(6) one can calculate the different asymmetries in the deuteron photodisintegration reaction. We are particularly interested in the recoil transverse P_y and longitudinal C_x, C_y polarizations for which the high energy data are becoming available from Jefferson Lab[12]. Based on Eq.(6) the predictions for these observables are as follows:

$$P_y = -2Im\left\{ [2(\phi_1(s,\tilde{t}) + \phi_2(s,\tilde{t})) + \phi_3(s,\tilde{t}) - \phi_4(s,\tilde{t})]\, \phi_5^\dagger(s,\tilde{t})\right\}/F(s,\tilde{t})$$
$$C_x = +2Re\left\{ [2(\phi_1(s,\tilde{t}) - \phi_2(s,\tilde{t})) + \phi_3(s,\tilde{t}) + \phi_4(s,\tilde{t})]\, \phi_5^\dagger(s,\tilde{t})\right\}/F(s,\tilde{t})$$
$$C_z = 0 \qquad (8)$$

where $F(s,\tilde{t}) = 2|\phi_1(s,\tilde{t})|^2 + 2|\phi_2(s,\tilde{t})|^2 + |\phi_3(s,\tilde{t})|^2 + 4|\phi_4(s,\tilde{t})|^2 + 6|\phi_5(s,\tilde{t})|^2$. In these derivations we neglected by the D wave contribution of the deuteron wave function, which is justified on the basis that the nucleon momenta which enter in the deuteron wave function in the integral of Eq.(6) is restricted, ($\leq 300 MeV/c$).

Based on the relation of Eq.(7) and the fact that for the on-shell amplitude $\phi_5 = 0$ at $\theta_{cm} = 90^0$ one can conclude in a rather model independent way that the P_y and C_x should be small at $\theta_{cm} = 90^0$ at $E_\gamma \geq 2GeV$. This result seems in qualitative agreement with the available data and the planned experiment at Jefferson Lab in $E_\gamma \geq 2GeV$ region will allow check these predictions in more detail.

It is interesting to note that the smallness of the polarization observables predicted in the model is not related to the commonly used assumption of the applicability of pQCD and related only to the fact that the considered asymmetries are determined by the small helicity component of NN amplitude, ϕ_5.

6. Summary and Outlook

The underlying hypothesis of hard rescattering model is that the dynamics of the photoproduction reaction is determined by the physics of high-momentum transfer contained in the hard scattering NN amplitude. As a result the short-distance aspect of the deuteron wave function is not important, which allows us to use the conventional deuteron wave function in numerical calculations. This hypothesis, if confirmed by additional studies, may suggest the existence of new type of calculable hard nuclear reactions in which the sum of the "infinite" number of quark interactions could be replaced by the hard amplitude of NN interaction.

The prediction of the model agrees reasonably well with the existing data on high energy photodisintegration of the deuteron at $\theta_{cm} = 90^0$. More data, especially with a two energetic proton final state in $\gamma+^3He \rightarrow pp$ (high p_t) + n ($p_t \approx 0$) reaction and a more detailed angular distribution would definitely allow to verify this hypothesis. The polarization measurement also will be crucial. It is very important to extend them in the energy region where anomalies are observed in the polarized hard pp scattering.

References

1. S.J. Brodsky and B.T. Chertok, Phys. Rev. Lett. **37**, 269 (1976).
2. R.J. Holt, Phys Rev. **C41**, 2400 (1990).
3. C. Bochna *et al.*, Phys. Rev. Lett. **81**, 4576 (1998).
4. J.E. Belz *et al.*, Phys. Rev. Lett. **74**, 646 (1995).
5. S.J. Brodsky and G.R. Farrar, Phys. Rev. Lett. **31**, 1153; V. Matveev, R.M. Muradyan and A.N. Tavkhelidze, Lett. Nuovo Cimento **7**, 719 (1973).
6. N. Isgur and C.H. Llewellyn Smith, Phys. Rev. Lett. **52**, (1984) 1080.
7. A. Radyushkin, Acta Phys. Pol. **B15**, 403 (1984).
8. R. Feynman, *Photon Hadron Interactions*, W.A. Benjamin Inc., 1972.
9. L.L. Frankfurt, G.A. Miller, M.M. Sargsian and M.I. Strikman, Phys. Rev. Lett. **84**, 3045 (2000).
10. J.P. Ralston and B. Pire, *Phys. Rev. Lett.* **61** 1823 (1988).
11. G. P. Ramsey and D. W. Sivers, Phys. Rev. D **45**, 79 (1992).
12. K. Wijesooriya *et al.* [Jefferson Lab Hall A Collaboration], Phys. Rev. Lett. **86**, 2975 (2001).

UNIFYING MESON PRODUCTION IN THE TWO NUCLEON SYSTEM WITH PHOTO-PRODUCTION ON THE NUCLEON

M. DILLIG

Institute for Theoretical Physics III,
University Erlangen-Nürnberg,
Staudtstr. 7, Erlangen,D-91058 , Germany
E-mail: mdillig@theorie3.physik.uni-erlangen.de

G. F. MARRANGHELLO, S. S. ROCHA, C. A. Z. VASCONCELLOS

Instituto de Física,
Universidade do Rio Grande do Sul,
Caixa Postal 15015, Porto Alegre RS, Brazil
E-mail: gfm@if.ufrgs.br

General aspects of production processes at high momentum transfers are discussed for the exclusive production of heavy mesons with masses up to 1 GeV near the corresponding production thresholds. Compared are the basic ingredients, ambiguities and results from effective meson-baryon models with findings in constituent quark and gluon models. For the example of the η meson, the intimate interplay between the exclusive production in the nucleon-nucleon system and the elctroproduction on the nucleon is discussed, emphasizing the need for a consistent analysis of hadron and photo-induced production processes at overlapping kinematics and final states.

1. Introduction

An important line in strong interaction physics at energies in the GeV region is the investigation of the structure of nucleons and nuclei at short distances as a testing ground for Quantum Chromodynamics (QCD) as the theory for strong interaction. The ultimate goal is to understand in detail the nature of various facets of QCD, in praticular its nonperturbative aspects as confinement and its transition to the deconfined phase at high density and/or temperature. For this end, various new facilities, which produce sophisticated data with very low cross sections with hadronic and electromagnetic probes came into operation just recently (complemented by heavy-ion machines with nuclei as complex projectiles). COSY as a cooled sychrotron with beams of protons, deuterons and light nuclei at energies up to 3.6 GeV [1] and JLab with an electron beam up to 4.2 GeV in its present and up to 12 GeV in its upgraded form [2] are characteristic examples for recent advances in accelerator technology towards the extraction of new physics at short distances.

2. Exclusive Meson Production Near Meson Thresholds

Presently, few aspects of strong interaction physics at short distances are quantiatively understood; frequently even qualitative insight in new phenomena is missing, in particular for processes which involve the conversion of a large kinetic energy into masses of produced particles. Consequently, with the advent of new facilities, experiments are focusing both on "simple" projectile - target combinations, together with kinematically complete final states, which reduce both the complexity of the input and the interpretation of the results and thus, hopefully, are able to map out the basic underlying physics.

2.1. *Kinematics and General Guides*

Modern accelerators, involving either electromagnetic or hadronic beams, are an ideal tool in testing this conjecture in detail. Exclusive meson production near the corresponding meson thresholds is a major part of their actual research programs [3],[4],[5]: in the most simple hadronic system, i.e., in nucleon-nucleon collisions, the threshold kinematics of heavy λ meson production

$$p + N \to p + B + \lambda \quad ; \quad \lambda = \eta, \eta^{'}, \omega, \Phi, K^{+}, a_0, \ ...$$

involves as a typical scale

$$\Delta q_0, \Delta q \sim m_\lambda \sim \sqrt{m_\lambda M} \sim 1 \, \text{GeV}$$

(above, m_λ, M denote the meson and the nucleon mass, respectively), which - together with phase space induced selectivity and the dominance of very few partial waves - maps out in detail hadronic degrees of freedom beyond the nucleon in its ground state. Unfortunately, even for this very simple kinematics, significant simplifications in the model assumptions are unavoidable in practical calculations. As both in meson-baryon framework and even more in a constituent quark model, the final hadronic state involves at least 3 or even 8 particles, rigorous Faddeev-type calculations are presently impractical. Instead one has to invoke factorization for the full production amplitude

$$M_{pp \to pB\lambda} = M_{ISI} * M_\lambda * M_{FSI} \ ,$$

where – in the spirit of the Distorted Wave Born Approximation – the "soft" inital (ISI) and final state interactions (FSI) are factorized out of the "hard" (meson or gluon exchange dominated) transition vertex. In practice ISI are related to the loss of flux from the pp channel due to open inelastic thresholds, while FSI are approximated in a scattering length and effective range expansion. Particularly around 1 GeV or at low GeV (even at JLab energies), factorization on a quantitative level is still an open issue.

2.2. Model Building

Having established the conceptional framework above, the following steps towards the production amplitude are canonical: starting from effective Lagrangians for mesons and baryons and constituent quarks and gluons as effective QCD inspired degrees of freedom, the full production vertex is formulated perturbatively. In the meson-baryon sector the exchange of mesons with masses below 1 GeV (in general $\pi, \eta, \sigma, \rho, \omega, \delta$ and K, K^* exchange in the strange sector) is included, together with the $N, \overline{N}$ and near-threshold Δ, N^* resonances as intermediate baryonic states, occasionally supplemented by uncorrelated $\pi\pi$ or $\pi\rho$ exchange. Symmetry relations or experimental information (partial decay widths, vector dominance, etc.) determine the various coupling constants on-shell; off-shell corrections are introduced in a basically phenomenological way). Similarly, to cope with the complexity of the formalism, quark models are based on effective exchange forces, such as the instanton induced interaction, the $^3P_0$ model, or (non)perturbative gluon-exchange: the dominant contribution involves the exchange of 2 gluons; 3-gluon exchange terms have to be supplemented to account for vector meson production (such as of the $\Phi(1020)$ meson as a pure $s\overline{s}$ state) or the excitation of intermediate N^* resonances via the colourless 2-gluon exchange (a Pomeron-like object; the 3rd gluon provides the appropriate momentum sharing). In practice, further approximations resort to effective zero-range exchange forces or to a quark-diquark representation of the interacting baryons. In both approaches the spectator-like direct meson production (direct radiation off the projectile or via one-gluon exchange) is dropped; already for pion production, being most favorable from momentum sharing, its contribution to the cross section is suppressed by typically one order of magnitude.

3. Interplay of Hadron and Electron Induced Meson Production

Within the N^* program at Jlab, in particular meson production into the η, Φ channels have been studied in detail; in our brief comparison for different probes we focus on the η meson final states [6],[7].

3.1. η Production on a Single Nucleon and in NN System

Focusing on η production in meson-exchange models (formulations in constitutent quark model follow conceptionally similar lines, though differ in important details of the parametrization of the production amplitude), the near-threshold regime is clearly dominated by the excitation of the negative parity S_{11} resonance $N^*(1535)$; furthermore it is understood that the ρ-meson provides the leading exchange contribution. Then assuming vector dominance for the virtial photon exchange together with the relation

$$L_{\rho NN^*}^v = g_{\rho NN^*} \overline{\Psi}_{N^*} \gamma_5 \, \gamma_\nu \Psi_N \, A^\nu \tag{1}$$

for the vector coupling of the ρ -meson at the NN^* vertex, together with

$$L_{\gamma \rho NN^*} = \frac{g_{\gamma \rho} g_{\rho NN^*}}{q^2 - m_\rho{}^2} \overline{\Psi}_{N^*} \gamma_5 \, \gamma_\nu \Psi_N \, A^\nu \, , \tag{2}$$

then the two amplitudes directly related via

$$M_{pp \to \eta pp}(q') \approx \frac{f_{\rho NN}}{2m_\rho} \, \overline{\Psi}_N \gamma_5 \, \sigma_{\mu\nu} \, q'^\mu \, \Psi_N \, M_{\gamma \rho p \to \eta N}^\nu(q) \, . \tag{3}$$

Thus, except of the effective momentum transfers q and q', the basic difference is reflected in the off-shell continuation of the initial ρpp vertex. An ideal experiment should minimize off-shell corrections: for threshold kinematics the constraints from equating both the three and four-momentum transfer in the pp and the ep systems, would be matched for an electron with momenta p_e, p_e' and a scattering angle $t_{e'}$

$$p_e' \approx p_e - \sqrt{M\,(m_\eta - M/2)}; \; t_{e'} \approx 1 - \frac{\sqrt{M\,(m_\eta - M/2)} + Mm_\eta}{p_e} \, . \tag{4}$$

It is evident, that such detail comparisons with respect to different observables would shed light on many problems hitherto unsolved, such as the quantitative nature of the strong $g_{\eta NN^*}$ coupling constant, the off-shell continuation at the ρNN^* vertex, the η scattering lenghts in the p and pp systems or add further information of the role of initial and final state interactions in the two-proton system.

3.2. η - Mesic Nuclei

The production of hidden new flavors as mentioned above can be extended to complex nuclei: exotic nuclear matter constitutes a major part of modern research programs in hadron physics. Guided by the success of unexpected structures in ongoing investigations of deeply bound pionic states in nuclei [8], the large attractive scattering length in the ηN system suggests the possiblity of quasi bound η-nucleus states. Given the dominance of the $N^*(1535)$ resonance in eta production near threshold, the recoilless production of A_η^* nuclei in

$$A(p,d)(A-1)_\eta^*; \ A(\gamma,N)(A-1)_\eta^* \tag{5}$$

at vanishing momentum transfers [9],[10] would be an ideal testing ground for the S_{11} dynamics in the nuclear environment. From comparison with detailed studies of the nonstatic $\Delta(1232)$ dynamics in nuclei in the framework of the isobar doorway model, the $N^*(1535)$ is expected to develop a much high degree of collectivity in multiple rescattering at production energies from below to above the η threshold. Focusing only on the leading η one-boson exchange term

$$V_{NN^*}(q_\eta^2) = -\frac{g_{\eta NN^*}^2}{\mathbf{q}_\eta^2 - (\omega_\eta^2 - m_\eta^2) - i\epsilon} \tag{6}$$

with $\omega_\eta \sim m_\eta$, i.e., the η-meson propagating close to its mass shell, the corresponding exchange potential exhibits a dramatic sensitivity both in its real and imaginary parts

$$V_{NN^*}(r) = -\frac{g_{\eta NN^*}^2}{4\pi}\frac{1}{r}\begin{cases} e^{-r\sqrt{m_\eta^2-\omega_\eta^2}} & m_\eta > \omega_\eta \\ \cdot \ 1 & m_\eta = \omega_\eta \\ e^{ir\sqrt{m_\eta^2-\omega_\eta^2}} & m_\eta < \omega_\eta \end{cases}$$

A rather similar significant variation of the imaginary part is expected for the quasielastic decay of the $N^* \to N\eta$ which exhausts at the resonance about 50% of the total N^* width $\Gamma^* \approx 95$ MeV. As, in addition, the leading nonmesic NN decay channel is strongly suppressed from the large momentum mismatch (final states with 3 or more particles are strongly quenched by phase space), a very different pattern of the energy dependence and relative strength of the $(N^*\overline{N})^{J^\pi}$ states compared to $\Delta(1232)$-hole doorway states in nuclei is to be expected. Keeping in mind the very different selectivity of the production mechanism in proton and photon induced reactions for the formation of η-mesic nuclei, their comparative study with different probes would allow a rather detailed dynamical picture of the modification of the S_{11} in the nuclear environment.

4. Conclusions

The new generation of accelerators, such as COSY or JLab, allow to test the structure and the dynamics of hadronic degrees of freedom in an unprecedented way; corresponding experiments stress and map out both unifying as well as different aspects of the same underlying pyhsics. It is just the unified description of these new and comprehensive data in a coherent microscopic model at appropriate kinematical ranges, which will lead to a fairly complete understanding of genuine aspects of the theory of strong interactions, bridging - hopefully - their formulation in terms of effective mesonic and baryonic degrees of freedom towards a more fundamental understanding within the framework of QCD. It is clear, that a satisfying picture, which is adequate both for the very light and in particular very heavy, dense hadronic systems, is still far ahead. We are, however, convinced that increasing systematical experimental evidence provides the key for a much deeper understanding of strong interaction physics in the very near future.

This work was supported by Forschungszentrum FZ Jülich, Contract COSY-062 41-445-386.

References

1. F. Rathman et al. *nucl-ex/0203008*
2. K. de Jager *Nucl. Phys.* **A699** 384 (2002)
3. P. Moskal et al. *J. Phys.* **G28** 1777 (2002)
4. T. Johannson *Invited talk QNP2002*, 10. - 14. July, Jülich, Germany (2002)
5. COSY-TOF Collaboration **nucl-ex/0205016** (2002)
6. V. D. Burkert *Nucl. Phys.* **A699** 261 (2002)
7. CLAS Collaboration (M. Ripani et al.) *Nucl. Phys.* **A699** 270 (2002)
8. H. Geissel et al. *Phys. Rev. Lett.* **88** 122301 (2002)
9. G. A. Sokol, A.I. L'vov, L. N. Pavlyuchenko **nucl-ex/0106005** (2002)
10. R. S. Hayano, S. Hirenzaki and A. Gilitzer *Eur. Phys. J.* **A6** 99 (1999)

PP COLLISIONS AT OR NEAR THE EXCLUSIVE LIMIT

S. HEPPELMANN*

Penn State University
University Park, Pa. 16802

The framework for interest in pp exclusive scattering, both for protons bound in a carbon nucleus and for free protons, is summarized. New results from BNL E850 are discussed in terms of color transparency. In addition, nearly exclusive processes, with additional π^0 production, are observed and characterized.

1. Exclusive PP and Color Transparency

In 1982 Brodsky[1] and Mueller[2] independently identified color transparency as an essential feature of QCD. Color transparency is the predicted tendency to form a non-interacting intermediate state when protons elastically scatter at large transverse momentum. The dimensional scaling laws for exclusive process energy dependence at large CM scattering angles[3] provide a very physical insight into the mechanism. The scaling law, $\frac{d\sigma}{dt} \propto s^{-10}$ for pp elastic scattering, reflects the price paid in cross section for a process that projects out only the spatially small (size inversely proportional to energy) component of the proton wave function. Scattering processes involving only the spatially small transverse components of the nucleon wave function are referred to as "point-like".

The point-like scattering process involves a rather unlikely configuration of the nucleon. Landshoff noted in 1974 that the phase space for point-like scattering was much smaller than that for independent quark scattering.[4] From phase space arguments, the independent scattering mechanism should dominate over the point-like mechanism. However, at large CM scattering angles and CM energies greater than a few GeV, the data clearly favor the point-like dimensional scaling model.

The realization that point-like scattering can compete in amplitude with the Landshoff process comes from an analysis of radiative effects.[5] Further study by Botts and Sterman[6] produced a clearer understanding of the

---

*for bnl e850 collaboration

mechanism. In the small angle region where the Landshoff picture seems to correctly predict the energy dependence of cross sections, QCD calculations showed that independent quark scattering occurred with the three valence quarks in the proton aligned out of the scattering plane.[7] Rather than projecting a point-like object from the normal proton state, a "cigar" shaped configuration is selected. They further find that as the energies and angles are increased, the cigar shape converges toward a point.

Even if the hard scattering is characterized by a singular point in space and time with a compact point-like wave packet, the question remains as to the persistence in time for the small system in comparison with the path length of a trajectory through a nucleus. The work by many theorists on unifying the quark-gluon based idea of color transparency with the traditional hadronic basis[8] has provided a framework for estimating this time evolution, complementing the purely quark based approaches.[9] In the view of many theorists, the observable transparency is most directly dependent on the time scale of recombinations.[10]

2. The Experimental Result

In 1988, the first measurement of the transparency of nuclei to the pp elastic scattering process at high momentum transfer was published.[11] The conclusion from that analysis was that there is a strong increase in transparency from Glauber levels at beam energy near 6 GeV toward a maximum transparency near 10 GeV and a reduction as the energy increased further toward 12 GeV. The results from the BNL experiments were followed with experiments at SLAC using electron beams. The 1994 publication of (e,e'p) data showed no significant energy dependence in transparency.[12]

Various interpretations of the structure observed for (p,2p) transparency were suggested. In one case, the entire transparency shape was attributed to a low energy effect associated with nuclear Fermi momentum distribution.[13]

Ralston and Pire observed that the variation of the energy dependence of transparency has a simple relationship with the experimental ratio of the energy dependence of the large angle pp cross section to the dimensional scaling prediction.[14] Connecting to the work of Mueller,[5] they introduced an energy dependent phase between two essential components of the scattering amplitude, the point-like component and the Landshoff contribution.

Taking the two component picture a step further, Brodsky and deTerramond proposed another mechanism for pp scattering. They proposed an "s" channel pp resonance near beam energies of 9-10 GeV and associated this channel with the open charm threshold,[15] It was argued that such a state would simultaneously explain the striking spin dependence of the pp cross section,[16] the transparency and the dimensional scaling violations of the cross section. They suggested that above 9 GeV, the resonance dominates the free pp cross section. However this contribution would tend to be filtered away by the nuclear medium. These models each lead to similar structure in the color transparency energy dependence, with different soft contributions to the scattering amplitude filtered away.

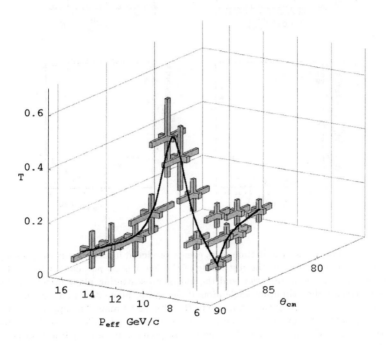

Figure 1. The dependence of carbon transparency on effective incident beam momentum (p_{eff}) and on center of mass scattering angle (θ_{cm}). The data are from the BNL E850 (p,2p) experiment.

2.1. The E850 Colar Transparency Experiment

Analysis of data from the E850 experiment[17] has been presented recently.[18,19] With the new data, the transparency of carbon is measured at larger beam energies, identifying quasi-elastic events with a better signal to background ratio than that of the 1988 measurement.

386

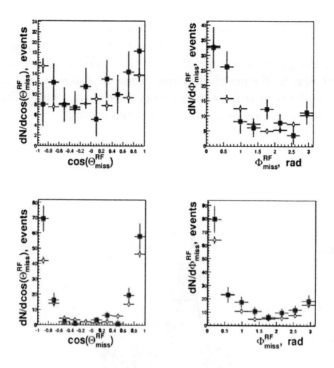

Figure 2. The angular distribution of π_0 events produced in near exclusive $pp \to p(p\pi^0)$ interactions as measured in the rest frame of the lower mass $p\pi^0$ combination. The Θ angle is between the π_0 direction and the out of pp scattering plane direction. The Φ angle represents the azimuthal angle around the out of plane direction with $\Phi = 0$ the direction of the boost from the lab frame to the π^0-p rest frame. The upper pair of plots represents data with low mass of the p π^0 system, $m_{p\pi} < 1.4\ GeV$. The lower graphs represent $1.4\ GeV < m_{p\pi} < 2\ GeV$. The square points represent hydrogen target data and the circle, the carbon target data.

The combined result, showing the energy and angular dependence of transparency measured by E850 is shown in Fig. 1. The falloff in transparency at high energy has been observed to an energy of about 15 GeV. The maximum transparency occurs at about 9 GeV near 90° CM. The general observation from 1988, that the scattering of protons in the carbon nuclear environment reduces the variance from dimensional scaling, remains true and is consistent with the Ralston-Pire idea. The complementary relationship with the single spin A_{NN} asymmetry, as discussed by Brodsky and deTerramond, remains very interesting.

2.2. *Additional* π^0 *Production*

Finally, we are motivated by the realization that in most of the models for exclusive pp scattering, there is a component filtered away by radiative processes. It is interesting to investigate the nature of the background under the quasi-elastic signal as a possible remnant of this radiative filtering process.[20]

The events that are considered here are background to the quasi-elastic event sample for the E850 data set with beam energy of 6 GeV. These events have a pair of large p_T protons (representing more than half of the kinematic limit for p_T in pp elastic scattering) in opposite azimuthal hemispheres. The proton momenta are measured in the charged track spectrometer. The missing momentum can be associated with π^0 production by applying a missing mass cut and by noting the absence of a corresponding charge track in the spectrometer.

If the background events are of the Landshoff type, with radiation showing up as an associated final state π_0, then the cigar shaped form of that interaction region could imply a non-trivial angular distribution for the final state pion. In Fig. 2 the angular distribution of a π^0 produced in the elastic background process $pp \rightarrow pp\pi^0$ is shown.

The choice for the reference frame used to display the angular distribution is a natural one for a $p\pi$ resonance decay, with the z direction defined by the boost from lab frame to the rest frame of the $p\pi$ system. In general the pion is produced along this z direction. In the low mass region, $M_{p\pi} < 1.5 GeV$, we see that the azimuthal distribution around the z axis is not exceptional. However, in the higher mass region, the azimuthal distribution is strongly peaked with a preference for the pion to be produced out of the scattering plane defined by the pp final state system.

This research was supported by the U.S. - Israel Binational Science Foundation, the Israel Science Foundation founded by the Israel Academy of Sciences and Humanities, NSF grants PHY-9804015, PHY-0072240 and the U.S. Department of Energy grant DEFG0290ER40553.

References

1. Stanley J. Brodsky. XIII International Symposium on Multi-particle Dynamics, page 963, 1982.
2. A. H. Mueller. XVII Rencontra de Moriand, Les Arcs, France, page 13, 1982.
3. Stanley J. Brodsky and Glennys R. Farrar. Scaling laws for large momentum transfer processes. *Phys. Rev.*, D11:1309, 1975.
4. P. V. Landshoff. Model for elastic scattering at wide angle. *Phys. Rev.*, D10:1024–1030, 1974.

388

5. Alfred H. Mueller. Perturbative qcd at high-energies. *Phys. Rept.*, 73:237, 1981.
6. James Botts and George Sterman. Hard elastic scattering in qcd: Leading behavior. *Nucl. Phys.*, B325:62, 1989.
7. Michael G. Sotiropoulos and George Sterman. Proton-proton near forward hard elastic scattering. *Nucl. Phys.*, B425:489–515, 1994.
8. B. K. Jennings and G. A. Miller. Realistic hadronic matrix element approach to color transparency. *Phys. Rev. Lett.*, 69:3619–3622, 1992.
9. G. R. Farrar, H. Liu, L. L. Frankfurt, and M. I. Strikman. Transparency in nuclear quasiexclusive processes with large momentum transfer. *Phys. Rev. Lett.*, 61:686–689, 1988.
10. L. L. Frankfurt, M. I. Strikman, and M. B. Zhalov. Pitfalls in looking for color transparency at intermediate- energies. *Phys. Rev.*, C50:2189–2197, 1994.
11. A. S. Carroll et al. Nuclear transparency to large angle p p elastic scattering. *Phys. Rev. Lett.*, 61:1698–1701, 1988.
12. N. Makins et al. Momentum transfer dependence of nuclear transparency from the quasielastic c-12 (e, e-prime p) reaction. *Phys. Rev. Lett.*, 72:1986–1989, 1994.
13. Byron K. Jennings and Boris Z. Kopeliovich. Color transparency and fermi motion. *Phys. Rev. Lett.*, 70:3384–3387, 1993.
14. John P. Ralston and Bernard Pire. Fluctuating proton size and oscillating nuclear transparency. *Phys. Rev. Lett.*, 61:1823, 1988.
15. Stanley J. Brodsky and G. F. de Teramond. Spin correlations, qcd color transparency and heavy quark thresholds in proton-proton scattering. *Phys. Rev. Lett.*, 60:1924, 1988.
16. E. A. Crosbie et al. Energy dependence of spin-spin effects in p p elastic scattering at 90-degrees center-of-mass. *Phys. Rev.*, D23:600, 1981.
17. J. Wu et al. The eva trigger: Transverse momentum selection in a solenoid. *Nucl. Instrum. Meth.*, A349:183–196, 1994.
18. I. Mardor et al. Nuclear transparency in large momentum transfer quasielastic scattering. *Phys. Rev. Lett.*, 81:5085–5088, 1998.
19. A. Leksanov et al. Energy dependence of nuclear transparency in c(p,2p) scattering. *Phys. Rev. Lett.*, 87:212301, 2001.
20. D. Zhalov, Studies Of The High Momentum Transfer Hard Quasi-Exclusive Proton-Proton Interactions With Eva Detector, UMI-30-36627.

STATUS OF COLOR TRANSPARENCY EXPERIMENTS

K. A. GRIFFIOEN

Department of Physics
College of William & Mary
PO Box 8795
Williamsburg, VA, USA
E-mail: griff@physics.wm.edu

Although there is growing evidence for color transparency in several high-energy reactions, the observation of its onset remains elusive. This report briefly summarizes the experiments, past, present and future.

1. Introduction

Color transparency (CT) is the phenomenon that small-sized, color-neutral quark wave packets have reduced cross sections in hadronic matter. The relevant questions are 1) How can one produce a small color singlet? and 2) How long does it take to expand to normal size? Brodsky and Mueller[1] introduced the concept of color transparency in 1982. By a decade later enough work had been done on the subject to merit two excellent reviews[2]. At that time there were some data, but no clear signs of CT. Now, in 2002, more data have been taken and new measurements planned. These pages briefly summarize the experimental situation.

2. Measurements of $(p, 2p)$

During the past 15 years several CT experiments were done at Brookhaven[4,5,6,7] using $(p, 2p)$ quasi-elastic reactions in nuclei. Fig. 1a shows schematically how the two colliding protons must be in small-size configurations in order to survive the scattering intact. This leads to reduced interactions with the surrounding nuclear material. The transparency T is defined as the ratio of the observed quasi-elastic $(p, 2p)$ cross section in nuclei to the corresponding off-shell elastic pp scattering cross section for equivalent kinematics. CT is signaled by a rise in this ratio with increasing incident beam momentum. The data show T rising up to an incident momentum of ≈ 8 GeV, but then falling again at higher momenta. This can

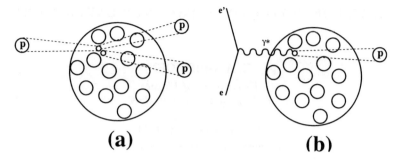

Figure 1.   Schematic of (a) the $(p, 2p)$ reaction and (b) the $(e, e'p)$ reaction.

be explained by a nuclear filtering process in which the nuclear medium reacts differently to the long-range and short-range scattering terms in the elastic cross section[2] or by the onset of charm production[3]. However, the data provide no clear evidence for CT up to incident momenta of 15 GeV/c.

3. Measurements of $(e, e'p)$

Several measurements[8,9,10,11,12] have been performed over the past decade using the $(e, e'p)$ quasi-elastic reaction within atomic nuclei. Fig. 1b shows the reaction schematically. High 4-momentum transfer squared (Q^2) elastic scattering is expected to proceed by photon absorption on a single quark when the two spectator quarks are close enough to be dragged along via single gluon exchanges. Therefore, the elastic scattering process selects the small components of the proton wavefunction. The propagating small-size object then has a reduced cross section with the surrounding nuclear medium. In plane-wave impulse approximation (PWIA), the cross section is given by $d^6\sigma/dE'_e d\Omega'_e d^3p' = \sigma_{ep} S(E_m, \vec{p_i})$ in which E'_e and Ω'_e correspond to the energy and solid angle of the outgoing electron, p' is the momentum of the outgoing proton, the spectral function $S(E_m, \vec{p})$ is the probability of finding a proton with initial momentum $\vec{p}$ and separation energy E_m, and σ_{ep} is the off-shell ep elastic cross section. The transparency $T \equiv \sigma_{\text{measured}}/\sigma_{\text{PWIA}}$ is less than unity, but should increase with Q^2 as the struck proton becomes color-transparent. The data, which extend to $Q^2 = 8$ GeV^2 are perfectly consistent with Glauber calculations for normal proton-nucleus interactions, and show no characteristic rise due to CT.

4. Measurements of ρ Production

Because CT should be a universal phenomenon, it must be observable for propagating mesons as well as baryons. The photo- or electro-production of

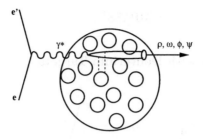

Figure 2.   Schematic of vector-meson production in nuclei.

vector mesons is a natural place to look, because the meson is expected to arise from a fluctuation of the photon into a $q\bar{q}$ pair that gets knocked on-shell by an elastically scattered nucleon. Since the $q\bar{q}$ pair originates from a point, and take some time to clothe itself with gluons and a quark sea, one hopes to see the effect of its reduced size on a surrounding nucleus. Two distance scales are important for these reactions, the coherence length (the distance traveled over the lifetime of a $q\bar{q}$ fluctuation), $\ell_c = \frac{2\nu}{Q^2+m_V^2}$, and the formation length (the distance traveled once the $q\bar{q}$ pair has interacted with a nucleon, during which it evolves from a small configuration to a normal hadron), $\ell_F = \frac{2\nu}{m_V'^2-m_V^2}$, in which ν is the photon (and likewise, nearly the vector meson) energy, Q^2 is the photon 4-momentum squared, m_V is the mass of the vector meson, and m'_V is the first excited state of that meson. These lengths can be derived from simple uncertainty-principle arguments. For the ρ^0, $\ell_c = 0.66\nu$ is greater than $\ell_F = 0.26\nu$ (ν is in GeV and $\ell_{c,F}$ are in fm). For a heavier meson like the J/ψ, $\ell_c = 0.04\nu$ is less than $\ell_F = 0.10\nu$.

Results from E665[13] at Fermilab show a significant rise in the A-dependence of the incoherent (elastic) ρ^0 production with increasing Q^2. The data for H, D, C, Ca and Pb were fit to $\sigma_A = \sigma_0 A^\alpha$ and the transparency was calculated as $T = \sigma_A/\sigma_0 A$. The large values of ν (typically about 120 GeV) resulting from a beam of 500 GeV muons, insured that both ℓ_c and ℓ_F were much larger than the nuclear radii. Although statistics are marginal, α increases from a small value for $Q^2 < 1$ to nearly 1 at $Q^2 = 5$ GeV^2, as expected for CT.

The results from HERMES[14,15] are similar. However, in this case, ν is much smaller and ℓ_c varies rapidly over the size of the nucleus. New and improved measurements[16] have been made which bin the transparency for fixed ℓ_c and fixed Q^2.

Additional measurements are planned using these techniques at

HERMES[16], JLab[17] and COMPASS[18].

5. Measurements with Deuterium Targets

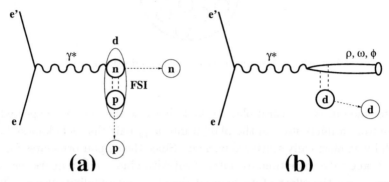

Figure 3.   Schematic of (a) the $d(e, e'p_s)n$ reaction and (b) coherent vector-meson production on the deuteron.

Early thinking about CT was geared toward measuring quasi-elastic cross sections on large nuclei as compared to on a single nucleon. However, since CT was not seen for protons at existing facilities, it became clear that the expansion distance was on the order of, or smaller than, the internucleon spacing. Therefore, once the struck proton has moved past its nearest neighbor, it has normal interactions with the nuclear medium thereafter, and a large nucleus actually dilutes any CT effect. Therefore, Frankfurt and Strikman proposed the counterintuitive step of looking for CT in deuterium, the smallest nucleus.

The first of two CLAS experiments on the deuteron at JLab detects spectator protons in the breakup of the target.[19] Fig. 3a shows a quasi-elastic reaction on the neutron in deuterium. The struck neutron is likely to make a grazing collision with the spectator proton through a final-state interaction (FSI). The resulting proton momentum perpendicular to the neutron's motion is enhanced. The Q^2-evolution of these FSI's will provide evidence for CT if the FSI's disappear faster than conventional calculations allow. With only a single nucleon in the neighborhood, one may be able to see CT in the range of Q^2 up to 6 GeV^2 where the $(e, e'p)$ experiments have seen nothing. The idea of this experiment is to form experimental ratios (unlike the ratios of experimental to theoretical cross sections in other searches). The cross section depends on both Born and FSI amplitudes. At high transverse momentum (p_T) of the proton the $|FSI|^2$ double scattering term enhances the cross section, whereas at moderate p_T the interference term Born·FSI (screening) reduces the cross section. Therefore, the ratio $\sigma(p_T = 400 \text{ MeV})/\sigma(p_T = 200 \text{ MeV})$ is most sensitive to FSI's. Model CT predictions show this ratio diverging from conventional calculations as Q^2 increases.

In a second experiment[20], coherent ρ^0 production from a deuteron is selected by detecting both the ρ^0 and the deuteron in CLAS at JLab. The reaction is shown schematically in Fig. 3b. The t- and x-dependence of the cross sections and the ℓ_c-dependence of the ratio $\sigma(t = 0.8 \text{ GeV})/\sigma(t = 0.4 \text{ GeV})$ will provide evidence for CT.

6. Measurements of Diffractive Dissociation

In normal pion-nucleus reactions, the pion-nucleon cross section is strong enough such that interactions occur preferentially with the surface nucleons. The internal nucleons are shadowed, and the resulting cross section goes as $\sigma \sim A^{2/3}$ (the nuclear radius $R = (1.2 \text{ fm})A^{1/3}$). However, if the nucleus becomes transparent, the amplitude for scattering is proportional to the total number of nucleons A in the nucleus, and the cross section goes as A^2. This is modified by the nuclear form factor $\int \exp(-\beta R^2 t)dt$ which contributes a factor of $A^{-2/3}$, in which β is proportional to the nuclear density. Therefore, CT implies an $A^{4/3}$ dependence of the cross section rather than an $A^{2/3}$ dependence. Measurements[21,22] of diffractive dissociation of 500 GeV/c pions from the Fermilab accelerator on nuclear targets yield fits to the dijet cross sections of the form $\sigma = \sigma_0 A^\alpha$ with $\alpha = 1.52$–1.64, consistent with the simplistic estimate of 4/3 as well as with more sophisticated CT calculations.

7. Measurements of J/ψ Production

In the photoproduction of J/ψ particles at Fermilab with 120 GeV photons[23], the coherent production (on a nucleus as a whole) should have the same A-dependence as diffractive dissociation. In this case the measured $\alpha = 1.40 \pm 0.06 \pm 0.04$, which is again consistent with $\alpha = 4/3$, is concrete evidence for CT. Similar evidence can also be found in muoproduction in the NMC experiment[24]. Both of these measurements have ℓ_c much longer than the size of the nucleus, and the combined data are sufficient to confirm CT in this limit. However, the onset of CT—short ℓ_c and varying ℓ_F—is yet to be observed.

The recently approved SLAC experiment E160[25], using a new polarized coherent bremsstrahlung facility for End Station A, will study the

394

A-dependence of J/ψ and ψ' photoproduction for beam energies of 15-45 GeV. In this region, ℓ_c is less than the internucleon spacing, but ℓ_F spans the sizes of target nuclei from Be to Pb.

8. Conclusions

Although CT is established experimentally for large ℓ_c, the details of its onset and its dependence on ℓ_c and ℓ_F remain unknown. A number of planned experiments at JLab, COMPASS, HERMES, and SLAC will shed new light on CT in the coming years.

References

1. S.J. Brodsky, G.T. Bodwin and G.P. Lepage, *Proceedings of the Thirteenth International Symposium on Multiparticle Dynamics,* Volendam, Netherlands, June 6–11, 1982, ed. W. Kittel, W. Metzger and A. Stergiou, (World Scientific, Singapore, 1983) p963. A.H. Mueller, *Proceedings of the Seventeenth Recontre de Moriond,* Les Arcs, France, 1982, ed. J. Tran Thanh Van (Editions Frontieres, Gil-sur-Yvette, France, 1982) p13.
2. L.L. Frankfurt, G.A. Miller and M. Strikman, Ann. Rev. Nucl. Part. Sci. **44** (1994)501. P. Jain, B. Pire and J. Ralston, Phys. Rep. **271** (1996) 67.
3. S. Brodsky and G. de Teramond, Phys. Rev. Lett. **60** (1988) 1924.
4. A.S. Carroll, *et al.,* Phys. Rev. Lett. **61** (1988) 1698.
5. I. Mardor, *et al.,* Phys. Rev. Lett. **81** (1998) 5085.
6. A. Leksanov, *et al.,* Phys. Rev. Lett. **87** (2001) 212301.
7. S. Heppelmann, these proceedings.
8. G. Garino, *et al.,* Phys. Rev. C **45** (1992) 780.
9. N.C.R. Makins, *et al.,* Phys. Rev. Lett. **72** (1994) 1986.
10. T.G. O'Neill, *et al.,* Phys. Lett. **B351** (1995) 87.
11. D. Abbott, *et al.,* Phys. Rev. Lett. **80** (1998) 5072.
12. K. Garrow, *et al.,* hep-ex/0109027
13. M.R. Adams, *et al.,* Phys. Rev. Lett. **74** (1995) 1525.
14. K. Ackerstaff, *et al.,* Phys. Rev. Lett. **82** (1999) 3025.
15. M. Hartig, Nucl. Phys. **A680** (2001) 264c.
16. A. Borissov, these proceedings.
17. K. Hafida, B. Mustapha and M. Holtrop, *et al.,* JLab Exp. E02-110.
18. A. Sandacz, M. Moinester, *et al.,* hep-ex/0106076; hep-ex/0109010.
19. K. Egiyan, K.A. Griffioen, M. Strikman, *et al.,* Jlab Exp. E94-019.
20. F. Klein, L. Kramer, and S. Stepanyan, *et al.,* JLab Exp. E02-012.
21. E.M. Aitala, *et al.,* Phys. Rev. Lett. **86** (2001) 4773.
22. D. Ashery, these proceedings.
23. M.D. Sokoloff, *et al.,* Phys. Rev. Lett. **57** (1986) 3003.
24. P. Amaudruz, *et al.,* Nucl. Phys. B **371** (1992) 553.
25. K. Griffioen, P. Bosted, D. Crabb, *et al.,* SLAC Exp. E160, http://www.slac.stanford.edu/exp/e160.

Participants List

Andrei Afanasev
Jefferson Laboratory
12000 Jefferson Ave., MS 12H
Newport News, VA 23606
(757) 269-7251; Fax (757) 269-7363
afanas@jlab.org

Daniel Ashery
School of Physics and Astronomy
Tel Aviv University
Tel Aviv 69978, Israel
(972) 3-640-7417; Fax (972) 3-640-6369
ashery@tauphy.tau.ac.il

Harut Avagyan
Jefferson Laboratory
12000 Jefferson Ave., MS 12H
Newport News, VA 23606
(757) 269-7764
avakian@jlab.org

Inna Aznauryan
Yerevan Physics Institute
Alikhanian Brothers St. 2
Yerevan 375036, Armenia
(7) 8852-341500
aznaur@jerewan1.YerPhI.AM

Marco Battaglieri
INFN
Via Dodecanes 33
Genova 16130, Italy
(39) 0103536458
battaglieri@ge.infn.it

Andrei Belitsky
Department of Physics
University of Maryland
College Park, MD 20742
(301) 405-6123
belitsky@physics.umd.edu

Pierre Bertin
Blaise PASCAL Univ./Jefferson Lab
12000 Jefferson Ave., MS 12H
Newport News, VA 23606
(757) 269-7544; Fax (757) 269-5703
bertin@jlab.org

Alexander Borissov
University of Michigan at HERMES
Doerpkamp 2
Hamburg 22527, Germany
(49) 40-8998-4664
Alexander.Borissov@desy.de

Will Brooks
Jefferson Laboratory
12000 Jefferson Ave., MS 12H
Newport News, VA 23606
(757) 269-6391; Fax (757) 269-5800
brooksw@jlab.org

Stanley Brodsky
Stanford Linear Accelerator Center
P.O. Box 4349
Stanford, CA 94305
(650) 926-2644; Fax (650) 926-2525
sjbth@slac.standford.edu

Matthias Burkardt
New Mexico State University
(505) 646-1928; Fax (505) 646-1934
burkardt@nmsu.edu

Carl Carlson
Physics Department
College of William and Mary
P.O. Box 8795
Williamsburg, VA 23187-8795
(757) 221-3509; Fax (757) 221-3540
carlson@physics.wm.edu

Jian-ping Chen
Jefferson Laboratory
12000 Jefferson Ave., MS 12H
Newport News, VA 23606
(757) 269-7413; Fax (757) 269-5703
jpchen@jlab.org

Victor Chernyak
Budker Institute of Nuclear Physics
630090 Novosibirsk, Russia
v.l.chernyak@inp.nsk.su

Eugene Chudakov
Jefferson Laboratory
12000 Jefferson Ave., MS 12H
Newport News, VA 23606
(757) 269-6959
gen@jlab.org

396

Stephen Cotanch
Department of Physics
North Carolina State University
Raleigh, NC 27695-8202
(919) 515-3316; Fax (919) 515-2471
cotanch@ncsu.edu

Dieter Cords
Jefferson Laboratory
12000 Jefferson Ave., MS 12H
Newport News, VA 23606
(757) 269-7526; Fax (757) 269-5800
cords@jlab.org

Kees de Jager
Jefferson Laboratory
12000 Jefferson Ave., MS 12H
Newport News, VA 23606
(757) 269-5254; Fax (757) 269-5703
kees@jlab.org

Martin DeWitt
North Carolina State University
1231-B Patrick Circle
Cary, NC 27511
(919) 515-4196
Madewitt@unity.ncsu.edu

Manfred Dillig
Institute for Theoretical Physics III
University of Erlangen
Staudtstr. 7, Erlangen, bav
D-91058, Germany
(49) 9131-852-8465
mdillig@theorie3.physik.uni-erlangen.de

Kim Egiyan
Yerevan Physics Institute
Alikhanian Brothers St. 2
Yerevan 375036 Armenia
7-8852364473
egiyan@jlab.org

Bruno El-Bennich
Rutgers University
618 Allison Road
Piscataway, NJ 08854
(732) 445-6886
bennich@physics.rutgers.edu

James Ely
University of Colorado
Campus Box 390
Boulder, CO 80216
(303) 735-5103
ely@ucsu.colorado.edu

Leonid Frankfurt
School of Physics
Tel Aviv University
Ramat Aviv, Tel Aviv 61390, Israel
(972) 3-640-7735
frankfur@lev.tau.ac.il

Haiyan Gao
Massachusetts Institute of Technology
77 Massachusetts Ave., Room 26-413
Cambridge, MA 02139
(617) 258-0256; Fax (617) 258-5440
haiyan@mit.edu

Anders Gardestig
Indiana Univ. Nuclear Theory Center
Indiana University
2401 Milo B Sampson Lane
Bloomington, IN 47408
(812) 855-5589
agardest@indiana.edu

David Gaskell
University of Colorado
UCB 390
Boulder, CO 80309
(303) 735-5103
gaskell@dilsey.colorado.edu

Shalev Gilad
Massachusetts Institute of Technology
77 Massachusetts Ave., Room 26-449
Cambridge, MA 02139
(617) 253-7785
sgilad@mitlns.mit.edu

Ronald Gilman
Rutgers University/Jefferson Lab.
12000 Jefferson Ave., MS 12H
Newport News, VA 23606
(757) 269-7011; Fax (757) 269-5703
gilman@jlab.org

Charles Glashaussser
Department of Physics
Rutgers University
136 Frelinghuysen Road
Piscataway, NJ 08854
(732) 445-2526
glashaus@physics.rutgers.edu

Jose Goity
Hampton University/Jefferson Lab
12000 Jefferson Ave., MS 12H2
Newport News, VA 23606
(757) 269-7345
goity@jlab.org

Javier Gomez
Jefferson Laboratory
12000 Jefferson Ave., MS 12H
Newport News, VA 23606
(757) 269-7498
Gomez@jlab.org

Keith Griffioen
Department of Physics
College of William and Mary
P.O. Box 8795
Williamsburg, VA 23187
(757) 221-3537
griff@physics.wm.edu

Ole Hansen
Jefferson Laboratory
12000 Jefferson Ave., MS 12H
Newport News, VA 23606
(757) 269-7627
ole@jlab.org

Steve Heppelmann
Pennsylvania State University
104 Davey Lab.
University Park, PA 16802
(814) 863-3157
heppel@psu.edu

Douglas Higinbotham
Jefferson Laboratory
12000 Jefferson Ave., MS 12H
Newport News, VA 23606
(757) 269-7627
doug@jlab.org

Roy J. Holt
Physics Division
Argonne National Laboratory
9700 South Cass Avenue
Argonne, IL 60439
(630) 252-4101; Fax (630) 252-3903
holt@anl.gov

Theodore Horton
North Carolina State University
821A Barringer Drive
Raleigh, NC 27606
(919) 858-0608
tuhorton@unity.ncsu.edu

Paul Hoyer
Nordita
Blegdamsvej 17
DK-2100 Copenhagen, Denmark
(45) 3532 5222; Fax (45) 3538 9157
hoyer@nordita.dk

Dae Sung Hwang
Stanford Linear Accelerator Center
MS 81
Stanford University
P.O. Box 4349
Stanford, CA 94309
(650) 926-4937; Fax (650) 926-2525
dshwang@slac.stanford.edu

Chueng Ji
Department of Physics
North Carolina State University
P.O. Box 8202
Raleigh, NC 27695-8202
(919) 515-3478; Fax (919) 515-2471
ji@ncsu.edu

Xiangdong Ji
Department of Physics
University of Maryland
College Park, MD 20742
(301) 405-7277
xji@physics.umd.edu

Xiaodong Jiang
Rutgers University/Jefferson Lab.
12000 Jefferson Ave., MS 12H
Newport News, VA 23606
(757) 269-7011
jiang@jlab.org

398

Mark Jones
Jefferson Laboratory
12000 Jefferson Ave., MS C126
Newport News, VA 23606
(757) 269-7733
jones@jlab.org

Kyungseon Joo
12000 Jefferson Ave., MS 12H
Newport News, VA 23606
(757) 269-7764; Fax (757) 269-5800
kjoo@jlab.org

Cynthia Keppel
Hampton University/Jefferson Lab
12000 Jefferson Ave., MS 12H
Hampton, VA 23606
(757) 269-7580; Fax (757) 269-7363
keppel@jlab.org

Valeri Koubarovski
Rensselaer Polytechnic Institute/Jefferson
Laboratory
12000 Jefferson Ave., MS 12H
Newport News, VA 23606
(757) 269-7196; Fax (757) 269-5800
vpk@jlab.org

Peter Kroll
Department of Physics
University of Wuppertal
Gaussstrasse 20
Wuppertal D-42097, Germany
(49) 202-439-2620; Fax (49) 202-439-3860
kroll@physik.uni-wuppertal.de

Jean-Marc Laget
CEA/SACLAY
DAPNIA/SPhN
GIF-SUR-YVETTE Cedex
France F91191
(33) 1 6908 7554; Fax (33) 1 6908 7584
jlaget@cea.ft

Jonathan Lenaghan
The Niels Bohr Institute
17 Blegdamsvej
Copenhagen 2100, Denmark
(45) 3532 5426; Fax (45) 3532 54 00
lenaghan@alf.nbi.dk

John LeRose
Jefferson Laboratory
12000 Jefferson Ave., MS 12H
Newport News, VA 23606
(757) 269-7624; Fax (757) 269-5703
lerose@jlab.org

Simonetta Liuti
Department of Physics
University of Virginia
382 McCormick Road
Charlottesville, VA 22901
(434) 924-4576; Fax (434) 924-4576
sl4y@virginia.edu

David Mack
Jefferson Laboratory
12000 Jefferson Ave., MS 12H
Newport News, VA 23606
(757) 269-7442; Fax (757) 269-5235
mack@jlab.org

Pieter Maris
Department of Physics
North Carolina State University
Box 8202
Raleigh, NC 27695
(919) 515-3441; Fax (919) 515-2471
pmaris@unity.ncsu.edu

Kathy McCormick
Rutgers University/Jefferson Laboratory
12000 Jefferson Ave., CC A123
Newport News, VA 23606
(757) 269-7011; Fax (757) 269-5703
mccormick@jlab.org

Wally Melnitchouk
Jefferson Laboratory
12000 Jefferson Ave., MS 12H2
Newport News, VA 23606
(757) 269-5854; Fax (757) 269-7002
melnitc@jlab.org

Gerald Miller
Department of Physics
University of Washington
Box 35-1560
Seattle, WA 98195-1560
(206) 543-2995; Fax (206) 685-9829
miller@phys.washington.edu

Igor Musatov
Old Dominion University/Jefferson Lab
12000 Jefferson Ave., MS 12H2
Newport News, VA 23606
(757) 269-7062; Fax (757) 269-7002
musatov@jlab.org

Sirish Nanda
Jefferson Laboratory
12000 Jefferson Ave., MS 12H
Newport News, VA 23606
(757) 269-7176
nanda@jlab.org

Alan Nathan
Department of Physics
Univ. of Illinois at Urbana-Champaign
1110 West Green Street
Urbana, IL 61801
(217) 333-0965; Fax (217) 333-1215
a-nathan@uiuc.edu

Bogdan Niczyporuk
Jefferson Laboratory
12000 Jefferson Ave., MS 12H
Newport News, VA 23606
(757) 269-7251; Fax (757) 269-5800
bogdan@jlab.org

Nikolai N. Nikolaev
Institut fuer Kernphysik
Forschungszentrum Juelich
Juelich, Germany D-52425
(49) 2461 616472; Fax (49) 2461 613930
n.nikolaev@fz-juelich.de

Tim Oppermann
Institute of Theoretical Physics
University of Regensburg
Universitaetsstr. 30
Regensburg D-93040, Germany
(49) 0941-943-2017
Tim.oppermann@physik.uni-regensburg.de

Kornelija Passek
Universitaet Wuppertal
Fachbereich Physik
Gaussstrasse 20, Wuppertal D-42097
Germany
(49) 202-439-2862; Fax (49) 202-439-3860
passek@theorie.physik.uni-wuppertal.de

Lubomir P. Pentchev
College of William and Mary/Jefferson
Laboratory
12000 Jefferson Ave., MS 16B
Newport News, VA 23606
(757) 269-7680
pentchev@jlab.org

Charles Perdrisat
Department of Physics
College of William and Mary
P.O. Box 8795
Williamsburg, VA 23187
(757) 269-5304
perdrisa@jlab.org

Ales Psaker
Department of Physics
Old Dominion University/Jefferson Lab
5115 Hampton Boulevard
Norfolk, VA 23529
(757) 683-4613; Fax (757) 269-7002
psaker@jlab.org

Vina A. Punjabi
Norfolk State Univ./Jefferson Laboratory
12000 Jefferson Ave., MS 16B
Newport News, VA 23606
(757) 269-5304 or 823-2472
Punjabi@jlab.org

Anatoly Radyushkin
Department of Physics
Old Dominion University/Jefferson Lab
5115 Hampton Boulevard
Norfolk, VA 23529
(757) 269-7377; Fax (757) 269-7002
radyush@jlab.org

John Ralston
Department of Physics
1082 Malott Hall
University of Kansas
Lawrence, KS 66045
(785) 864-4626; Fax (785) 864-5262
ralston@ukans.edu

400

Bodo Reitz
Jefferson Laboratory
12000 Jefferson Ave., MS 12H
Newport News, VA 23606
(757) 269-5064
reitz@jlab.org

Michael Roedelbronn
University of Illinois
312 North James Street
Champaign, IL 61821
(217) 356-8216
roedelbr@jlab.org

Franck Sabatie
CEA SACLAY
DAPNIA SPhN
Gif sur Yvette, 91191, France
(33) 1-6908-3206
sabatie@jlab.org

Arun Saha
Jefferson Laboratory
12000 Jefferson Ave., MS 12H
Newport News, VA 23606
(757) 269-7605; Fax (757) 269-5703
saha@jlab.org

Elton Smith
Jefferson Laboratory
12000 Jefferson Ave., MS 12H
Newport News, VA 23606
(757) 269-8790; Fax (757) 269-5800
elton@jlab.org

Paul Stoler
Department of Physics
Rensselaer Polytechnic Institute
Troy, NY 12180
(518) 276-8388; Fax (518) 276-6680
stoler@rpi.edu

Mark Strikman
Pennsylvania State University
104 Davey Laboratory
University Park, PA 16803
strikman@phys.psu.edu

Brian C. Tiburzi
Department of Physics
University of Washington
Box 351560
Seattle, WA 98185
(206) 652-5281; Fax (206) 685-0635
bctiburz@u.washington.edu

Maurizio Ungaro
Jefferson Laboratory
12000 Jefferson Ave., MS 16B
Newport News, VA 23606
(757) 269-5328
ungaro@jlab.org

Marc Vanderhaeghen
University of Mainz
Institut fuer Kernphysik
J.J. Becher Weg 45
Mainz, Germany D-55099
(49) 6131 3924277
marcvdh@kph.uni-mainz.de

Carsten Vogt
Nordita
Blegdamsvej 17
Copenhagen 2100, Denmark
(45) 353-25501; Fax (45) 353-89157
cvogt@nordita.dk

Christian Weiss
Institut fuer Theoretische Physik
Regensburg University
Regensburg D-93053, Germany
(49) 941-943-2187; Fax (49) 941-943-3887
Christian.weiss@physik.uni-regensburg.de

Krishni Wijesoorya
Argonne National Lab/Jefferson Lab
12000 Jefferson Avenue
Newport News, VA 23606
(757) 269-5852
krishniw@jlab.org

Bogdan Wojtsekhowski
Jefferson Laboratory
12000 Jefferson Ave., MS 12H
Newport News, VA 23606
(757) 269-7191; Fax (757) 269-5703
bogdan@jlab.org

Feng Yuan
Department of Physics/TQHN
University of Maryland
College Park, MD 20770
(301) 405-6124
fyuan@physics.umd.edu